Springer Series in
MATERIALS SCIENCE 70

Springer Series in
MATERIALS SCIENCE

Editors: R. Hull   R. M. Osgood, Jr.   J. Parisi

The Springer Series in Materials Science covers the complete spectrum of materials physics, including fundamental principles, physical properties, materials theory and design. Recognizing the increasing importance of materials science in future device technologies, the book titles in this series reflect the state of the art in understanding and controlling the structure and properties of all important classes of materials.

61 **Fatigue in Ferroelectric Ceramics and Related Issues**
By D.C. Lupascu

62 **Epitaxy**
Physical Principles and Technical Implementation
By M.A. Herman, W. Richter, and H. Sitter

63 **Fundamentals of Ion-Irradiated Polymers**
By D. Fink

64 **Morphology Control of Materials and Nanoparticles**
Advanced Materials Processing and Characterization
Editors: Y. Waseda and A. Muramatsu

65 **Transport Processes in Ion-Irradiated Polymers**
By D. Fink

66 **Multiphased Ceramic Materials**
Processing and Potential
Editors: W.-H. Tuan and J.-K. Guo

67 **Nondestructive Materials Characterization**
With Applications to Aerospace Materials
Editors: N.G.H. Meyendorf, P.B. Nagy, and S.I. Rokhlin

68 **Diffraction Analysis of the Microstructure of Materials**
Editors: E.J. Mittemeijer and P. Scardi

69 **Chemical–Mechanical Planarization of Semiconductor Materials**
Editor: M.R. Oliver

70 **Applications of the Isotopic Effect in Solids**
By V.G. Plekhanov

71 **Dissipative Phenomena in Condensed Matter**
Some Applications
By S. Dattagupta and S. Puri

72 **Predictive Simulation of Semiconductor Processing**
Status and Challenges
Editors: J. Dabrowski and E.R. Weber

73 **SiC Power Materials**
Devices and Applications
Editor: Z.C. Feng

74 **Plastic Deformation in Nanocrystalline Materials**
By M.Yu. Gutkin and I.A. Ovid'ko

75 **Wafer Bonding**
Applications and Technology
Editors: M. Alexe and U. Gösele

76 **Spirally Anisotropic Composites**
By G.E. Freger, V.N. Kestelman, and D.G. Freger

77 **Impurities Confined in Quantum Structures**
By P.O. Holtz and Q. Zhao

78 **Macromolecular Nanostructured Materials**
Editors: N. Ueyama and A. Harada

Volumes 10–60 are listed at the end of the book.

V.G. Plekhanov

Applications of the Isotopic Effect in Solids

With 151 Figures and 24 Tables

Springer

Professor Vladimir G. Plekhanov
Computer Science College
Erika Street 7a, Tallinn
10416 Estonia
E-mail: vgplekhanov@hotmail.com

Series Editors:

Professor Robert Hull
University of Virginia
Dept. of Materials Science and Engineering
Thornton Hall
Charlottesville, VA 22903-2442, USA

Professor R. M. Osgood, Jr.
Microelectronics Science Laboratory
Department of Electrical Engineering
Columbia University
Seeley W. Mudd Building
New York, NY 10027, USA

Professor Jürgen Parisi
Universität Oldenburg, Fachbereich Physik
Abt. Energie- und Halbleiterforschung
Carl-von-Ossietzky-Strasse 9–11
26129 Oldenburg, Germany

Professor Hans Warlimont
Institut für Festkörper-
und Werkstofforschung,
Helmholtzstrasse 20
01069 Dresden, Germany

ISSN 0933-033X

ISBN 3-540-40809-6 Springer Berlin Heidelberg New York

Library of Congress Cataloging-in-Publication Data:

Plekhanov, Vladimir G.
Applications of the isotopic effect in solids / V.G. Plekhanov. p. cm. – (Springer series in materials science, ISSN 0933-033X ; 70) Includes bibliographical references and index. ISBN 3-540-40809-6 (acid-free paper)
1. Solid-state physics. 2. Isotopes. I. Title. II. Springer series in materials science; v. 70)
QC176.P54 2004
530.4'1–dc22 2004040798

Springer is a part of Springer Science+Business Media.

springeronline.com

Printed in Germany

Final processing by PTP-Berlin Protago-TeX-Production GmbH, Germany
Cover concept: eStudio Calamar Steinen
Cover production: *design & production* GmbH, Heidelberg

Printed on acid-free paper SPIN: 10825876 57/3141/YU 5 4 3 2 1 0

To the memory of my parents

Preface

This is the first book in world literature devoted to applications of the isotopic effect in solids. It is accessible to physicists, chemists, electronic engineers, and material scientists alike. This book is intended both as tutorial and as reference. Readers seeking to learn the basics of application of the isotopic effect in solids should start by reading the first few overview chapters, and then dig into descriptions of specific applications to see how they really work. It is hoped that it will be useful to undergraduate and graduate students of physics and optics as well as engineers, physicists, material scientists, and medical doctors who are interested in investigation or applications of the isotopic effect in solids. This book is a state-of-the-art introduction to very recent activity in solid-state physics which has developed in the main during the last half century and promises a new technology of isotopic engineering. Important applications are to be expected for information storage and developing materials for computer memory, quantum computers, isotopic fibers, isotopic optoelectronics, and quantum electronics as well as UV lasers.

The references I cite are those with which I am most familiar and which have helped us understand the subject as presented here. There has been no attempt to give credit to each contributor, but I have tried to cite the original papers, which brought new and important results (methods) to applications of the isotopic effect in solids, covered in this text.

I wish to thank all the authors cited throughout the text for their permission to use the figures from their work and also the publishers concerned for allowing reproduction. I would like to express thanks to my many students who have contributed to this investigations in various ways over the years. Last but by no means least I owe a great debt to my wife and my children for their patience and encouragement over the long period spent in preparing this book and I thank their most sincerely.

I thank my many friends for their help stimulation and the Production and Editorial Staff of Springer-Verlag, especially Dr. C. Asheron and Ms. A. Duhm, and also the staff of PTP-Berlin for their continual and very long assistance in preparing the copy for publication of my manuscript.

Tallinn, May 2004 *V.G. Plekhanov*

Contents

1 Introduction

The availability of isotopically pure crystals with low carrier and impurity concentrations has allowed the investigation of isotopic effects on lattice dynamic and electronic properties of solids in the last three decades [23]. Due to zero-point motion, the atoms in a solid feel the anharmonicity [28] of the interatomic potential, even at low temperatures. Therefore, the lattice parameters of two chemically identical crystals formed by different isotopes do not coincide [23, 106, 428]; heavier isotopes have smaller zero-point delocalization (as expected in a harmonic approximation) and smaller lattice parameters (an anharmonic effect). Moreover, phonon-related properties such as thermal conductivity, thermal expansion, and melting temperature are expected to depend on isotope mass.

We start with the fact that phonon frequency is directly affected by changes in the average mass of the whole crystal or its sublattice (the VCA-model), even if we look upon them as noninteracting particles, i.e., as harmonic oscillators. The direct influence of the isotopic mass on the frequencies of coupled phonon modes may be used to determine their eigenvectors. Second, the mean square amplitude $\langle u^2 \rangle$ of phonons depends on isotopic masses only at low temperature, and they are determined only by the temperature T, once T becomes larger than the Debye temperature. A refinement of these effects must take place when taking interactions among phonons into account. These interactions lead to finite phonon lifetimes and additional frequency renormalization. The underlying processes can be divided into two classes:

(1) anharmonic interactions in which a zone center phonon decays into two phonons or more with wave-vector and energy conservation and
(2) elastic scattering in which a phonon scatters into phonons of similar energies but different wave vectors.

The former processes arise from cubic and quartic terms in the expansion of lattice potential [28], whereas the latter are due to the relaxed wave-vector conservation rule in samples that are isotopically disordered and thus not strictly translationally invariant. Since the vast majority of compounds derives from elements that have more than one stable isotope, it is clear that both processes are present most of the time. Unfortunately, their absolute sizes and relative importance cannot be predicted easily. However, isotopic enrichment allows one to suppress the elastic scattering induced by isotopic disorder. In contrast, anharmonic phonon–phonon interaction cannot be suppressed, so that isotopic-disorder-induced effects can be studied only against a background contribution from anharmonic processes. However, if one assumes that the two processes are independent of each other, one can measure the disorder-induced renormalization by comparing the phonon energies and line-widths of isotopically pure samples with those gained from disordered ones.

The isotopic composition affects the band gaps through electron–phonon coupling and through the change in volume with isotopic mass. Although the electronic properties of different isotopes of a given atom are, to a very good approximation, the same, isotopic substitution in a crystal modifies the phonon spectrum [23] which, in turn, modifies the electron energy bands through electron–phonon interaction. Measuring the energy gaps in samples with different isotopic composition then yields the difference in the changes in the valence- and conduction-band renormalization. The reason for the changes lies in the fundamental quantum mechanical concept of zero-point motion – the vibrational energy that the atoms in the crystal have, even at low temperatures. If we excite an electron from one electronic state to another, we actually excite the whole crystal. In other words, we move the crystal from a ground state made up of low-energy electrons plus zero-point vibrations to an excited state in which there is one excited electron plus the zero-point vibrations of the crystal. The values of zero-point energy in the two electronic states are slightly different because the vibrational frequencies depend on chemical bonding, which is changed by exciting an electron. If the average mass of the vibrating atoms is increased, then the vibrational frequencies will be reduced. As a result, the difference in zero-point motions will be smaller, and the transition energy will therefore increase with increasing mass.

In this book we present the results of experimental and theoretical studies of the different applications of the isotopic effect in solids. The first three chapters are devoted to the fundamental properties of the isotopic effect and its manifestation in the elastic, thermal, and vibrational properties of different materials. The first chapter describes how phonon spectra of solids are a good indicator of their isotopic purity. It is shown that the simplest defects in a crystalline lattice that distort its translational symmetry are isotopes of the elements forming a crystal. Moreover, a change in atomic mass results in the appearance of three features in the phonon spectrum [7, 11]: localized, gap, and quasi-localized (resonant) modes. Localized mode frequencies lie above the maximum frequency of the phonon spectrum and are good indicator of the isotopic purity of solids studied. The second chapter describes the dependence of thermal conductivity on the isotopic impurities of crystals. It is shown that isotopically enhanced Ge, Si, and diamond crystals increase in thermal conductivity by about one order of magnitude. The third chapter analyzes the electronic properties of isotopic pure and mixed crystals. The results presented in this book indicate that substituting a light isotope with a heavy one leads to an increase in the band-to-band transitions with a nonlinear dependence on the concentration of isotopes. The replacement of the light isotope with a heavier one causes a nonlinear increase in the binding energy of Wannier–Mott excitons and an increase in the energy of longitudinal–transverse splitting (for details, see [23]).

The last five chapters are devoted to different applications of the isotopic effect in solids. In Chap. 5 we analyze the process of self-diffusion in isotopically pure materials and heterostructures. Interest in diffusion in solids is as old as metallurgy and ceramics, but the scientific study of the phenomenon may probably be dated to some six to seven decades ago. As is well known, a measured diffusion coefficient depends on the chemistry and structure of the sample on which it is measured. We have organized the chapter around general principles that are applicable to all materials. We briefly discuss the SIMS (secondary ion mass spectrometry) technique, in which a the sample is bombarded by reactive ions and the sputtered molecules are ionized in a plasma and fed into a mass spectrometer. Self-diffusion is the migration of constituent atoms in materials. This process is mediated by native defects in solids and thus can be used to study the dynamics and kinetics of these defects. The knowledge obtained in these studies is pivotal for understanding many important mass transport processes such as impurity diffusion in solids.

Chapter 6 describes a new reactor technology – neutron transmutative doping (NTD). The capture of thermal neutrons by isotopic nuclei followed by nuclear decay produces new elements, resulting in a large number of possibilities for isotope-selective doping of solids. Different facilities which use this reactor technology are presented. The feasibility of constructing reactors dedicated to the production of NTD silicon, germanium, and other compounds is analyzed in terms of technical and economic viability, and the practicality of such a proposal is examined. The importance of this technology for studies of semiconductor doping as well as metal–insulator transitions and neutral impurity scattering process is underlined. The introduction of particle irradiation into processing of semiconductor materials and devices creates a new need for additional understanding of atomic-displacement-produced defects in semiconductors. Measurement of decay rates of induced radioactivity and the system of clearance and certification that allows the solids to be internationally transported as "exempt material," as defined in IAEA Regulations, are dealt with.

The use of isotopes in the theory and technology of optical fibers is considered in Chap. 7. This chapter is addressed to readers who wish to learn about fiber communications systems and, particularly, about the properties of optical fibers. In this chapter we describe very briefly the Maxwell equations as well as wave electromagnetic equations. In this chapter we also describe the properties of optical fibers and the materials for optical fiber and fiber technology. The influence of isotopes on the properties of the optical fibers is also shown.

Chapter 9 is devoted to the application of the isotopic effect in laser physics. There is a short description of the theory and practice of semiconductor lasers. The discovery of the linear luminescence of free excitons observed over a wide temperature range has placed lithium hydride [23], as well as crystals of diamond [604] in line as prospective sources of coherent radiation in the UV spectral range. Emission has also been shown for LiH isotope tuning of the exciton [23, 230].

The last chapter of this book is devoted to a description of other unexplored applications of isotopic engineering. First, we consider materials for information storage in modern personal computers as well as in biology. In this chapter, it is shown that isotopic substitution has made it possible to produce objects of research that earlier were simply inaccessible (except for the LiH–LiD system). The use of such objects allows the investigation of isotopic effects in lattice dynamics (elastic, thermal, and vibrational properties) and also the influence of such effects on electronic states (the renormalization of the band-to-band transition energy E_g, exciton binding energy E_B, and the size of longitudinal–transverse splitting Δ_{LT} [23]. We should add here that the strength of the hyperfine interaction is proportional to the probability density of the electron wave-function in the nucleus. In semiconductors, the electron wave-function extends over large distances through the crystal lattice. Two nuclear spins can consequently interact with the same electron, leading to electron-mediated or indirect nuclear spin coupling. Because the electron is sensitive to externally applied electric fields, the hyperfine interaction and electron-mediated nuclear spin interaction can be controlled by voltages applied to metallic gates in a semiconductor device, enabling external manipulation of nuclear spin dynamics that is necessary for quantum computation in quantum computers (for details, see [624]).

A brief summary is presented in the conclusion. The difficult and unsolvable problems of isotopic effects in solids are considered there. The main aim of this book is to familiarize readers with present and some future applications in isotopic science and engineering.

2 Phonon Spectra of Solids: Indicator of Their Isotope Purity

2.1 Theory of Lattice Dynamics

We begin by considering a static crystal lattice. The position of an atom at equilibrium is given by

$$\boldsymbol{x}\begin{pmatrix} l \\ k \end{pmatrix} = \boldsymbol{x}_{\mathrm{L}} + \boldsymbol{x}_k, \tag{2.1}$$

where

$$\boldsymbol{x}_{\mathrm{L}} = l_1\boldsymbol{a}_1 + l_2\boldsymbol{a}_2 + l_3\boldsymbol{a}_3, \qquad -N_i \leq l_i \leq N_i, \tag{2.2}$$

is a lattice vector and

$$\boldsymbol{x}_k = k_1\boldsymbol{a}_1 + k_2\boldsymbol{a}_2 + k_3\boldsymbol{a}_3, \qquad 0 \leq k_i \leq 1, \tag{2.3}$$

are basis vectors. The three vectors $\boldsymbol{a}_1$, $\boldsymbol{a}_2$, $\boldsymbol{a}_3$ are primitive vectors. For the rock-salt-structure crystals, the primitive vectors are

$$\boldsymbol{a}_1 = \frac{a}{2}(\hat{y} + \hat{z}), \quad \boldsymbol{a}_2 = \frac{a}{2}(\hat{x} + \hat{z}), \quad \boldsymbol{a}_3 = \frac{a}{2}(\hat{x} + \hat{y}), \tag{2.4}$$

where a is the lattice constant. There is a reciprocal lattice associated with this real-space crystal lattice. Every $\boldsymbol{k}$–vector in this reciprocal lattice is given by

$$\boldsymbol{k} = \frac{n_1}{N_1}\boldsymbol{b}_1 + \frac{n_2}{N_2}\boldsymbol{b}_2 + \frac{n_3}{N_3}\boldsymbol{b}_3, \qquad n_i - \text{integers}, \tag{2.5}$$

where N_i's are integers giving the number of cells $N = 8N_1N_2N_3$ and the vectors $\boldsymbol{b}_1, \boldsymbol{b}_2, \boldsymbol{b}_3$, the primitive vectors for this reciprocal space, are defined by

$$\boldsymbol{a}_i \cdot \boldsymbol{b}_j = 2\pi\delta_{ij}. \tag{2.6}$$

For a rock-salt-structure crystal, the reciprocal lattice vectors are

$$\boldsymbol{b}_1 = \frac{2\pi}{a}(\hat{y} + \hat{z} - \hat{x}), \quad \boldsymbol{b}_2 = \frac{2\pi}{a}(\hat{x} + \hat{z} - \hat{y}), \quad \boldsymbol{b}_3 = \frac{2\pi}{a}(\hat{x} + \hat{y} - \hat{z}). \tag{2.7}$$

The total kinetic energy of a crystal can be written as (see, e.g., [1])

$$T = T_{\mathrm{N}} + T_{\mathrm{E}} = \sum_i \left(\frac{\boldsymbol{P}_i^2}{2M_i}\right) + \sum_j \left(\frac{\boldsymbol{p}_j^2}{2m}\right), \tag{2.8}$$

where T_{N} and T_{E} are the nuclear kinetic energy and the electronic kinetic energy, respectively. The mutual interactive Coulomb potential between charged particles is given by

$$U(\boldsymbol{X}, \boldsymbol{x}) = \frac{1}{2}\sum_{i,j} \frac{Z_i Z_j e^2}{|\boldsymbol{X}_i - \boldsymbol{X}_j|} + \frac{1}{2}\sum_{i,j} \frac{e^2}{|\boldsymbol{x}_i - \boldsymbol{x}_j|} - \sum_{i,j} \frac{Z_i e^2}{|\boldsymbol{X}_i - \boldsymbol{x}|}, \tag{2.9}$$

where $\boldsymbol{X}$ and $\boldsymbol{x}$ refer to the nuclear and electronic coordinates, respectively.

Given the crystal Hamiltonian, H, the quantum mechanical many-body Schrödinger equation for a crystal can now be constructed:

$$H(\boldsymbol{P}, \boldsymbol{X}; \boldsymbol{p}, \boldsymbol{x})\psi(\boldsymbol{X}, \boldsymbol{x}) = E\psi(\boldsymbol{X}, \boldsymbol{x}), \tag{2.10}$$

where H is the quantum mechanical operator,

$$H = T_{\mathrm{N}} + T_{\mathrm{E}} + U(\boldsymbol{X}, \boldsymbol{x}) \equiv H_0 + T_{\mathrm{N}}, \tag{2.11}$$

and $\psi(\boldsymbol{X}, \boldsymbol{x})$ is the many-body eigenfunction. In (2.11), H_0 is the Hamiltonian for all electrons moving in the field of nuclei whose instantaneous positions are given as the set $\{\boldsymbol{X}\}$.

We will now apply the Born–Oppenheimer [2] adiabatic approximation to the many-body Schrödinger equation. The object of this work is to show that, if one chooses the basis functions appropriately, it is possible, to a good approximation, to decouple the various terms in this series into groups of terms, each of which is a set of vibrational levels that belongs to a single electronic state. However, the electronic state does depend on the nuclear configuration, $\{\phi_\nu(\boldsymbol{X}, \boldsymbol{x})\}$. By solving the electronic problem,

$$H_0\phi_\nu(\boldsymbol{X}, \boldsymbol{x}) = W_\nu(\boldsymbol{X})\phi_\nu(\boldsymbol{X}, \boldsymbol{x}), \tag{2.12}$$

where the eigenfunctions $\phi_\nu(\boldsymbol{X}, \boldsymbol{x})$ and the eigenenergy $W_\nu(\boldsymbol{X})$ depend parametrically on $\{\boldsymbol{X}\}$ and the quantum number ν specifies the electronic state, the many-body eigenfunction can be expanded as

$$\psi(\boldsymbol{X}, \boldsymbol{x}) = \sum_\nu \phi_\nu(\boldsymbol{X}, \boldsymbol{x})\chi_\nu(\boldsymbol{X}), \tag{2.13}$$

where the expansion coefficients, $\chi_\nu(\boldsymbol{X})$, are functions only of the nuclear coordinates. Since the set of electronic eigenfunctions $\{\phi_\nu(\boldsymbol{X}, \boldsymbol{x})\}$ is complete, we have the orthonormality condition

$$\int \phi_\mu^*(\boldsymbol{X}, \boldsymbol{x})\phi_\nu(\boldsymbol{X}, \boldsymbol{x})\,\mathrm{d}x = \delta_{\mu\nu} \tag{2.14}$$

where (2.14) is independent of nuclear coordinate $\boldsymbol{X}$. Substituting (2.13) in (2.10) and exploiting the orthonormality condition, we find an equation for $\chi_\nu(\boldsymbol{X})$:

$$[T_{\mathrm{N}}(\boldsymbol{X}) + W_\nu(\boldsymbol{X}) - E]\,\chi_\nu(\boldsymbol{X}) + \sum_{\mu\nu} C_{\mu\nu}(\boldsymbol{X}, \boldsymbol{P})\,\chi_\nu(\boldsymbol{X}) = 0, \tag{2.15}$$

where

$$C_{\mu\nu}(\boldsymbol{X}, \boldsymbol{P}) = \sum_k M_k^{-1}\left[A_{\mu\nu}^k(\boldsymbol{X})\,\boldsymbol{P}_k + \boldsymbol{B}_{\mu\nu}^k(\boldsymbol{X})\right] \tag{2.16}$$

with $\boldsymbol{P}_k$ the nuclear momentum operator and

$$\begin{aligned} A_{\mu\nu}^k(\boldsymbol{X}) &= \int \phi_\mu^*(\boldsymbol{X}, \boldsymbol{x})\,\boldsymbol{P}_k \phi_\nu(\boldsymbol{X}, \boldsymbol{x})\,\mathrm{d}x, \\ \boldsymbol{B}_{\mu\nu}^k(\boldsymbol{X}) &= \frac{1}{2}\int \phi_\mu^*(\boldsymbol{X}, \boldsymbol{x})\,\boldsymbol{P}_k^2 \phi_\nu(\boldsymbol{X}, \boldsymbol{x})\,\mathrm{d}x. \end{aligned} \tag{2.17}$$

Assuming that the different electronic states are not coupled, then the off-diagonal elements $C_{\mu\nu}(\boldsymbol{X}, \boldsymbol{P}) = 0$ for $\mu \neq \nu$. For $\mu = \nu$ and for stationary states of the crystal Hamiltonian, $\boldsymbol{A}_{\mu\nu}^k(\boldsymbol{x}) = 0$. Therefore, (2.15) can be written as

$$[T_{\mathrm{N}}(\boldsymbol{X}) + U_\nu(\boldsymbol{X}) - E]\,\chi_\nu(\boldsymbol{X}) = 0, \tag{2.18}$$

where a redefined complete potential is

$$U_\nu(\boldsymbol{X}) = W_\nu(X) + \sum_k M_k^{-1}\boldsymbol{B}_{\nu\nu}^k(\boldsymbol{X}). \tag{2.19}$$

Equations (2.18) and (2.19) are complete equations for nuclear motion under the supposition that coupling between electronic states ($\mu \neq \nu$) can either be ignored or vanishes. As pointed out by Hardy and Karo [3], justifying these assumptions is not particularly easy with any degree of rigor, and there is something of a gap between the Born–Oppenheimer arguments and those used in developing potential functions for the various model calculations. However, the Born–Oppenheimer approximation provides the basic justification for computing the effective potential function by evaluating of the change in energy of the whole system for the static, but distorted, lattice. In the work that follows, we will treat T_{N} and $\Phi \equiv U_\nu(\boldsymbol{X})$ as functions of Cartesian displacement variables (see also [4]).

Now, we will derive the equations of motion in the harmonic approximation. Let the displacement vector from the equilibrium position at $\boldsymbol{x}\begin{pmatrix} l \\ k \end{pmatrix}$ be denoted $\boldsymbol{u}\begin{pmatrix} l \\ k \end{pmatrix}$, where $\boldsymbol{u}\begin{pmatrix} l \\ k \end{pmatrix}$ are considered independent dynamic variables. Assuming that the magnitudes of the displacements are small, the potential energy of the crystal can be expanded about its equilibrium value in terms of $\boldsymbol{u}\begin{pmatrix} l \\ k \end{pmatrix}$. The first three terms in the expansion of $\Phi(\boldsymbol{u})$ with respect to displacement from equilibrium are

$$\Phi(\boldsymbol{u}) = \Phi_0 + \sum_{l,k,\alpha} \Phi_\alpha \begin{pmatrix} l \\ k \end{pmatrix} u_\alpha \begin{pmatrix} l \\ k \end{pmatrix}$$
$$+\frac{1}{2} \sum_{l,k\alpha} \sum_{l'k'} {}_{,\alpha'} \mathrm{u}_\alpha \begin{pmatrix} l \\ k \end{pmatrix} \Phi_{\alpha\beta} \begin{pmatrix} l\ l' \\ k\ k' \end{pmatrix} u_\beta \begin{pmatrix} l' \\ k' \end{pmatrix}, \tag{2.20}$$

where

$$\Phi_\alpha \begin{pmatrix} l \\ k \end{pmatrix} = \left[\frac{\partial \Phi}{\partial u_\alpha} \begin{pmatrix} l \\ k \end{pmatrix} \right] \Bigg|_{u=0} \tag{2.21}$$

and the generalized force constant matrix

$$\Phi_{\alpha\beta} \begin{pmatrix} l\ \mathrm{l}' \\ k\ \mathrm{k}' \end{pmatrix} = \left[\frac{\partial^2 \Phi}{\partial u_\alpha} \begin{pmatrix} l \\ k \end{pmatrix} \partial u_\beta \begin{pmatrix} l' \\ k' \end{pmatrix} \right] \Bigg|_{u=0} . \tag{2.22}$$

Here, the derivatives are evaluated at the equilibrium configuration. The second term vanishes by definition because it is assumed that the expansion is made about the equilibrium position defined by the criterion that such first-order vanishes. Hence, in the harmonic approximation,

$$\Phi(\boldsymbol{u}) = \Phi_0 + \frac{1}{2} \sum_{l,k,\alpha} \sum_{l',k',\alpha'} u_\alpha \begin{pmatrix} l \\ k \end{pmatrix} \Phi_{\alpha\beta} \begin{pmatrix} l\ l' \\ k\ k' \end{pmatrix} u_\beta \begin{pmatrix} l' \\ k' \end{pmatrix} . \tag{2.23}$$

The first term in (2.23) is the potential energy evaluated at equilibrium positions, thus it represents the reference potential energy. The lattice dynamics is then contained in the second term in this harmonic approximation. Since the order in which the differentiation is carried out is immaterial, the force constant matrix satisfies the relation

$$\Phi_{\alpha\beta} \begin{pmatrix} l\ l' \\ k\ k' \end{pmatrix} = \Phi_{\beta\alpha} \begin{pmatrix} l'\ l \\ k'\ k \end{pmatrix} . \tag{2.24}$$

There are $3Nr$ equations of motion since there is one for each of the three components of the displacement of the r basis ions. By applying Newton's equations of motion, we obtain

$$M_k \ddot{u}_\alpha \begin{pmatrix} l \\ k \end{pmatrix} = -\frac{\partial \Phi}{\partial u_\alpha} \begin{pmatrix} l \\ k \end{pmatrix} = -\sum_{l',k',\beta} \Phi_{\alpha\beta} \begin{pmatrix} l\ l' \\ k\ k' \end{pmatrix} u_\beta \begin{pmatrix} l' \\ k' \end{pmatrix} . \tag{2.25}$$

Recalling that plane waves form a complete basis, we have the following completeness and orthonormality relations:

$$\sum_{k}\left[\frac{\exp\left(\mathrm{i}\boldsymbol{k}\boldsymbol{x}_{\mathrm{L}}\right)}{\sqrt{N}}\right]\left[\frac{\exp\left(\mathrm{i}\boldsymbol{k}\boldsymbol{x}_{\mathrm{L}'}\right)}{\sqrt{N}}\right]^{*}=\delta_{\mathrm{L},\mathrm{L}'}, \tag{2.26}$$

$$\sum_{\boldsymbol{x}_{\mathrm{L}}}\left[\frac{\exp\left(\mathrm{i}\boldsymbol{k}\cdot\boldsymbol{x}_{\mathrm{L}}\right)}{\sqrt{N}}\right]\left[\frac{\exp\left(\mathrm{i}\boldsymbol{k}'\cdot\boldsymbol{x}_{\mathrm{L}}\right)}{\sqrt{N}}\right]^{*}=\delta_{\mathrm{k},\mathrm{k}'}. \tag{2.27}$$

Taking the plane wave expansion of the Cartesian components of the displacements,

$$u_{\alpha}\begin{pmatrix} l \\ k \end{pmatrix}=\left(NM_{k}\right)^{-\frac{1}{2}}\sum_{k}w_{\alpha}\left(\frac{k}{\boldsymbol{k}}\right)\exp\left\{\mathrm{i}\left[\boldsymbol{k}\cdot\boldsymbol{x}\begin{pmatrix} l \\ k \end{pmatrix}-\omega\left(\boldsymbol{k}\right)t\right]\right\}. \tag{2.28}$$

Clearly, $w_{\alpha}\left(\frac{k}{\boldsymbol{k}}\right)$ are complex variables, and we may take

$$w_{\alpha}^{*}\left(\frac{k}{\boldsymbol{k}}\right)=\pm w_{\alpha}\left(\frac{k}{-\boldsymbol{k}}\right) \tag{2.29}$$

so that $\boldsymbol{u}\begin{pmatrix} l \\ k \end{pmatrix}$ are real. The negative sign is chosen for a crystal with inversion symmetry. Upon substituting (2.28) in (2.25), we obtain an eigenvalue equation,

$$\omega^{2}\left(\boldsymbol{k}\right)w_{\alpha}\left(k\mid\boldsymbol{k}\right)=\sum_{k',\beta}D_{\alpha\beta}\left(kk'\mid\boldsymbol{k}\right)w_{\beta}\left(\frac{k'}{\boldsymbol{k}}\right), \tag{2.30}$$

where the Fourier transform of the mass-weighted force constant matrix

$$D_{\alpha\beta}\left(kk'\mid\boldsymbol{k}\right)\equiv\left(M_{k}M_{\mathrm{k}'}\right)^{-\frac{1}{2}}\times\sum_{l,l'}\Phi_{\alpha\beta}\begin{pmatrix} l & l' \\ k & k' \end{pmatrix}\exp\left\{\mathrm{i}\boldsymbol{k}\left[\boldsymbol{x}\begin{pmatrix} l \\ k \end{pmatrix}-\boldsymbol{x}\begin{pmatrix} l' \\ k' \end{pmatrix}\right]\right\} \tag{2.31}$$

is known as the dynamic matrix of the crystal. It is a $3r\times3r$ matrix, here r is the number of atoms in the basis. Thus, for alkali halides, it will be a 6×6 matrix. The dynamic matrix is Hermitian

$$D_{\alpha\beta}\left(kk'\mid\boldsymbol{k}\right)=D_{\beta\alpha}^{*}\left(k'k\mid\boldsymbol{k}\right). \tag{2.32}$$

Moreover, we see from (2.31) that

$$D_{\alpha\beta}\left(kk'\mid-\boldsymbol{k}\right)=D_{\alpha\beta}^{*}\left(kk'\mid\boldsymbol{k}\right). \tag{2.33}$$

The eigenfrequencies and eigenvectors are obtained by diagonalizing the dynamic matrix of (2.30),

$$\parallel D_{\alpha\beta}\left(kk'\mid\boldsymbol{k}\right)-\omega^{2}\left(\boldsymbol{k}\right)\delta_{\alpha\beta}\delta_{\mathrm{kk}'}\parallel=0, \tag{2.34}$$

which possesses $3Nr$ roots $\omega_j(\boldsymbol{k})$; $j = 1, 2, \ldots, 3r$ for $N\boldsymbol{k}'$ s. There is an associated eigenvector for each eigenfrequency. These eigenvectors are the solutions of (2.30) which can now be formally written as

$$\sum_{k',\beta} D_{\alpha\beta}(kk' \mid \boldsymbol{k})\, e_\beta \begin{pmatrix} k' \mid \boldsymbol{k} \\ \mid j \end{pmatrix} = \omega^2 \begin{pmatrix} \boldsymbol{k} \\ j \end{pmatrix} e_\alpha \begin{pmatrix} k \mid \boldsymbol{k} \\ \mid j \end{pmatrix}. \tag{2.35}$$

The eigenvectors are orthonormalized and satisfy completeness relation:

$$\sum_{k\alpha} e_\alpha^* \begin{pmatrix} k \mid \boldsymbol{k} \\ \mid j \end{pmatrix} e_\alpha \begin{pmatrix} k \mid \boldsymbol{k} \\ \mid j' \end{pmatrix} = \delta_{jj'}, \tag{2.36}$$

$$\sum_{j} e_\beta^* \begin{pmatrix} k' \mid \boldsymbol{k} \\ \mid j \end{pmatrix} e_\alpha \begin{pmatrix} k \mid \boldsymbol{k} \\ \mid j \end{pmatrix} = \delta_{\alpha\beta}\delta_{\mathrm{kk}'}. \tag{2.37}$$

If we replace $\boldsymbol{k}$ by $-\boldsymbol{k}$ in (2.35), take the complex conjugate of the resulting equation, and use (2.32) and (2.33), we obtain the result that

$$\omega^2 \begin{pmatrix} -\boldsymbol{k} \\ j \end{pmatrix} = \omega^2 \begin{pmatrix} \boldsymbol{k} \\ j \end{pmatrix}. \tag{2.38}$$

In a later section, this result will be applied to calculate two-phonon dispersion curves when we discuss second-order Raman scattering.

Now we introduce a set of complex normal coordinates $Q\begin{pmatrix} \boldsymbol{k} \\ j \end{pmatrix}$ by writing the amplitude $w_\alpha(k \mid \boldsymbol{k})$ of (2.30) as

$$w_\alpha(k \mid \boldsymbol{k}) = \sum_{j=1}^{3r} Q\begin{pmatrix} \boldsymbol{k} \\ j \end{pmatrix} e_\alpha \begin{pmatrix} k \mid \boldsymbol{k} \\ \mid j \end{pmatrix}. \tag{2.39}$$

This can be inverted to solve for $Q\begin{pmatrix} \boldsymbol{k} \\ j \end{pmatrix}$ by using (2.40)

$$Q\begin{pmatrix} \boldsymbol{k} \\ j \end{pmatrix} = \sum_{k\alpha} w_\alpha(k \mid \boldsymbol{k})\, e_\alpha^* \begin{pmatrix} k \mid \boldsymbol{k} \\ \mid j \end{pmatrix}. \tag{2.40}$$

Substituting (2.30) and (2.39) in the expression for kinetic energy, we obtain

$$T = \frac{1}{2}\sum_{l,k\alpha} M_k \left[\dot{u}_\alpha \begin{pmatrix} l \\ \boldsymbol{k} \end{pmatrix}\right]^2 = \frac{1}{2}\sum_{k,j} \dot{Q}^* \begin{pmatrix} \boldsymbol{k} \\ j \end{pmatrix} \dot{Q}\begin{pmatrix} \boldsymbol{k} \\ j \end{pmatrix}. \tag{2.41}$$

Similarly, by substituting (2.39) in (2.30) and using (2.35), we obtain harmonic approximation,

$$\Phi = \frac{1}{2}\sum_{k,j} \omega^2 \begin{pmatrix} \boldsymbol{k} \\ j \end{pmatrix} \dot{Q}^* \begin{pmatrix} \boldsymbol{k} \\ j \end{pmatrix} \dot{Q} \begin{pmatrix} \boldsymbol{k} \\ j \end{pmatrix}, \tag{2.42}$$

where the reference term Φ_0 is neglected.

For many applications and, in particular, to implement the transition from classical mechanics to quantum mechanics, it is preferable to have real dynamic variables. It is possible to accomplish this by any number of canonical transformations. However, as Birman pointed out, the conventional quantization procedure based on real normal coordinates is not consistent with preserving the space group symmetry properties of normal coordinates. In view of this apparent inconsistency, we will follow Birman's lead by making the ansatz that a set of normal coordinates $q \begin{pmatrix} \boldsymbol{k} \\ j \end{pmatrix}$ that preserves the symmetry can be quantized in the conventional manner.

The transformed Hamiltonian assumes the form of a set of noninteracting simple harmonic oscillators,

$$H = \frac{1}{2}\sum_{k,j} \left[p^2 \begin{pmatrix} \boldsymbol{k} \\ j \end{pmatrix} + \omega^2 \begin{pmatrix} \boldsymbol{k} \\ j \end{pmatrix} q^2 \begin{pmatrix} \boldsymbol{k} \\ j \end{pmatrix} \right], \tag{2.43}$$

where the canonical momenta $p \begin{pmatrix} \boldsymbol{k} \\ j \end{pmatrix}$ conjugate to $q \begin{pmatrix} \boldsymbol{k} \\ j \end{pmatrix}$ are given by the Lagrangian prescription:

$$p \begin{pmatrix} \boldsymbol{k} \\ j \end{pmatrix} = \frac{\partial L}{\partial \dot{q}} \begin{pmatrix} \boldsymbol{k} \\ j \end{pmatrix} = \dot{q} \begin{pmatrix} \boldsymbol{k} \\ j \end{pmatrix}. \tag{2.44}$$

Now, we make the transition to quantum mechanics by imposing the commutator relation

$$\left[q \begin{pmatrix} \boldsymbol{k} \\ j \end{pmatrix}, p \begin{pmatrix} \boldsymbol{k}' \\ j' \end{pmatrix} \right] = \mathrm{i}\hbar \Delta \left(\boldsymbol{k} - \boldsymbol{k}' \right) \delta_{jj'}, \tag{2.45}$$

and all commutators involving other combinations vanish. The Δ function in (2.45) denotes modulo reciprocal lattice vectors, i.e., they differ by a reciprocal lattice vector. We will now follow the conventional procedure by defining the creation and annihilation operators a^+_{kj} and a_{kj}:

$$q \begin{pmatrix} \boldsymbol{k} \\ j \end{pmatrix} \equiv \left[\frac{\hbar}{2\omega} \begin{pmatrix} \boldsymbol{k} \\ j \end{pmatrix} \right]^{\frac{1}{2}} \left(a^+_{-kj} + a_{kj} \right), \tag{2.46}$$

$$p \begin{pmatrix} \boldsymbol{k} \\ j \end{pmatrix} \equiv \mathrm{i} \left[\hbar\omega \frac{\begin{pmatrix} \boldsymbol{k} \\ j \end{pmatrix}}{2} \right]^{\frac{1}{2}} \left(a^+_{-kj} - a_{kj} \right). \tag{2.47}$$

If we express the creation and annihilation operators in terms of $q\begin{pmatrix}\boldsymbol{k}\\ j\end{pmatrix}$ and $p\begin{pmatrix}\boldsymbol{k}\\ j\end{pmatrix}$,

$$a_{kj} = \left[\frac{\omega\begin{pmatrix}\boldsymbol{k}\\ j\end{pmatrix}}{2\hbar}\right]^{\frac{1}{2}} q\begin{pmatrix}\boldsymbol{k}\\ j\end{pmatrix} + \mathrm{i}\left[2\hbar\omega\begin{pmatrix}\boldsymbol{k}\\ j\end{pmatrix}\right]^{-\frac{1}{2}} p\begin{pmatrix}\boldsymbol{k}\\ j\end{pmatrix}, \tag{2.48}$$

$$a_{kj}^{+} = \left[\frac{\omega\begin{pmatrix}\boldsymbol{k}\\ j\end{pmatrix}}{2\hbar}\right]^{\frac{1}{2}} q\begin{pmatrix}\boldsymbol{k}\\ j\end{pmatrix} - \mathrm{i}\left[2\hbar\omega\begin{pmatrix}\boldsymbol{k}\\ j\end{pmatrix}\right]^{-\frac{1}{2}} p\begin{pmatrix}\boldsymbol{k}\\ j\end{pmatrix} \tag{2.49}$$

and apply the commutation relations (2.46–2.47), we obtain

$$\left[a_{kj}, a_{k'j'}^{+}\right] = \Delta\left(\boldsymbol{k} - \boldsymbol{k}'\right)\delta_{jj'}, \tag{2.50}$$

and all commutators involving other combinations vanish. Therefore, these operators are Bose operators and the particles are bosons. The quanta of this particle field, which were obtained by second-quantizing the lattice vibrations in the harmonic approximation, are called phonons. The Hamiltonian of the phonons takes on a particularly simple form in terms of the creation and annihilation operators,

$$H = \sum_{k,j} \hbar\omega\begin{pmatrix}\boldsymbol{k}\\ j\end{pmatrix}\left(a_{kj}^{+}a_{kj} + \frac{1}{2}\right). \tag{2.51}$$

This Hamiltonian has a very simple interpretation. By defining the number operator N_{kj}

$$\begin{aligned} a_{kj}^{+}a_{kj} \mid \cdots n_{kj-1}n_{kj}n_{kj+1}\cdots\rangle &= N_{kj} \mid \cdots n_{kj-1}n_{kj}n_{kj+1}\cdots\rangle \\ &= \mid \cdots n_{kj-1}n_{kj}n_{kj+1}\cdots\rangle n_{kj} \end{aligned} \tag{2.52}$$

where $\mid n_{kj}\rangle$ is the eigenket or the number state of the phonons in the $\mid \boldsymbol{k}, j\rangle$ state in the system. We see that the total energy of the system is just the energy of the phonon, $\hbar\omega\begin{pmatrix}\boldsymbol{k}\\ j\end{pmatrix}$, multiplied by the number of the phonons in the system. The second term is known as the zero-point energy of the system, which is the minimal energy possessed by the crystal lattice. The number of phonons is evidently not conserved because they can be created or annihilated (see also Reissland [5, 6]).

Local modes

The lattice dynamics of perfect crystals was analyzed in the preceding section. In an imperfect lattice, both masses and force constants may be perturbed from their perfect counterparts (Maradudin et al. [7], Barker and Sievers [8], and Taylor [9]). We will denote the imperfect mass as $M'\begin{pmatrix} l \\ k \end{pmatrix}$ and the force constants as $\Theta_{\alpha\beta}\begin{pmatrix} l\ l' \\ k\ k' \end{pmatrix}$ at site $\boldsymbol{x}\begin{pmatrix} l \\ k \end{pmatrix}$. The equations of motion in the harmonic approximation for this imperfect mass can then be written as

$$M'\begin{pmatrix} l \\ k \end{pmatrix}\ddot{u}_\alpha\begin{pmatrix} l \\ k \end{pmatrix}+\sum_{l'k',\beta}\Theta_{\alpha\beta}\begin{pmatrix} l\ l' \\ k\ k' \end{pmatrix}u_\beta\begin{pmatrix} l' \\ k' \end{pmatrix}=0, \tag{2.53}$$

where $\boldsymbol{u}\begin{pmatrix} l \\ k \end{pmatrix}$ is the perturbed displacement at the site of imperfection. Assuming that the displacement has a harmonic time dependence, (2.53) becomes

$$-\omega^2 M'\begin{pmatrix} l \\ k \end{pmatrix}u_\alpha\begin{pmatrix} l \\ k \end{pmatrix}+\sum_{l',k',\beta}\Theta_{\alpha\beta}\begin{pmatrix} l\ l' \\ k\ k' \end{pmatrix}u_\beta\begin{pmatrix} l' \\ k' \end{pmatrix}=0, \tag{2.54}$$

where ω is the eigenfrequency of the local mode at $\boldsymbol{x}\begin{pmatrix} l \\ k \end{pmatrix}$.

Equation (2.54) can be put in a form that explicitly shows the perturbation due to th e imperfect lattice. We begin by defining a matrix

$$C_{\alpha\beta}\begin{pmatrix} l\ l' \\ k\ k' \end{pmatrix}\equiv -M_{\mathrm{k}'}\omega^2\varepsilon\begin{pmatrix} l' \\ k' \end{pmatrix}\delta_{\alpha\beta}\delta_{\mathrm{ll}'}\delta_{\mathrm{kk}'}+\Phi_{\alpha\beta}\begin{pmatrix} l\ l' \\ k\ k' \end{pmatrix}-\Theta_{\alpha\beta}\begin{pmatrix} l\ l' \\ k\ k' \end{pmatrix}, \tag{2.55}$$

where

$$\varepsilon\begin{pmatrix} l \\ k \end{pmatrix}=\frac{\left[M_k-M'\begin{pmatrix} l \\ k \end{pmatrix}\right]}{M_k} \tag{2.56}$$

is the fractional change in the mass. It follows that (2.54) becomes

$$-M_k\omega^2 u_\alpha\begin{pmatrix} l \\ k \end{pmatrix}+\sum_{l',k',\beta}\Phi_{\alpha\beta}\begin{pmatrix} l\ l' \\ k\ k' \end{pmatrix}u_\beta\begin{pmatrix} l' \\ k' \end{pmatrix}=\sum_{l',k',\beta}C_{\alpha\beta}\begin{pmatrix} l\ l' \\ k\ k' \end{pmatrix}u_\beta\begin{pmatrix} l' \\ k' \end{pmatrix}. \tag{2.57}$$

This is an inhomogeneous system of equations.

Solutions of (2.57) can obtained by the method of Green's function. Green's function is defined by (see, e.g., Economou [10])

$$-M_k\omega^2 G_{\alpha\gamma}\begin{pmatrix} l\ l' \\ k\ k' \end{vmatrix}\omega\Bigr) + \sum_{l',k',\beta} \Phi_{\alpha\beta}\begin{pmatrix} l\ l' \\ k\ k' \end{pmatrix} G_{\beta\gamma}\begin{pmatrix} l'\ l'' \\ k'\ k'' \end{vmatrix}\omega\Bigr) = \delta_{\alpha\gamma}\delta_{\mathrm{ll'}}\delta_{\mathrm{kk'}}. \tag{2.58}$$

Green's function is found by expanding $G_{\alpha\beta}\begin{pmatrix} l\ l' \\ k\ k' \end{vmatrix}\omega\Bigr)$ in terms of the eigenvectors of the unperturbed system. It can be shown that

$$G_{\alpha\beta}\begin{pmatrix} l\ l' \\ k\ k' \end{vmatrix}\omega\Bigr) = \sum_{\boldsymbol{k},j}\left(\frac{e_\alpha^*\begin{pmatrix} k \mid \boldsymbol{k} \\ \mid j \end{pmatrix} e_\beta\begin{pmatrix} k' \mid \boldsymbol{k} \\ \mid j \end{pmatrix} \exp\left\{\mathrm{i}\boldsymbol{k}\left[\boldsymbol{x}\begin{pmatrix} l \\ k \end{pmatrix} - \boldsymbol{x}\begin{pmatrix} l' \\ k' \end{pmatrix}\right]\right\}}{N\left(M_k M_{\mathrm{k'}}\right)^{\frac{1}{2}}\left[\omega^2\begin{pmatrix} \boldsymbol{k} \\ j \end{pmatrix} - \omega^2\right]}\right) \tag{2.59}$$

which can be readily verified by direct substitution in (2.58). The displacement then is given by

$$u_\alpha\begin{pmatrix} l \\ k \end{pmatrix} = \sum_{l',k',\beta}\sum_{l''k''\gamma} G_{\alpha\beta}\begin{pmatrix} l\ l' \\ k\ k' \end{vmatrix}\omega\Bigr) C_{\alpha\beta}\begin{pmatrix} l'\ l'' \\ k'\ k'' \end{pmatrix} u_\gamma\begin{pmatrix} l'' \\ k'' \end{pmatrix}. \tag{2.60}$$

Again this can be verified by direct substitution in (2.57). By factoring $u_\gamma\begin{pmatrix} l'' \\ k'' \end{pmatrix}$, (2.60) can be written as

$$\sum_{l'',k'',\gamma}\left\{\sum_{l',k',\beta}\left[G_{\alpha\beta}\begin{pmatrix} l\ l' \\ k\ k' \end{vmatrix}\omega\Bigr) C_{\beta\gamma}\begin{pmatrix} l'\ l'' \\ k'\ k'' \end{pmatrix} - \delta_{\alpha\gamma}\delta_{\mathrm{ll'}}\delta_{\mathrm{kk'}}\right]\right\} u_\gamma\begin{pmatrix} l'' \\ k'' \end{pmatrix}, \tag{2.61}$$

and if the effect of the perturbation can be restricted to the nearest neighboring cells, i.e., a localized mode, then the condition for a solution of (2.61) is given by

$$\left\|\sum_{l',k',\beta}\left[G_{\alpha\beta}\begin{pmatrix} l\ l' \\ k\ k' \end{vmatrix}\omega\Bigr) C_{\beta\gamma}\begin{pmatrix} l'\ l'' \\ k'\ k'' \end{pmatrix} - \delta_{\alpha\gamma}\delta_{\mathrm{ll'}}\delta_{\mathrm{kk'}}\right]\right\| = 0. \tag{2.62}$$

For the given perturbation matrix C, this equation has solutions only at the eigenfrequencies of the perturbed problem.

We will simplify (2.62) further by considering the mass defect without force constants (see, however, below), giving

$$C_{\alpha\beta}\begin{pmatrix} l & l' \\ k & k' \end{pmatrix} = -M_{k'}\omega^2\varepsilon\begin{pmatrix} l' \\ k' \end{pmatrix}\delta_{\alpha\beta}\delta_{ll'}\delta_{kk'}. \tag{2.63}$$

Upon substitution (2.63) in (2.62), we obtain a 3×3 secular equation involving only x, y, z components:

$$\left\| M_k\omega^2\varepsilon\begin{pmatrix} l \\ k \end{pmatrix} G_{\alpha\beta}\begin{pmatrix} l & l \\ k & k \end{pmatrix}\omega\Big) + \delta_{\alpha\beta} \right\| = 0. \tag{2.64}$$

In a cubic crystal, Green's function can be evaluated exactly from (2.59)

$$G_{\alpha\beta}\left(\begin{matrix} l & l \\ k & k \end{matrix}\middle|\, \omega\right) = -\frac{1}{NM_k}\sum_{\boldsymbol{k},j}\frac{e^*_\alpha\left(\begin{matrix} k & \boldsymbol{k} \\ & j \end{matrix}\right)e_\beta\left(\begin{matrix} k & \boldsymbol{k} \\ & j \end{matrix}\right)}{\omega^2 - \omega^2\begin{pmatrix} \boldsymbol{k} \\ j \end{pmatrix}}. \tag{2.65}$$

For a cubic crystal, G transforms under rotation as a second-rank symmetrical tensor and has only one independent nonzero component:

$$G_{11}\left(\begin{matrix} l & l \\ k & l \end{matrix}\middle|\, \omega\right) = G_{22}\left(\begin{matrix} l & l \\ k & k \end{matrix}\middle|\, \omega\right) = G_{33}\left(\begin{matrix} l & l \\ k & k \end{matrix}\middle|\, \omega\right). \tag{2.66}$$

It follows that

$$G_{\alpha\alpha}\left(\begin{matrix} l & l \\ k & k \end{matrix}\middle|\, \omega\right) = -\frac{1}{3NM_k}\sum_{\boldsymbol{k},j}\frac{e_\alpha\left(\begin{matrix} k & \boldsymbol{k} \\ & j \end{matrix}\right)^2}{\omega^2 - \omega^2\begin{pmatrix} \boldsymbol{k} \\ j \end{pmatrix}}. \tag{2.67}$$

With this simplified Green's function, the secular equation (2.63) reduces to a product of three identical terms, giving a root of three-fold degeneracy:

$$1 - (3N)^{-1}\omega^2\varepsilon\begin{pmatrix} l \\ k \end{pmatrix}\sum_{\boldsymbol{k},j}\frac{e_\alpha\left(\begin{matrix} k & \boldsymbol{k} \\ & j \end{matrix}\right)^2}{\omega^2 - \omega^2\begin{pmatrix} \boldsymbol{k} \\ j \end{pmatrix}} = 0. \tag{2.68}$$

This equation gives the eigenfrequency of the local mode (for details, see, e.g., [11]).

A detailed understanding of the basic mechanisms that determine vibrational properties is useful for the current technologically most important semiconductors and insulators. Even though applications (see below) are based on electronic properties in the majority of cases (naturally exluding thermal conductivity), technology may also benefit in the future from a predictive knowledge of vibrational properties. A detailed understanding of the dynamic lattice properties of bulk crystals also forms the basis for interpreting vibrational effects in low-dimensional systems, which nowadays attract much interest.

2.2 Elastic Properties

2.2.1 Theoretical Background of Elastic Constant Measurements

The velocity v of an elastic wave in a crystalline solid depends on the density ρ and the adiabatic elastic constants c_{ij}. The velocities of the three modes of propagation are given by the solution of the Christoffel equation:

$$\left|\Gamma_{ik} - \delta_{ik}\rho v^2\right| = 0, \tag{2.69}$$

where

$$\Gamma_{ik} = mc_{ijkl}\, n_i n_l \tag{2.70}$$

and n_i are the direction cosines of the direction of propagation (see Musgrave [12]). Materials with cubic symmetry have three independent elastic constants (see also Nye [13]), and for any of the $\langle 100\rangle$ directions, the solutions for the velocities of the three modes of propagation are

$$v_1 = \left(\frac{c_{11}}{\rho}\right)^{\frac{1}{2}}, \tag{2.71}$$

$$v_2 = v_3 \left(\frac{c_{44}}{\rho}\right)^{\frac{1}{2}}, \tag{2.72}$$

where v_1 is the longitudinal wave velocity and v_2 are transverse wave velocities. In the $\langle 100\rangle$ direction, substitution of the appropriate Γ_{ik} in (2.70) leads to

$$v_4 = \left\{\frac{\left[\frac{1}{2}(c_{11}+c_{12})+c_{44}\right]}{\rho}\right\}^{\frac{1}{2}}, \tag{2.73}$$

$$v_5 = \left(\frac{c_{44}}{\rho}\right)^{\frac{1}{2}}, \tag{2.74}$$

$$v_6 = \left[\frac{(c_{11}-c_{121})}{\rho}\right]^{\frac{1}{2}}, \tag{2.75}$$

where v_4 is the longitudinal wave velocity, v_5 is the transverse wave velocity with particle motion in the $\langle 001\rangle$ direction and v_6 is the transverse wave velocity with particle motion in the $\langle 1\bar{1}0\rangle$ direction.

Provided that suitably oriented specimens are available elastic constants c_{11}, c_{12} and c_{44} may be obtained by using (2.71)–(2.75) together with the measured values of an appropriate set of velocities. For most directions of propagation in a crystal, the solution of (2.69) does not lead to a simple relation between the elastic constants and the velocities. Consequently, direct evaluation of the elastic constants is not possible for general orientations of specimens. However, Neighbors and Schacher [14] have shown that a perturbation method may be used to determine the elastic constants from the measured velocities of ultrasonic waves propagated in arbitrary directions. Gluyas et al. [15] have used this method to evaluate the elastic constants of thallium chloride, where it was not possible to obtain the $\langle 100 \rangle$ and $\langle 110 \rangle$ faces with sufficient accuracy to use (2.71)–(2.75), according to the criteria of Watherman [16], who examined the errors introduced into the velocity due to misoriented specimens. After that, this method was used for determination of the elastic moduli of LiH crystals and its isotopic analog–LiD crystals (James and Kherandish [17]).

Another method for determining elastic moduli in the solid state is Brillouin scattering of light. There are many excellent accounts of the theory of Brillouinn scattering (see, e.g., Born and Huang [1]; Hayes and Loudon [18]). In the context of our narration, it is sufficient to recall that the phonons, satisfying wave vector conservation in a given Brillouin scattering geometry in a cubic and hence optically isotropic material, have wave vectors given by

$$\boldsymbol{q} = \pm\left(\overrightarrow{k_i} - \overrightarrow{k_\mathrm{s}}\right); \quad |\boldsymbol{q}| = \pm 2\left|\overrightarrow{k_i}\right| \sin\left(\frac{\theta}{2}\right), \tag{2.76}$$

where $\boldsymbol{q}$, $\boldsymbol{k}_i$ and $\boldsymbol{k}_\mathrm{s}$ are the wave vectors of the phonon, and incident and scattered light, respectively, and θ is the angle subtended by $\boldsymbol{k}_i$ and $\boldsymbol{k}_\mathrm{s}$. Note that the k's and θ defined inside the material so that refractive effects at the surfaces are not yet included in (2.76).

In an isotropic medium, the velocity v of an elastic wave is deduced from the frequency shift $\delta\nu$ of the Brillouin line by the usual relation

$$\frac{\delta\nu}{\nu_0} = 2n\left(\frac{v}{c}\right)\sin\frac{\theta}{2}, \tag{2.77}$$

where c is the velocity of light, n the refractive index of the medium for the radiation of frequency ν_0, θ is the scattering angle inside the crystal, and $\delta\nu$ is equal to the frequency ν_e of the hypersonic waves involved (see also Hayes and Loudon [18]).

2.2.2 Experimental Results and Interpretation

In this section, we will follow in the main the results of James and Kherandish [17]. Values of c_{11} and c_{44} of LiH crystals were obtained from measured velocities of longitudinal and transverse waves propagation in the $\langle 100 \rangle$ direction. The elastic constants c_{11} and c_{44} were calculated from measured transit times by using (2.71) and (2.72) together with values of specimen length and density corrected for thermal expansion. The densities of ^{7}LiH and ^{7}LiD at room temperature, are 783 kgm^{-3} and 891 kgm^{-3} respectively, as quoted by Guinan and Cline [19] and by Gerlich and Smith [20]. The elastic constant c_{12}, corrected for thermal expansion, was obtained at 10 K intervals from values of v_1, v_2, and v_6^1 by using a computer program that carried out the perturbation calculations suggested by Neighbours and Schacher [14]. According to the results of the authors cited, the absolute uncertainties of c_{11} and c_{44} are due to uncertainties in the measurements of the length of the specimen (±0.1%), the transit time (±0.7%), and the density estimated at (±0.1%). However, the relative uncertainties arise only from the measurement of transit time. The relative uncertainties in the values of c_{12} are much higher than those in c_{11} and c_{44} because of two velocities of almost the same value. Thus, together with uncertainties in the measurement of sample length, transit time, and density, the results in absolute uncertainty in c_{12} are ±5%.

The room temperature elastic constants of ^{7}LiH and ^{7}LiD have also been measured by Haussuhl and Skorczyk [21], Guinan and Cline [19] and those of LiH by Gerlich and Smith [20] and by Laplaze et al. [22]. The values of elastic constants of different authors are given in [23, Table 1]. The uncertainties quoted in this table are absolute uncertainties. The results for ^{7}LiH are in good agreement. We should note the large discrepancy in the value of c_{12} of different authors. The reason for this discrepancy is not well understood.

The date presented in [23, Table 1] demonstrate a great difference in elastic constants c_{11} and c_{44} for both crystals. It is well known (see, for example, Leibfried [24]) that for crystals with the inversion center in the model of central forces, which act along the line connecting neighboring atoms, Cauchy's relation should be valid, i.e., $c_{11} = c_{44}$. In the microscopic description of lattice dynamics, Cauchy's relation follows from the spherical distribution of ionic charges. Cauchy's relation is quite satisfactorily obeyed in many ionic crystals (for details, see James and Kherandish [17]). The greatest deviation from Cauchy's relation is observed for LiF, where $\frac{c_{44}}{c_{12}} = 1.53$, and the effective Szigetti charge $e_s^* = 0.80$ (see, for example, Plekhanov [25]). One can see from Table 1 that for LiH (LiD), this ratio exceeds three. This means that large deviations from Cauchy's relation suggest that noncentral forces play an important role in a LiH crystal and its isotopic analog (see also Wilson and Johnson [26] and details below). The isotopic effect, which is manifested in the fact that elastic constants c_{11} and c_{44} in LiD are greater (whereas c_{12} is smaller) than those in LiH will be described below. Note also that the direct calculations of the elastic constants in LiH by Gerlich and Smith show that the c_{44} value depends strongly on the interaction between nearest neighbors in the second coordination sphere (for details, see, e.g., [27]).

The temperature dependences of the elastic constants of LiH crystals (Fig. 2.1) were studied by the ultrasonic method in the temperature range between 4.2 and 300 (James and Kheyrandish [17]) and by Brillouin scattering in the temperature range between 100 and 300 (Laplaze et al. [22]). It was shown that constants c_{11} and c_{44} increased with decreasing temperature, whereas c_{12} decreased (see also [23, Table 1]). The temperature dependence of the bulk modulus proves to be very weak, which is most likely explained primarily by the partial compensation of contributions from c_{11} and c_{12}.

In the papers of James and Kherandish [17] the method of Leibfried and Ludvig [28] was used to predict the temperature variation of the elastic constants of ^{7}LiH and ^{7}LiD, which are compared with the corresponding values observed experimentally, to evaluate the degree of success of this extrapolative technique. This technique has also been applied to calculate values of c_{12} for ^{7}LiH in the temperature range 0–160 K, where the measurements of the appropriate wave velocities were not possible. The method involved extrapolation proportional to the thermal energy so that each of the slopes of the curves of c_{11}, c_{12} and c_{44} versus temperature are proportional to heat capacity for both ^{7}LiH and ^{7}LiD. The elastic constants and the initial slopes used were the appropriate c_{ij} and $\frac{\partial c_{ij}}{\partial T}$ at room temperature obtained for the experimental data. The heat capacity data used were from Yates et al. [29]. The temperature variation of the elastic constants of ^{7}LiH and ^{7}LiD calculated in this way from the room temperature values of the elastic constants are shown in Fig. 2.1 (for more details, see [27]).

The strong, tetrahedrally coordinated, covalent bonds between nearest neighbors and the light mass of constituent atoms lead to many striking and unique properties of diamond, e.g., the largest elastic moduli (c_{ij}) known

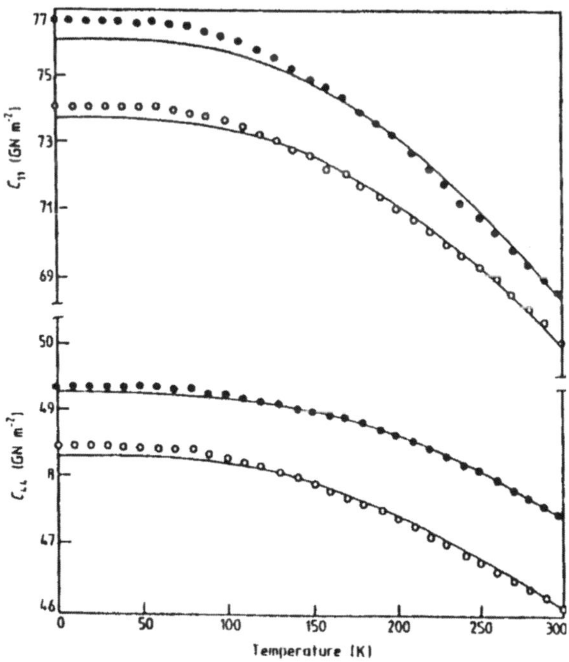

Fig. 2.1. The elastic constants c_{11} and c_{44} of ^{7}LiH and ^{7}LiD (*smoothed values*: O – ^{7}LiH; • – ^{7}LiD; *full curve* is theoretical calculations). (After [17])

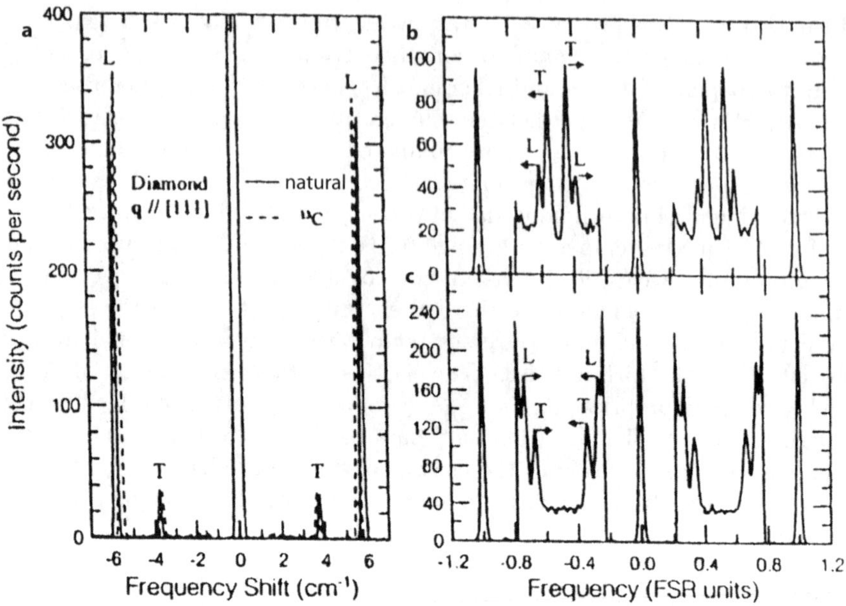

Fig. 2.2. Brillouin spectra of natural and isotopically enriched diamonds. **(a)** The spectra were recorded in backscattering geometry for 5145 Å radiation incident along ⟨111⟩ and the backscattered light analyzed with a (5 + 4) tandem Fabry–Perot interferometer. The phonon wave vector is along ⟨111⟩. Here L and T denote longitudinal and transverse, respectively. **(b)** Brillouin spectrum for ^{13}C diamond in the same geometry as for **(a)** but analyzed with a five-pass interferometer with a FRS of 0.67067 cm^{-1} and with an analyzer in the scattered beam to reduce the intensity of the longitudinal peaks. The longitudinal L and transverse T peaks are ≈ 8.4 and ≈ 5.4 orders from their parent laser line as indicated by the arrows. **(c)** Same as **(b)** but for ^{12}C diamond; the L and T peaks are now ≈ 8.7 and ≈ 5.7 orders from their parent line, respectively. (After [33])

for any material and correspondingly, the largest sound velocities (Grimsditch and Ramdas [30])and a very large Debye temperature, making it, like LiH [25], a "quantum" crystal, even at room temperature. Crimsditch and Ramdas made a comprehensive study of Brillouin scattering in natural diamond. On the basis of their investigations, they deduced the frequencies of the phonons with wave vectors corresponding to a large number of critical points of the phonon dispersion curves spanning the entire Brillouin zone and of the elastic constants with high precision. Determination of ultrasonic velocities from the round trip transit time is, as is well-known, an alternative technique which is equally precise. However, Brillouin scattering offers special advantages: it is a "contactless" technique; oriented specimens as small as a few millimeters are entirely satisfactory and can be conveniently placed in a high-temperature environment; the incident light is brought to the samples, and the scattered light is collected with appropriate optical windows. In contrast, the ultrasonic technique is advantageous for precision, large specimens, and a suitable bond to the transducer. Finally, Brillouin scattering geometries can be suitably devised that obviate the need for knowledge of the refractive index in deducing the elastic constants from the Brillouin shifts (see, also Sandercock [31]). The dependence of the elastic constants of diamond on the isotopic effect was investigated. In the papers of Ramdas et al. [32], Vogelgesang et al. [33], and Zouboulis et al. [34] the method of Brillouin scattering of light was used, whereas in the paper of Hurley et al. [35], the elastic moduli were measured using the method of ultrasound waves. The Brillouin experiments were performed in the backscattering geometry for phonon propagation vectors $\boldsymbol{q}$ along $\langle 001 \rangle$, $\langle 111 \rangle$, and $\langle 110 \rangle$. The incident beam was normal to the sample surface. Since the facets produced during the growth of natural and synthetic diamonds were exactly normal to the desired direction, single measurements are sufficient to determine the Brillouin shift for each particular wave vector (see, e.g., Fig. 2.2). For diamond surfaces which have been prepared by polishing and, hence were possibly misoriented by a few degrees, the Brillouin shift was measured in the paper of Vogelsang et al. for the various propagation directions close to the surface normal. As noted in the paper cited, this does not change the scattering angle θ. In this paper was formulated the theory for the x dependence ($^{12}C_x^{13}C_{1-x}$) of $\hbar\omega_0$, lattice constant a, and elastic moduli c_{ij} in terms of a dynamic lattice description for the zone center optical F_{2g} phonon and the third-order bulk modulus, incorporating anharmonicity as well as zero-point motion (see also Ramdas et al. [32], Plekhanov [25]). Dynamic lattice theories for crystals with a diamond structure have been given by Musgrave and Pople [36], by Keating [37], and by Martin [38] based on models in which the strain energy of a crystal is expressed in terms of the change in the bond length and the bond angles. In this approach, called the valence to force-field method (for detail, see also [23, Chapter 111]), the potential is manifestly invariant under both translation and rotation (Lax [39]). Anharmonic forces were used by Keating in a model involving two harmonic and three anharmonic force constants. The paper of Vogelgesang et al. [33] considered a simple model containing four parameters, two harmonic and two anharmonic. The harmonic part was identical to that employed by Musgrave [12]. Of the two anharmonic force constants, only one is of significance for bond stretching energy.

Let Δr_{in} be the change in distance between the central ion and its tetrahedral surrounding ions and let $\Delta\theta_{ij\mathbf{n}}$ be the change in the angle between the bonds of two adjacent tetrahedral ions. The energy required to stretch (or contact) a bond by Δr_i is

$$\frac{1}{2}k_1 (\Delta r_i)^2 - \frac{1}{6}g_1 (\Delta r_i)^3 + \cdots , \tag{2.78}$$

and that required to alter the angle is

$$\frac{1}{2}k_2\left(R\Delta\theta_{ij}\right)^2-\frac{1}{6}g_2\left(R\Delta\theta_{ij}\right)^3+\cdots \tag{2.79}$$

Here k_1 and k_2 are the force constants for the harmonic part of the potential, whereas g_1 and g_2 define the lowest order anharmonic contributions. Neglecting interactions other than those between nearest neighbor atoms, the deformation energy is

$$\Delta U=\sum_n \Delta U_{\mathrm{n}}=\sum_n\left[\frac{1}{2}k_1\sum_i\left(\Delta r_{i\mathbf{n}}\right)^2+\frac{1}{2}k_2\sum_{i\prec j}\left(R\Delta\theta_{ij\mathbf{n}}\right)^2\right.$$
$$\left.-\frac{1}{6}g_1\sum_i\left(\Delta r_{i\mathbf{n}}\right)^3-\frac{1}{6}g_2\sum_{i\prec j}\left(R\Delta\theta_{ij\mathbf{n}}\right)^3+\cdots\right]. \tag{2.80}$$

Bulk Modulus

The energy of deformation in which all nearest neighbor distances change from R to $R+u$ ($u \ll R$) without change in symmetry, i.e., $\Delta r_{i\mathbf{n}}=u$ and $\Delta\theta_{ij\mathbf{n}}=0$ is

$$\Delta U=N\left(2k_1u^2-\frac{2}{3}g_1u^3+\cdots\right), \tag{2.81}$$

where N is the number of primitive cells in the crystal. The change in volume is $\Delta V=\frac{3Vu}{R}$, so that

$$\Delta U=\frac{k_1\left(\Delta V\right)^2}{6aV}-\frac{g_1\left(\Delta V\right)^3}{72\sqrt{3}\,V^2}+\cdots \tag{2.82}$$

Recall the change in free energy F with volume

$$\Delta F=\frac{1}{2}\left(\Delta V\right)^2\left(\frac{\partial^2F}{\partial V^2}\right)_0+\cdots\approx\frac{\left(\Delta V\right)^2B}{2V}, \tag{2.83}$$

where B is the bulk modulus. A comparison (2.81) and (2.82) at zero temperature yields

$$B=\frac{k_1}{3a}. \tag{2.84}$$

Zone Center Optical Mode

Consider a motion in which the two face-centered cubic (fcc) sublattices of a structure experience a relative displacement u with respect to each other, as in the F_{2g} zone center optical mode, and let u be along $\langle 111 \rangle$. Noting that the terms in g_2 cancel exactly, the deformation energy due to this relative displacement is

$$\Delta U = N \left[\frac{2}{3} (k_1 + 4k_2) u^2 - \frac{4g_1 u^3}{27} + \cdots \right]. \tag{2.85}$$

The kinetic energy per primitive cell for the relative motion of the fcc sublattices is $\frac{P^2}{M}$ where P is the momentum canonically conjugate which u and M the mass of the atom ($\frac{M}{2}$ is the reduced mass for diamond). Thus, the Hamiltonian per primitive cell associated with motion is

$$H = \frac{P^2}{M} + \frac{2}{3} (k_1 + 4k_2) u^2 - \frac{4g_1 u^3}{27} + \cdots \tag{2.86}$$

Neglecting the cubic term, the equation of motion for u is harmonic with the angular frequency of the F_{2g} zone center optical phonon

$$\Omega_0 = \left[\frac{8 (k_1 + 4k_2)}{3M} \right]^{\frac{1}{2}}. \tag{2.87}$$

To estimate the effect of zero-point motion, the paper cited, quantized the Hamiltonian in (2.86) and investigated the nature of the ground state of such a system by a variational procedure. Choosing, after Vogelgesang et al. [33], a displaced Gaussian distribution characterized by the normalized wave function

$$\Psi (u) = \left(\frac{\alpha}{\sqrt{\pi}} \right)^{\frac{1}{2}} \exp \left[-\frac{\alpha^2}{2} (u - \delta)^2 \right], \tag{2.88}$$

where α and δ are the variational parameters, one obtains the expectation value of H in state $\Psi (u)$,

$$E(\alpha, \delta) = \frac{\hbar^2 \alpha^2}{2M} + \frac{2}{3} (k_1 + 4k_2) \left(\delta^2 + \frac{1}{2\alpha^2} \right) - \frac{4g_1}{27} \left(\delta^3 + \frac{3\delta}{2\alpha^2} \right) + \cdots. \tag{2.89}$$

For a given δ, the minimum $E(\alpha, \delta)$ occurs for

$$\alpha = \left(\frac{M\Omega_0}{2\hbar} \right)^{\frac{1}{2}} \left(1 - \frac{16 g_1 \delta}{9M\Omega_0^2} \right)^{\frac{1}{4}} \tag{2.90}$$

and

$$E(\delta) = \frac{\hbar\Omega_0}{2} + \frac{1}{4} M\omega_0^2 \left(\delta - \frac{8\hbar g_1}{9M^2\Omega_0^3} \right)^2 - \frac{16\hbar^2 g_1}{81 M^3 \Omega_0^4} + \cdots \tag{2.91}$$

with

$$\omega_0 = \Omega_0 \left(1 - \frac{32\hbar g_1^2}{81 M^3 \Omega_0^5}\right) . \tag{2.92}$$

We note that minimum total energy corresponds to a displacement

$$\delta = \left(\frac{8\hbar g_1}{9M^2 \Omega_0^3}\right) \tag{2.93}$$

with respect to the classical value obtained by neglecting the kinetic energy arising from zero-point motion of the atoms. As a result, the zone center optical phonon frequency ω_0 is the renormalized value given by (2.92).

Now we shall briefly discuss the effect of zero-point motion on the value of the bulk modulus. Consider uniform expansion of a crystal, without altering its symmetry, but including the effect of zero-point motion. By a procedure similar to that employed above for the zone center optical phonon, one can relate additional macroscopic parameters to the microscopic parameters k_1 and g_1. For a uniform change in volume, the Hamiltonian per primitive cell is

$$H = \frac{P^2}{M} + 2k_1 u^2 - \frac{2}{3} g_1 u^3 + \cdots . \tag{2.94}$$

Taking a variational wave function of the form defined in (2.88) and minimizing the expectation value of the Hamiltonian given in (2.94) with respect to the parameter α, one obtains

$$E(\delta) = \hbar \left(\frac{2k_1}{M}\right)^{\frac{1}{2}} + 2K_1 (\delta - \delta_0)^2 + \cdots , \tag{2.95}$$

where K_1 is the renormalized stiffness constant,

$$K_1 = k_1 \left(1 - \frac{\hbar g_1^2}{8\sqrt{2} k_1^{\frac{5}{2}} M^{\frac{1}{2}}}\right) , \tag{2.96}$$

and

$$\delta_0 = \frac{\hbar g_1 \sqrt{2}}{8 k_1^{\frac{3}{2}} M^{\frac{1}{2}}} . \tag{2.97}$$

The renormalized bulk modulus, T is then given by

$$T = \frac{K_1}{3a} . \tag{2.98}$$

The lattice parameter was obtained from $R = R_\infty + \delta_0$, where R_∞ is the nearest neighbor distance if the atomic mass were infinitely large and hence the zero-point motion negligible. Taking it into account, we obtain

Table 2.1. Elastic moduli c_{ij} and bulk modulus T of diamond in units of 10^{12} dyn/cm^2. (After Vogelgesang et al. [33])

x	sample	c_{11}	c_{12}	c_{44}	T
0.0	D 29	10.799(5)	1.248(10)	5.783(5)	4.432(8)
0.01105	D1, D2 and D17	10.804(5)	1.270(10)	5.766(5)	4.448(8)
0.992	D 30	10.792(7)	1.248 (14)	5.776(7)	4.429(12)

$$a_0 = a_\infty + \frac{\hbar g_1}{\sqrt{6} k_1^{\frac{3}{2}} M^{\frac{1}{2}}}. \qquad (2.99)$$

The results of the Brillouin measurements (Vogelgesang et al. [33]) for the four directions of phonon wave vectors were used to obtain the elastic constants of diamond. The squares of the velocity of sound clearly show a decrease with increasing average mass (for details, see Table 2.1). The corresponding elastic moduli X were calculated according to the equation:

$$X(x) = \frac{c^2}{4\omega_{\mathrm{L}}^2} \frac{\rho(x)}{n^2(x)} \Delta\omega^2(x), \qquad (2.100)$$

where the ^{13}C concentration-dependent mass density ρ expressed by $\rho(x) = \frac{8M_x}{a^3(x)}$ and $a(x)$ was determined by Holloway et al. [40] as

$$a(x) = (3.56715 - 0.00053\,x). \qquad (2.101)$$

The calculated $\rho(x_{\mathrm{nat}}) = 3.5152\,\mathrm{mg/cm^3}$ based on (2.101) compares very well with $3.5153\,\mathrm{g/cm^3}$ quoted by Mykolajewycz et al. [41]. From the data obtained by Vogelgesang et al., it was possible to determine c_{11}, c_{12}, and c_{44} separately for $x = 0.0$, $x = 0.01105$, and $x = 0.992$. The results are displayed in Table 2.1. From these elastic constants, one can deduce the bulk modulus, the values given in the last column of Table 2.1. The theoretical prediction for $T(x)$ is

$$\frac{T(x)}{T(0)} = \frac{a(0)}{a(x)} \left\{ 1 + \frac{\hbar g_1^2}{8\,(2k_1^5 M_{12})^{\frac{1}{2}}} \left[1 - \left(\frac{M_{12}}{M_x} \right)^{\frac{1}{2}} \right] \right\} = 1 + 0.0012\,x, \qquad (2.102)$$

where the values k_1 and g_1 equal, respectively, $4.76 \cdot 10^5$ dyn/cm and $(4.5 \pm 0.4) \cdot 10^{14}\,\mathrm{erg/cm^3}$.

The third-order bulk modulus T' can be expressed in terms of the anharmonicity parameter g_1. According to the results of Vogelgesang et al.

$$T' = \frac{dT}{dP} = -\frac{V}{T} \left(\frac{\partial T}{\partial g_1} \right) \left(\frac{\partial V}{\partial g_1} \right)^{-1} = \frac{g_1}{12\sqrt{3}T} = \frac{g_1 a}{4\sqrt{3}k_1} = 4.9. \qquad (2.103)$$

Assuming that the bulk modulus is a linear function of pressure, one obtains the Mungham equation of state,

$$\frac{V_0}{V} = \left(1 + P\frac{T_0'}{T_0}\right)^{\frac{1}{T_0^1}}, \tag{2.104}$$

where T_0^1 is the derivative of the bulk modulus with respect to pressure, evaluated at $P = 0$. With the experimentally determined value (McSkimin and Andreatch [42]) $T_0^1 = 4.03$, one can conclude that the molar volume of ^{12}C diamond equals that of ^{13}C diamond at zero pressure when $P = 0.2\,\text{GPa}$. One can conclude that

$$\frac{T(^{13}\text{C}) - T(^{12}\text{C})}{T(^{12}\text{C})} = 1.8 \cdot 10^{-3}, \tag{2.105}$$

in reasonable agreement with previous estimates using (2.102).

We should add here that the results of Vogelgesang et al. are consistent with the data obtained by Muinov et al. [43].

Based on experimental values of a for LiH and LiD, the value of $g_1 = 3.5 \cdot 10^{15}\,\text{erg/cm}^3$ was obtained in the paper of Plekhanov [25]. Inserting (2.96) into (2.100), one can obtain the relation described by the elastic constants renormalized upon isotopic substitution, which is to (see also Cerdeira and Cardona [44])

$$\Delta c_{44} = \left(\frac{k_1}{8a}\right)\left[1 - \left(\frac{\hbar g_1}{8K^2}\right)^2 \frac{1}{\sqrt{K\mu}}\right], \tag{2.106}$$

where μ is the reduced mass of a unit cell. According to the experimental data (Gerlich and Smith [20]), the relative change δc_{44} in c_{44} upon isotopic substitution amounts to 2.2–3.4% (see [23, Table 1]), whereas the calculation based on (2.106) yields only 1.8% (see also Plekhanov [25]). In concluding of this section, note that poor agreement, in our opinion, between theoretical values and the experimental data of the elastic constant of LiH (LiD) could suggest that consideration of the lowest order anharmonicity is insufficient. We should add in this connection that some reason for this discrepancy may be linked with the theoretical model used.

Zoubolis et al. [34] studied the temperature dependence of the elastic constants of diamond in the temperature range 300 to 1600 K. Since the frequency (ω_0) of the F_{2g} zone center optical mode is determined by the bond stretching (k_1) and bond bending (k_2) force constants which also determine the elastic constants (see above also), a combination of Raman (see below) and Brillouin results allow extracting the temperature dependence of the microscopic force constants (for more details, see also [23]).

2.3 Vibrational Properties

2.3.1 Phonon Dispersion and Density of Phonon States

In view of the obvious mass dependence of phonon frequencies, dynamic lattice properties have been studied intensively, mainly by Raman scattering. In addition to changes in the average atomic mass, mass fluctuation due to isotopic disorder also affects phonon frequencies and line-widths. At low temperatures, isotopic substitution allows one to tune the zero-point vibrational amplitudes of the nuclei which, via anharmonicity, cause changes in the lattice constant (see, e.g., [25]). This effect and, even more important, the electron–phonon interaction (for more details, see Sect. 4) lead to renormalization of band gaps which therefore exhibit isotopic shifts [23].

In a phenomenological approach to lattice dynamics, the force constants are viewed as parameters to be determined from experimental data. To obtain a meaningful fit, the underlying model must not involve too many parameters, and they must have a definite physical meaning. Different models have been developed for the various systems (see, e.g., [3, 6, 45, 46]). Especially the lattice dynamics of insulators attracted much interest in the twentieth century. In these materials, an account of the polarization of ions and an explanation of the effective ionic charges observed are very challenging problems. A phenomenological approach to lattice dynamics of insulators (ionic crystals) is the subject of many books. Alternative approaches are the shell model, the deformation dipole model, and the bond charge model (see, e.g., [3,4,6,23,45]).

In the shell model, the ions are regarded as cores, to which massless shells, representing the outer valence electrons are bound by isotropic springs. In this model, there is only one type of polarization, the relative displacement of the core and the shell, irrespective of whether this is brought about by an electric field or by the influence of short-range forces. By allowing for additional radial symmetrical deformability of the ions, i.e., so-called breathing deformability, the fit of experimental phonon dispersion curves (see below) by the shell model can be substantially improved.

In the deformation dipole model, there two possible types of polarization, which are assumed to be independent: (i) electric field induced and (ii) overlap induced polarization. The latter polarization originates from the distortion of the special symmetry of the charge clouds around ions by the overlap with the charge distributions of its neighboring ions (for details, see [3]). Accordingly, a lattice wave produces a distortion of the charge cloud about a given ion which is equivalent to a point dipole at the undisplaced position of the ion.

The applicability of the shell model to covalent crystals has been called into question because it is unphysical to divide the electrons in a bond between two atoms, when they are shared between the two. The bond charge model seems to provide a more adequate description of these materials. In this model, the incomplete screening of an ion in a covalent crystal by a diagonal dielectric matrix is compensated for by bond charges placed midway between neighboring ions. The bond charges yield effective noncentral forces between the ions and thus produce the stability of a diamond-type lattice to shear. Accordingly, they represent the covalent character of the bonding (see also [47]).

By allowing for adiabatic motion of bond charges, the agreement of the calculated phonon spectrum with that experimentally determined can be improved. At first sight, there is no relationship between the different phenomenological models used for studying lattice vibrations in the various systems. The microscopic theory of phonons provides a tool for examining a possible relationship between the different models because it represents a unified treatment of lattice vibrations in conducting as well as in nonconducting crystals (for details see, e.g., [5,6]).

The phonon spectrum is characterized by the dispersion law $\omega(\boldsymbol{q})$ and the frequency distribution function $g(\omega)$ (see, e.g., Krivoglaz [48]). Both functions are commonly determined from experiments on scattering of thermal neutrons, if coherent and incoherent scattering can be separated. Thermal neutrons play an important role in the study of lattice dynamics because their energy ($k_{\mathrm{B}}T \sim 10^{-1} - 10^{-2}\,\mathrm{eV}$) is of the same order of magnitude as the phonon energy. At the same time, their de Broglie wavelength is comparable with the interatomic distance in a crystal. In this respect, neutrons have an advantage over electromagnetic waves, for which matching can be obtained only in energy (in the IR range) or wavelength (gamma rays). The features of scattering of thermal neutrons by a lattice are determined by the following main factors:

1. Because the neutron wavelength greatly exceeds the size of the nucleus, the scattering is isotropic and independent of the neutron energy.
2. Scattering of neutrons by the lth nucleus is described with good accuracy by the Fermi pseudopotential, which is proportional to the delta function
$$U_{\mathrm{l}}(\boldsymbol{r}) = \left(2\pi\hbar^2 \frac{b_l}{m_{\mathrm{n}}}\right) \delta(\boldsymbol{r} - \boldsymbol{R}), \tag{2.107}$$
where m_{n} is the neutron mass and b_l is the so-called scattering length.
3. The total scattering cross section by an ensemble of nuclei forming a crystal is determined by the summation over individual nuclei, taking properly into account the phase relations for scattered waves. If nuclei are vibrating, the neutron scattering can be both elastic and inelastic, i.e., it can be accompanied by creation and annihilation of a phonon or several phonons.
4. The scattering length b_l can greatly differ for different isotopes of the same element.

In addition, when the nucleus spin is not zero, b_l depends on the mutual orientation of the neutron and nucleus spins. It is these factors that cause incoherent scattering, i.e., the scattering during which the law of conservation of momentum is not satisfied because of the violation of translational symmetry (Dolling [49]). The principle of measuring phonon dispersion by means of one-phonon coherent scattering can be easily understood from the laws of conservation of energy and momentum:

$$E_{\mathrm{i}} - E_{\mathrm{s}} = \frac{\hbar^2}{2m_{\mathrm{n}}} \left(\boldsymbol{k}_i^2 - \boldsymbol{k}_{\mathrm{s}}^2 \right) = \hbar\omega\left(\boldsymbol{q}\right), \tag{2.108}$$

$$\boldsymbol{Q} = \boldsymbol{k}_i - \boldsymbol{k}_{\mathrm{s}} = \boldsymbol{q} + \boldsymbol{G}, \tag{2.109}$$

where E_{i} and E_{s} are energies of the incident and scattered neutrons, respectively; $\hbar\boldsymbol{k}$ are their momenta; $\boldsymbol{G}$ is the reciprocal lattice vector; $\boldsymbol{q}$ is the phonon wave vector; and j is the number of the phonon spectrum branch. The frequency of phonons involved in scattering is found from (2.108) by measuring neutron velocities before and after the interaction. If the scattering angle is also measured, the phonon wave vector $\boldsymbol{q}$ can be found from (2.109). Thus, the dispersion law at a point in the Brillouin zone is totally determined. What actually happens is that the situation is complicated by the presence of several frequency values for each $\boldsymbol{q}$; in other words, for each value of the transmitted momentum $\boldsymbol{Q}$, there are several groups of neutrons scattered by each branch (Dolling [49]). In incoherent scattering, only the law of conservation of energy (2.108) is fulfilled, so that the scattering spectrum proves to be continuous within $E_{\mathrm{i}} \pm \hbar\omega_{\mathrm{max}}$. The scattering cross section is proportional to the frequency distribution function $g(\omega)$ (see above) and also contains a frequency factor, which depends on the polarization vectors of phonons $\boldsymbol{e}^s(\boldsymbol{q})$ (Krivoglaz [48]). For this reason, function $g(\omega)$, reconstructed from experiments, reflects to some degree the model that was used for determining polarization vectors. In lithium hydride crystals, neutron scattering strongly depends on isotopic composition. The nonzero spin of the hydrogen nucleus results in very strong incoherent scattering. For this reason, in lithium hydride, only the frequency distribution function can be measured, but not the dispersion law. In contrast, in LiD (containing 100% D), only coherent scattering takes place (for details, see [25]). However, an addition of a small amount of hydrogen allows one to observe incoherent scattering as well. Scattering of thermal neutrons by LiH(LiD) crystals was investigated in papers of Zemlianov et al. [50] and Verble et al. [51]. Zemlianov et al. studied incoherent scattering. Figure 2.3 presents dispersion curves for lithium deuteride, measured by Veble et al. The accuracy of these data, according to the authors, is 3–4%. Despite great efforts, Verble et al. failed to observe scattering by longitudinal optical phonons. They explained this fact by the very short lifetime of LO phonons (see also Plekhanov [52]).

Very accurate experiments on the dispersion relation were performed for diamond (Fig. 2.4) by Warren et al. [53], for silicon by Dolling [54], and for germanium by Nilsson and Nelin [55]. The numerous calculations (Warren et al. [53], Tubino et al. [56], Bilz et al. [46], and Baroni et al. [57]) of the dispersion relation for these semiconductors show rather good accordance with the experimental ones (see, also Dolling and Cowley [68]), excluding diamond (for details, see [23]).

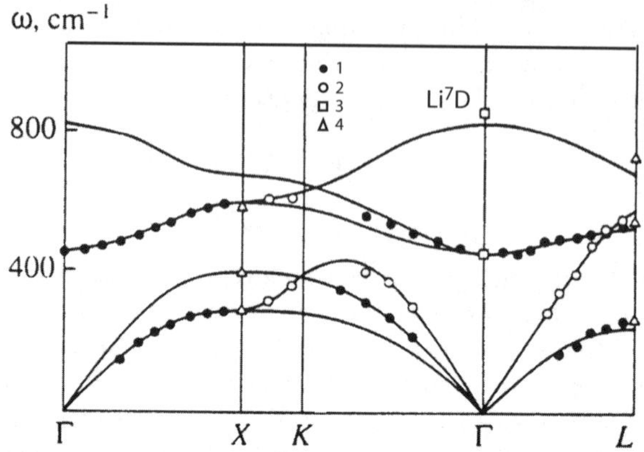

Fig. 2.3. Dispersion of phonons in a ^{7}LiD crystal. (1) and (2) measurements of inelastic neutron scattering by TO and LO phonons, respectively (after Verble et al. [51]); (3) IR measurements (after Brodsky and Burstein [232] and Laplaze [64]); and (4) data on Raman scattering (after Anderson and Luty [77], Tyutyunnik and Tyutyunnik [77], and Plekhanov [27]). *Solid lines* correspond to calculations according to the shell model of Verble et al. [51]. (After [27])

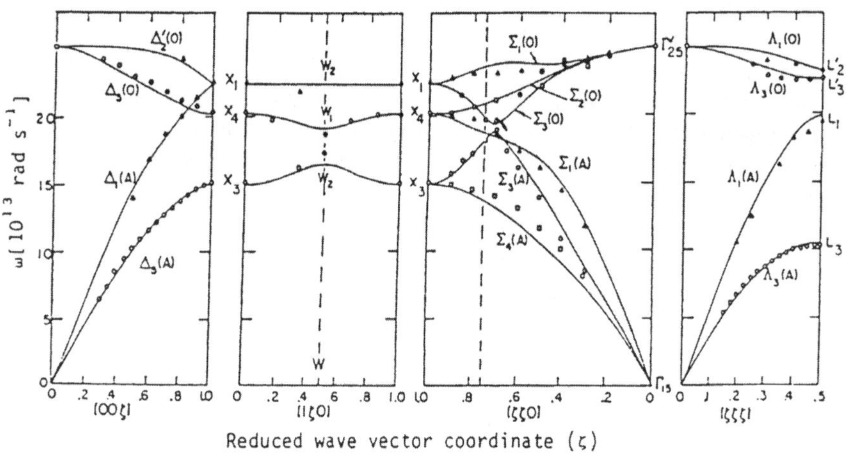

Fig. 2.4. The dispersion relation for the normal modes of vibration of diamond in the principal symmetry directions of diamond at 296 K (after Warren et al. [53]). The full curves represent a shell-model fit to the data points. (After Bilz and Kress [45])

The phonon density of states $g(\omega)$ of a crystalline solid is defined as follows. The function $g(\omega)\mathrm{d}\omega$ signifies the fraction of the total number of phonon states in the frequency interval $(\omega, \omega + \mathrm{d}\omega)$ if $\int g(\omega)\mathrm{d}\omega$ is normalized to 1. Knowledge of $g(\omega)$ is essential for understanding thermodynamic properties, as well as for extraction of the electron-phonon coupling, etc. If is difficult

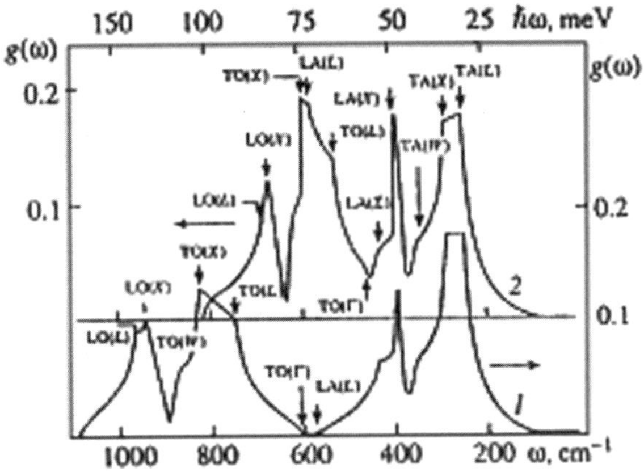

Fig. 2.5. Density of phonon states in (1) LiH and (2) LiD crystals. (After Plekhanov [25])

to make good direct measurements of the density of states, and few spectra actually (e.g., LiH) have been studied. Usually, measurements of incoherent scattering of neutrons reveal only the main peaks. Theoreticians have produced mathematically accurate $g(\omega)$ functions for lattice models in about four decades, and one of the most striking results of these calculations is the existence of kinks in $g(\omega)$ or discontinuities in $\frac{\mathrm{d}g}{\mathrm{d}\omega}$.

A direct comparison between Raman spectra and overtone density of states cannot be done with great precision because the Raman scattering efficiency is a combination of overtone phonon density of states weighted by electron–phonon matrix elements and energy denominators (see, e.g., Cardona [59]). For tetrahedrally coordinated semiconductors, these matrix elements in second-order spectra involving phonon overtones dominate the scattering efficiency and, away from resonance, vary weakly with the Raman frequency shift, thus second-order spectra are a good measure of the overtone density of states. It must be kept in mind, however, that resonant intermediate states will in general distort the shape of Raman spectra away from that of the density of states.

Figure 2.5 shows frequency distribution functions $g(\omega)$ calculated in the shell model for lithium hydride and lithium deuteride crystals (see, e.g., Plekhanov [27]). One can see that spectral distributions of phonons in LiH and LiD are virtually identical in the region of acoustic vibrations ($\leq 450\,\mathrm{cm}^{-1}$). At the same time, the optical band in LiH is broader by a factor of $\sqrt{2}$ than that in LiD. This is explained by the fact that in a crystal lattice containing ions with substantially different masses, heavy ions primarily take part in acoustic vibrations and light ions are involved in optical vibrations. In addition, there is a narrow gap in $g(\omega)$ between 583 and $608\,\mathrm{cm}^{-1}$ in LiH crystals, whereas in LiD, such a gap is absent. Note also that the frequency

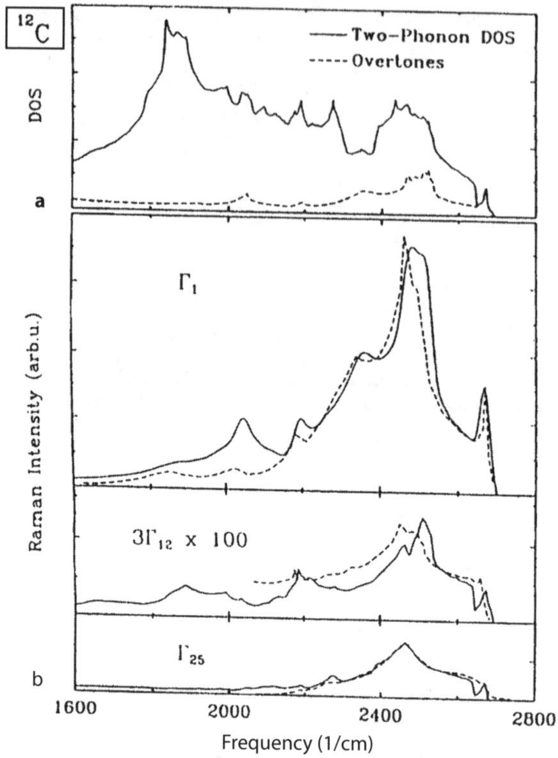

Fig. 2.6. (a) Two-phonon and overtone density of states of diamond^{12}C. **(b)** Second-order Raman spectrum of diamond ^{12}C in the high-frequency region. The *solid lines* represent calculated spectra; the *dashed lines* are experimental from Solin and Ramdas (Solin and Ramdas [99]). (After Windl et al. [66])

of a transverse optical phonon in LiH lies at the band boundary, where the density of states is very low.

As noted above from the appropriate models for different materials, it is relatively straightforward to compare the frequency distribution function $g(\omega)$ for the normal modes of vibrations. As we can see in Figs. 2.6 and 2.7, the curves clearly display sharp critical points arising from regions where the dispersion curves have zero gradient. Many of these can be correlated with features of $\omega(\boldsymbol{q})$ for wave vectors in the principal symmetry directions, although some must be from nonsymmetry direction modes. The two-phonon absorption spectra (Raman spectra) of semiconducting crystals (C; Ge; Si; α-Sn) were compared with the density of the combined vibrational states in the paper of Tubino and co-workers (Tubino et al. [56]). In that paper, it was shown that the comparison holds only in what concerns the peak position and not the intensity ratios, since the joint density of states has not been modified by a thermal factor and the frequency dependence of the coupling

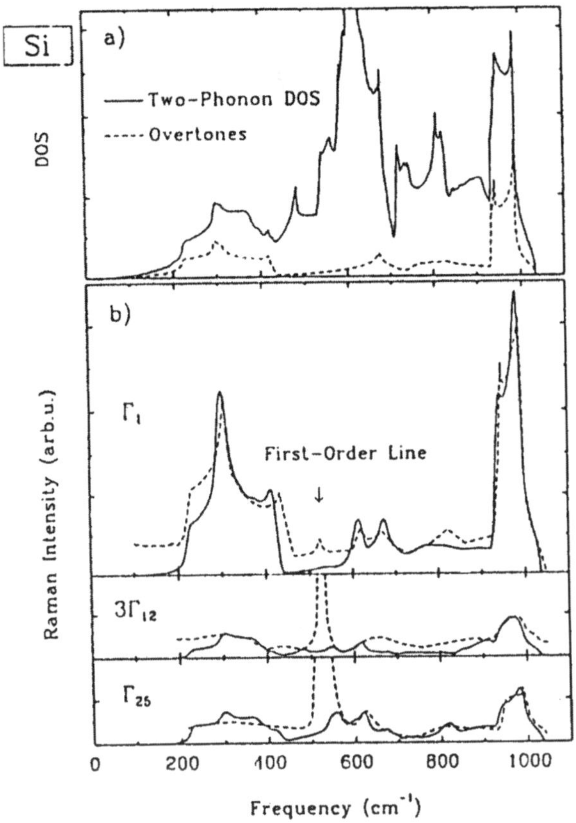

Fig. 2.7. (a) Two-phonon and overtone density of states of silicon. **(b)** Second-order Raman spectrum of silicon. The *solid lines* represent calculated spectra; the *dashed lines* are experimental. (After Windl et al. [66])

Hamiltonian has not been taken into account because of the intrinsic difficulty of a priori knowledge of the second-order dipole moment (see also Lax and Burstein [60]).

In studying Raman scattering due to phonons, one is concerned with the changes in the electronic polarizability of a crystal due to lattice vibrations (Loudon [61]). For small oscillations, the polarizability tensor can be expanded in terms of the ionic displacement about the equilibrium position of the ions. Since a cubic crystal (with NaCl structure; e.g., LiH) has inversion symmetry about every ion, the first-order terms in the polarizability expansion are zero, and hence this crystal does not produce any one-phonon Raman scattering. It should be noted that diamondlike crystals (C, Si, Ge, etc.) have no inversion symmetry and possess first-order Raman scattering [60]. The second-order terms in the expansion give rise to continuous two-phonon

Raman scattering due to all phonons pairs with wave vectors $+\boldsymbol{q}$ and $-\boldsymbol{q}$ allowed by the symmetry of the crystal.

The intensity of Raman scattered light from the system of phonons is given by Born and Huang [1]:

$$I(\omega_{\mathrm{f}}) = \frac{\omega_i^4}{2\pi c^3} \sum_{k=1}^{2} \sum_{\alpha\beta\gamma\delta} \boldsymbol{n}_\alpha^k \boldsymbol{n}_\beta^k i_{\alpha\beta\gamma\delta}(\omega) E_\gamma^- E_\lambda^+ \tag{2.110}$$

with

$$i_{\alpha\beta\gamma\delta}(\omega) = \sum_{v'} \left\{ \langle' | \boldsymbol{P}_{\alpha\gamma}^{\bullet} | v \rangle \langle v | \boldsymbol{P}_{\beta\lambda} | v' \rangle \right\}_{\mathrm{Av}}, \tag{2.111}$$

where $\omega = \omega_i - \omega_{\mathrm{f}}$, with ω_i and ω_{f} the frequencies of incident and scattered light respectively; $\boldsymbol{n}^1$ and $\boldsymbol{n}^2$ are two mutually perpendicular unit vectors in a plane perpendicular to the direction of scattering; E_γ^- and E_λ^+ are the components of the amplitudes of the incident electric field; v and v' respectively, refer to the initial and final vibrational states of system; c is the velocity of light; $\{\}_{\mathrm{Av}}$ means thermal average over the initial states; $\boldsymbol{P}_{\alpha\beta}$ is the electronic polarizability tensor; the subscripts α, β, etc. refer to the three axes of our coordinate system; and ε_{v} is the vibrational energy of the state v of the system. In (2.111), ω_i actually should be ω_{f}, but the small change in ω_i corresponding to the change in the vibrational energy of the system is negligible. By choosing polarization of the exciting light parallel to one of the axes (x, y, z), one can avoid the summation in (2.111). Then, the intensity of the scattered light is essentially determined by $i_{\alpha\beta\gamma\delta}(\omega)$.

The polarizability tensor is expanded in terms of the displacement of the ions about their equilibrium positions. The zeroth-order term gives rise to Raleigh scattering, and the first term is zero because of the inversion symmetry about every ion (excluding crystals with a diamondlike structure). When the displacements are Fourier analyzed in terms of normal coordinates, the second order term in the polarizability expansion becomes

$$\boldsymbol{P}_{\alpha\beta}^{(2)} = \frac{1}{2} \sum_{gjj'} P_{\alpha\beta} \begin{pmatrix} -\boldsymbol{q}\boldsymbol{q} \\ j'j \end{pmatrix} Q \begin{pmatrix} \boldsymbol{q} \\ j \end{pmatrix} Q \begin{pmatrix} -\boldsymbol{q} \\ j' \end{pmatrix} \tag{2.112}$$

with

$$\begin{aligned} \boldsymbol{P}_{\alpha\beta} \begin{pmatrix} -\boldsymbol{q}\boldsymbol{q} \\ j'j \end{pmatrix} &= \sum_{l'kk'} \boldsymbol{P}_{\alpha\beta\gamma\delta} \begin{pmatrix} l & -l' \\ k & k' \end{pmatrix} \sigma_\gamma (k \mid \boldsymbol{q}j) \\ &\times \sigma(k' \mid -\boldsymbol{q}j') (M_k M_{\mathrm{k}'})^{-\frac{1}{2}} \exp\left[2\pi \mathrm{i} \boldsymbol{q}\boldsymbol{a} \begin{pmatrix} l & -l' \\ k & k' \end{pmatrix} \right], \end{aligned} \tag{2.113}$$

where $\mathrm{Q}\begin{pmatrix} \boldsymbol{q} \\ j \end{pmatrix}$ is the normal coordinate corresponding to wave vector $\boldsymbol{q}$ and branch j of the phonon dispersion curves and

$$P_{\alpha\beta\gamma\delta}\begin{pmatrix} l & -l' \\ k & k' \end{pmatrix} = \frac{\partial^2 \boldsymbol{P}_{\alpha\beta}}{\partial \boldsymbol{u}_\gamma (lk)\, \partial \boldsymbol{u}_\delta (l'k')\, |_o} \tag{2.114}$$

is the coefficient of the second-order term in the polarizability expansion with $u_\alpha \begin{pmatrix} l \\ k \end{pmatrix}$ is the α-component of the kth ion in the lth cell; $\boldsymbol{\sigma}(k \mid \boldsymbol{q}j)$ is the eigenvector; M_k is the mass of the kth type of ion; and $\boldsymbol{a}\begin{pmatrix} l & -l' \\ k & k' \end{pmatrix}$ is the position vector of ion $(l'k')$ with respect to ion (lk). When (2.112) is substituted in (2.111) and the thermal average is taken, one gets

$$\begin{aligned} i_{\alpha\beta\gamma\delta}(\omega) = \left(\frac{\hbar^2}{8}\right) \sum_{\boldsymbol{q}jj'} \left[\omega_j(\boldsymbol{q})\, \omega_{j'}(-\boldsymbol{q})\right]^{-1} \cdot \boldsymbol{P}_{\alpha\gamma}\begin{pmatrix} -\boldsymbol{q}\boldsymbol{q} \\ j'j \end{pmatrix} \boldsymbol{P}_{\beta\delta}\begin{pmatrix} -\boldsymbol{q}\boldsymbol{q} \\ j'j \end{pmatrix} f\left[\omega_j(\boldsymbol{q})\right] \\ \times\ f\left[\omega_{j'}(-\boldsymbol{q})\right] \delta\left[\omega \pm \omega_j(\boldsymbol{q}) \pm \omega_{j'}(-\boldsymbol{q})\right], \end{aligned} \tag{2.115}$$

where $\omega_j(\boldsymbol{q})$ is the phonon frequency; $f[\omega_j(\boldsymbol{q})] = n[\omega_j(\boldsymbol{q})] + 1$ for the creation of a phonon and $f[\omega_j(\boldsymbol{q})] = n[\omega_j(\boldsymbol{q})]$ when a phonon is destroyed where $n[\omega_j(\boldsymbol{q})] = \left\{\exp\left[\frac{\hbar\omega_j(\boldsymbol{q})}{kT}\right] - 1\right\}^{-1}$ is the usual Bose factor; and + and − signs in the δ function correspond to the destruction and creation of a phonon, respectively. One can see from (2.115) that the second-order Raman spectrum is a weighted two-phonon DOS (density optical states). For positive and negative values of ω, the spectra are called Stokes and anti-Stokes, respectively. In this book, we consider only Stokes spectra.

If we assume that the polarizability tensor is symmetrical, then the tensor $i_{\alpha\beta\gamma\delta}(\omega)$ has the same transformation properties as the elastic constant tensor (see above, Sect. 2.1 of this chapter). Cubic LiH(LiD) crystals possess three different components of scattering tensor, $i_{11} = i_{1111}$, $i_{12} = i_{1212}$, and $i_{44} = i_{1122}$. It is known from group theory (see, e.g., Birman [62]) that components of $\boldsymbol{P}_{\alpha\beta}$ of the second-rank tensor are transformed over three irreducible representations, A_{1g}, E_g, and T_{2g} of the O_h^5 group. In nonpolarized scattering, all three components are present with certain weights. However, by choosing the experimental geometry of scattering propery, one can separate the E_g and T_{2g} components of the scattering tensor. The intensities of these components are described by the expressions (see also review [25])

$$\begin{aligned} I(A_{1g}) &\sim \left(\frac{A_{xx} + A_{yy} + A_{zz}}{\sqrt{3}}\right)^2, \\ I(E_g) &\sim \left[\left(\frac{A_{xx} + A_{yy} - 2A_{zz}}{\sqrt{6}}\right)^2 + \left(\frac{A_{xx} + A_{yy}}{\sqrt{2}}\right)^2\right], \\ I(T_{2g}) &\sim \frac{(A_{xy}^2 + A_{yz}^2 + A_{zx}^2)}{3}. \end{aligned} \tag{2.116}$$

In this case,

$$
\begin{aligned}
i_{1111}(\omega) &= \frac{\left[I(A_{1g}) + \frac{I(E_g)}{4} 3I(T_{2g})\right]}{3} = AET, \\
i_{1212}(\omega) &= \frac{I(E_g)}{4} = E_g, \\
i_{1313}(\omega) &= I(T_{2g}) = T_{2g}, \\
i_{3333}(\omega) &= \frac{[I(A_{1g}) + I(E_g)]}{3} = AE.
\end{aligned}
\tag{2.117}
$$

One can see from (2.117) that theoretical calculation of the Raman spectrum consists of the calculation of three tensor components i_{11}, i_{12}, and i_{44}. Such calculations were done in the paper of Jaswal et al. [63]. In the paper cited, the dynamic lattice model used for LiD and LiH was a deformation dipole model. This model is based on neutron scattering data for ^{7}LiD (Verble et al. [51]) and experimental second-order Raman spectra of LiD. Jaswal et al. [63] showed that if the contributions to polarizability from nearest and next-to-nearest neighbors are taken into account, one can select the values of tensor components $\boldsymbol{P}_{\alpha\gamma}$ that allow one to reproduce Raman spectra with good accuracy for different polarizations (see also [27, 64]).

The DOS of diamond and Si was calculated in papers of Pavone et al. [65] and Windl et al. [66]. The eigenvectors and eigenvalues that are necessary to evaluate $\boldsymbol{P}_{\alpha\beta}$, (2.113) have been calculated in the papers cited using the local-density approximation (LDA) plane-wave pseudopotential method (for details, see also [67]). The results for second-order Raman spectra for diamond ^{12}C and silicon are shown together with experimental curves in Figs. 2.6 and 2.7.

To show that the origin of the sharp peak at the high-frequency cutoff is not an effect of the matrix elements, Windl et al. [66] also fitted the second-order Raman spectrum in its three representations for silicon. The results of these calculations are shown in Fig. 2.7. Again, Pavone et al. [65] found good agreement between calculation and experiment for all representations. The positions of the calculated peaks with experimental values are within generally less than 2% (see also [23]). As expected from phonon dispersion (see Fig. 2.4) and the high-frequency density of states, there is no sharp peak at the high-frequency cutoff. So, these results give strong evidence that the long-standing controversy seems to be resolved.

2.3.2 Low Concentrations: Localized, Resonant, and Gap Modes

Studies of vibrational properties of crystals containing impurities (defects of various types) were described in detail in a number of reviews (Maradudin et al. [7], Dean [68], Bell [69], Taylor [9], Barker and Sievers [8], Chang and Mitra [70], Elliott et al. [71], Taylor [72], and Lifshitz [11]). These studies are traditionally classified under two regions corresponding to low ($x < 10\%$) and high ($x \geq 10\%$) concentrations of defects (impurities). The latter region, which corresponds to so-called mixed crystals, will be considered below. Here, we will analyze the results obtained in the region of low concentrations, when the interaction between defects (impurities) can be neglected in the first approximation (see, e.g., [8]). The simplest defects in a crystalline lattice that distort its translational symmetry are isotopes of the elements that form a crystal. In most cases, the effect of isotopes on vibrational spectrum is small and can be neglected in calculations of the dispersion curves (this picture is changed when we analyze thermal conductivity). However, in the LiH–LiD system, the situation is different. The great change in mass upon replacement of H by D results in a shift of the entire optical vibrational band by $\sim 250\,\mathrm{cm}^{-1}$. In low concentrations, the main effects consist of the appearance of various processes induced by defects. For example, in the harmonic approximation, the IR absorption spectrum of an ideal lattice consists of a single delta-shaped peak at a frequency of ω_{TO} (Γ), and the introduction of defects lifts the selection rule over the wave vector, which results in the appearance of additional absorption in the entire frequency region (whose intensity is proportional to the concentration of defects). Lifshitz [73] developed the theory for extremly low concentrations, which does not contain free parameters an isotopic defect. The method developed by Lifshitz is based on the technique of Green's functions. This method is especially convenient because it allows one to uniformly write the solution of a vibrational equation in the presence of point defects of different types. A change in the atomic mass results in the appearance of three features in the phonon spectrum [73, 74]: localized, gap, and quasi-localized (resonant) modes. Localized modes appear upon a sufficiently strong decrease in atomic (ion) mass, and their frequencies lie above the maximum frequency of the phonon spectrum [see (2.68)]. Frequencies of gap modes lie in the gap between optical and acoustic bands. These modes can appear both upon decreasing and increasing atomic mass. Resonant modes appear if the defect mass greatly exceeds the mass of the substituted ion. These modes are manifested as sharp maxima of the frequency distribution function in the $\omega \ll \omega_{\max}$ region.

For lithium deuteride, a condition of the appearance of a localized mode upon replacement of D^- ions by H^- ions has the form $\frac{M'}{M} = 0.66$ [25] (for the phonon spectrum in the shell model). Therefore, the introduction of H^- ions into a LiD crystal should result in the appearance of the localized mode. Figure 2.8 shows the second-order Raman spectrum of a pure LiD crystal at room temperature (Plekhanov and Veltri [75]). First of all, note that this spectrum agrees well with the spectra measured earlier (Jaswal et al. [76] and Anderson and Luty [77]). Although, according to a nomogram of excitonic states (see, e.g., [78]), this crystal is a pure LiD crystal, its Raman spectrum nevertheless exhibits a high-frequency peak at $1850\,\mathrm{cm}^{-1}$ (that makes this method, on the other hand, very useful for estimating isotopic impurities in

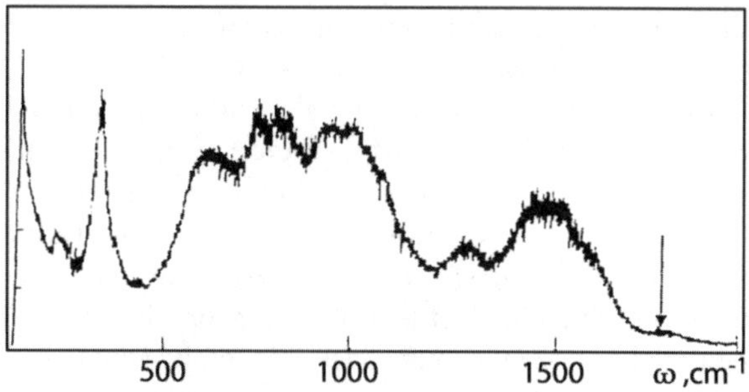

Fig. 2.8. Second-order Raman spectrum of a LiD crystal excited at $\lambda = 532\,\text{nm}$ at room temperature. (After Plekhanov and Veltri [75])

crystals). This peak has no analog in the Raman spectra of a pure LiH crystal. Although the concentration of H^- in the LiD crystals under study is very low (according to [78], the position of the n = 1S line in the reflection and luminescence spectra corresponds to a pure LiD crystal), it is nevertheless sufficient for observing of the localized mode in the second-order Raman spectra (Fig. 2.8). The intensity of the peak of the localized mode increases with increasing concentration of hydrogen ions (the concentration was estimated by Plekhanov and Veltri from the position of the ground-state level of excitons in the reflection spectra), and it slightly [78] shifts to the blue (Fig. 2.9).

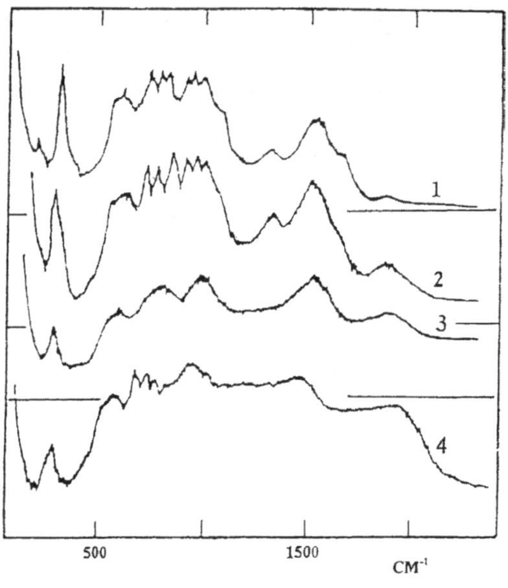

Fig. 2.9. Second-order Raman spectra of LiH_xD_{1-x} crystals at room temperature: $x = 0(1); 0.05(2); 0.12(3)$; and $0.35(4)$. (after Plekhanov [25])

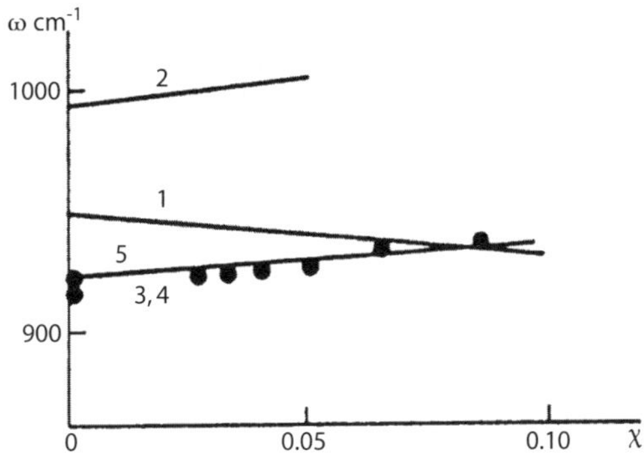

Fig. 2.10. Dependence of the localized mode frequency on the concentration of hydrogen ions. Data from Elliott et al. [71] – (1); Jaswal et al. [76] – (2); Jaswal et al. [63] – (3) and Anderson and Luty [77] – (4) Experimental data from Plekhanov and Veltri [75] and Plekhanov [25]. (After [27])

This behavior is retained up to a concentration of hydrogen ions $x \leq 10\%$. Note here that the study of the concentration dependence of the shape of this peak allows one to examine the percolation threshold, which separates the local states from continuous phonon states (Anderson [79], Belitz and Kirkpatrick [80], and Plekhanov [81]). Figure 2.10 shows the dependence of the frequency of the localized mode on the concentration of hydrogen ions obtained from the Raman spectra presented in Figs. 2.8 and 2.9 and spectra of other crystals (Plekhanov and Veltri [75] and Plekhanov [81]). Figure 2.11 also shows the theoretical dependence $\omega_{\text{loc}} \sim f(x)$ calculated in the liter-

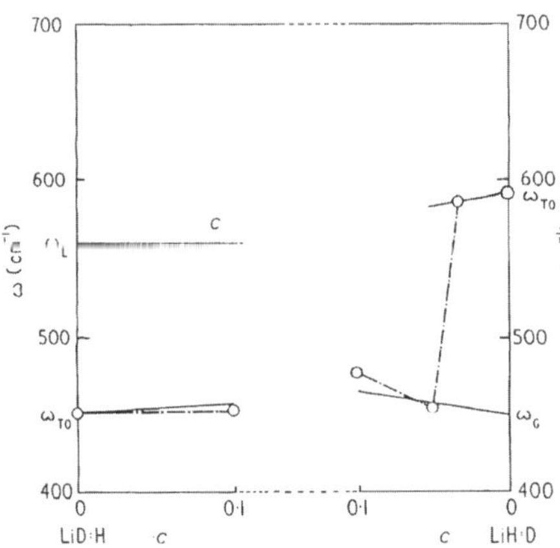

Fig. 2.11. The calculated concentration dependence (*full lines*) of ω_{TO}, ω_{g}, and ω_{L} together with experimental data (*chain lines*) for LiH–LiD systems. (After Behera and Tripathi [82])

Table 2.2. Frequencies of optical and acoustic phonons (in cm^{-1}) calculated at different points of the Brillouin zone (Verble et al. [51] and Plekhanov [25])

Substances	^{7}LiH					^{7}LiD				
Phonon branches	Γ	X	W	L	K	Γ	X	W	L	K
LO	1085	944	889	968	913	880	672	633	689	651
TO	608	825	847	749	808	444	589	604	535	578
LA	0	399	366	580	387	0	285	338	251	278
TA	0	291	336	248	279	0	285	338	251	278

ature (for details, see [27]). These data show that the model of deformable dipoles quite adequately describes the behavior of the localized mode, which is related to vibrations of hydrogen ions in lithium deuteride crystals.

Upon the the introduction of deuterium ions into a LiH crystal, the gap mode appears, whose frequency, in the model of deformable dipoles, is $\omega_{\mathrm{g}} = 576\,\mathrm{cm}^{-1}$. The shell model yields almost the same value of ω_{g}. One can see from Table 2.2 that this frequency is close to the IR-active frequency ω_{TO} in LiH. For this reason, it is unlikely that the gap mode can be observed at very low concentrations of deuterium ions. However, the gap mode has a number of interesting properties that should be manifested with an increasing concentration of deuterium ions. The consideration of a change in the force constant in the calculation of the localized mode upon introducing D into LiH allowed Behera and Tripathi [82] to achieve agreement with experimental data (Montgomery and Yeung [83]) in the region of low concentrations ($c_{\mathrm{D}} \leq 0.1$). For this purpose, ω_{TO} was renormalized according to the equation

$$\omega_{\mathrm{TO}} = \omega_{\mathrm{TO}}^{0}\left(1 + \frac{C}{2\pi}\frac{4\lambda\tau}{[\tau - \lambda]}\right), \tag{2.118}$$

where $\lambda = (M' - M/M')$ is the mass defect parameter and $\tau = \frac{(\Phi - \Phi_0)}{\Phi_0} = \frac{\Delta\Phi}{\Phi_0}$ is the parameter describing the change in the force constant between the defect and lattice atoms. The agreement between the theory and observation of low c_{D} (LiH:D) was obtained for $\lambda = 0.5$ and $\tau = 0.21$ (Fig. 2.11); LiH:D, these parameters were 1.0 and 0.21, respectively. This shows that an actual change in the force constant upon replacing the hydrogen ion by the deuterium ion, which is smaller, is quite large.

2.3.3 Phonon Spectra of Isotopically Mixed Crystals

If the impurity concentration in a crystal is high enough that the interaction between impurity atoms (ions) plays an important role, such a system is called a mixed crystal with various degrees of disorder. There are two types of disordered systems: disordered alloys (isotopic mixtures) or mixed crystals and glassy substances, which possess a more pronounced spatial disorder than configurational disorder. The first theoretical dynamic model of mixed crystals was a linear chain, which represented the development of the virtual model (Nordheim; Pant and Joshi [84] Ipatova [85]). Despite its simplicity, this model adequately described general features of lattice dynamics of mixed alkali halide crystals. This model uses two independent force constants, f_0 and f_0', which are obtained, as a rule, from the observed frequencies of LO phonons in pure substances, according to the expression $f = \omega^2 mM/2\,(m+M)$, where m and $M(M')$ are masses of crystal-forming particles. The dependence of the force constants on concentration was described by the equation $F = f_0'x - (f_0' - f_0)x$ by assuming a linear dependence of $f_0\,(f_0')$ on concentration x (see also Chang and Mitra [70]). A more complex concentration dependence of the force constant was considered in detail in comprehensive reviews (Barker and Sievers [8], Elliott et al. [71], Ipatova [85], and Taylor [72]), where the cluster model and isodisplacement model in lattice dynamics, based on the CPA (coherent potential approximation) (Taylor [86] and Ehrenreich and Schwartz [87]) or averaging of the T-matrix (Taylor [9, 72]), were also described.

Elemental semiconductors (C, Si, Ge, α-Sn) with diamondlike structures are ideal objects to study isotopic effects by the method of Raman scattering. At the present time, high-quality isotopically enriched, labeled crystals are also available. In this section, we describe our understanding of first-order Raman spectra of isotopally-mixed elements.

The materials with a diamond structure discussed here have a set of threefold-degenerate phonons (frequency ω_0) at the center ($\boldsymbol{k} = 0$, Γ-point) of the Brillouin zone (BZ). These phonons are Raman active but infrared inactive (Lax and Burstein [60]). Let us consider Ge with five isotopes (Agekyan et al. [88] and Fuchs et al. [89]). The uninitiated will ask whether one should see five phonons [or more knowing that there are two atoms per primitive cell (PC)] corresponding to the five different masses, or only one corresponding to the average mass. The reason that the Raman spectrum (see Fig. 2.12) of natural Ge does not show the local modes of the individual isotopes is that the scattering potentials for the phonons due to mass defects (mass fluctuations) are too small to induce bound states, i.e., Anderson localization of the phonons (Anderson [79]). In a three-dimensional crystal, fluctuations in the parameters of the secular equation lead to localization if these fluctuation (measured in units of frequency, i.e., $\left(\frac{\Delta M}{M}\right)\omega_0$) are larger than the bandwidth of the corresponding excitations. For optical phonons in Ge, this bandwidth is $\sim 100\,\mathrm{cm}^{-1}$ (see, e.g., Etchegoin et al. [91]) whereas $\left(\frac{\Delta M}{M}\right)\omega_0 \le 0.4 \cdot 300 = 12\,\mathrm{cm}^{-1}$. Hence, no phonon localization (with lines corresponding to all pairs of masses) is expected, in agreement with the observation of only one line at $304\,\mathrm{cm}^{-1}$ ($\sim 80\,\mathrm{K}$) for natural Ge.

Figure 2.12 shows superimposed Raman lines of natural and of isotopically enriched Ge single crystals recorded near LNT (liquid nitrogen temperature) (Fuchs et al. [90]). The lines in these spectra are fully resolved instrumentally (the experimental resolution was better than $0.1\,\mathrm{cm}^{-1}$ (Cardona et al. [92]),

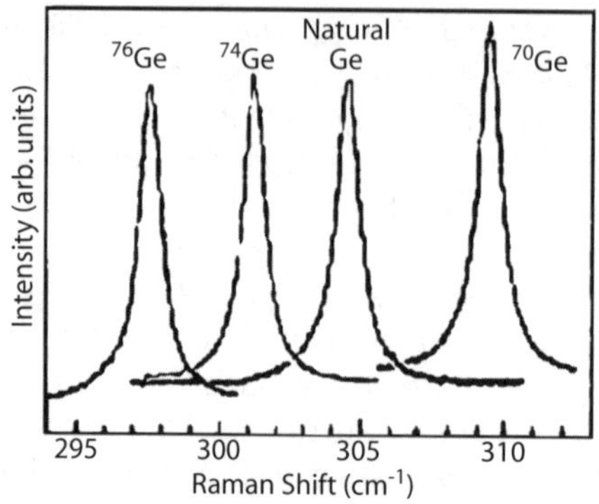

Fig. 2.12. First-order Raman spectra of the L phonon of several Ge isotopes and natural Ge at 80 K. (After Fuchs et al. [90])

and their widths are caused by homogeneous broadening. The centroid of the Raman line shifts follows the relation $\omega_0 \sim M^{-\frac{1}{2}}$. This behavior is expected within harmonic approximation. Additional frequency shifts are observed (Zhang et al. [93]) for natural and alloy samples which arise from their isotope mass disorder. This additional shift is $0.34 \pm 0.04\,\mathrm{cm}^{-1}$ in natural Ge and $1.06 \pm 0.04\,\mathrm{cm}^{-1}$ (Fig. 2.13) in the $^{70/76}$Ge alloy sample, which has nearly the maximum isotopic disorder possible with natural isotopes.

As is well known, natural diamond exhibits a single first-order Raman peak at $\omega_{\mathrm{LTO}}(\Gamma) = 1332.5\,\mathrm{cm}^{-1}$. Figure 2.14 shows the first-order Stokes

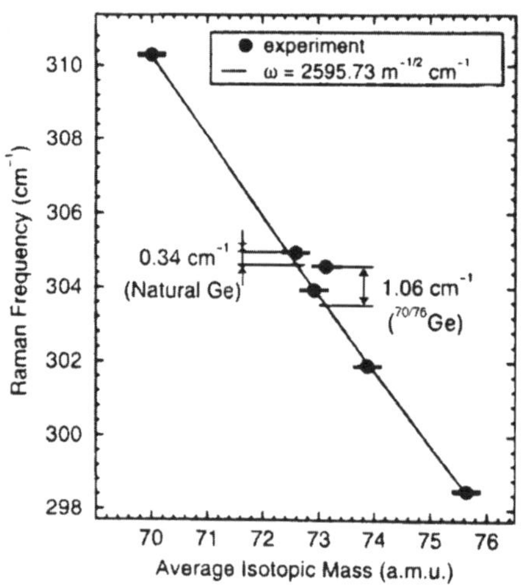

Fig. 2.13. Raman frequency as a function of average mass, measured at 10 K, for isotopically enriched and disordered Ge samples. The *solid line* is a calculation with $\omega = \frac{2595.73}{\sqrt{M}}\,\mathrm{cm}^{-1}$. (After Zhang et al. [93])

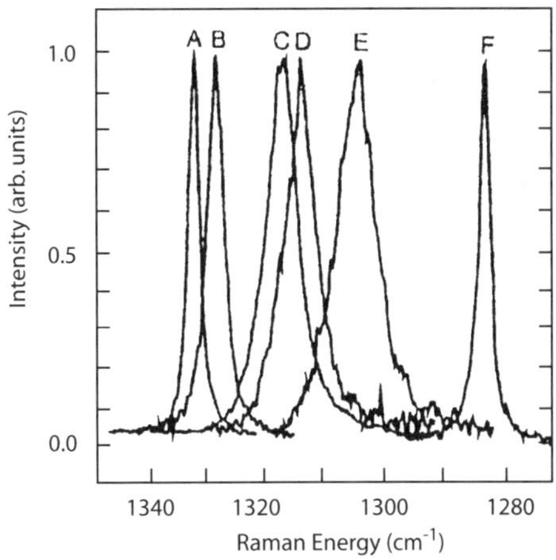

Fig. 2.14. First-order Raman spectra of $^{12}C_x^{13}C_{1-x}$ diamonds with different isotopic compositions. The labels A, B, C, D, E, and F correspond to $x = 0.989$, 0.90, 0.60, 0.50, 0.30, and 0.01, respectively. The intensity is normalized at each peak. (After Hanzawa et al. [94])

Raman spectra for several samples with different isotopic ratios (Hanzawa et al. [94]). The Raman energy increases continuously, but nonlinearly, with decreasing x. The energy difference between the extreme compositions is $52.3\,\mathrm{cm}^{-1}$, which is consistent with the isotopic mass ratio. Additional Raman data on isotopically mixed polycrystalline diamond films (Chu et al. [95]) are not discussed here because of the numerous extrinsic factors (e.g., stress, impurities) (Knight and White [96]) that can obscure the intrinsic composition dependence in such films. For example, the linear Raman variation assumed by Chu et al., suggests that the x values determined in that work may be in error by as much as 0.1 (for details, see Hass et al. [97]). The Raman spectra and their explanation of the other elemental and compound semiconductors can be found in published reviews (see, e.g., [27]). Thus, depicted in the figures are experimental results that show the nonlinear dependence frequency shift of the first Raman lines on the isotopic concentration.

The second-order Raman spectra for natural and isotopically-mixed crystals of diamond were investigated by Hass et al. [97] and Chrenko [98]. Second-order Raman spectra for synthetic diamonds are shown in Fig. 2.15. The second-order spectra were measured by Hass et al. with slightly lower resolution ($\sim 4\,\mathrm{cm}^{-1}$) than the first-order spectra because of the much lower count rate. The results of Hass et al. for 1.1 atom% ^{13}C agree well with previous measurements for natural diamond (Solin and Ramdas [99]). The spectra for 0.07 and 99 atom% ^{13}C also look similar, if one ignores the shifts that occur as a result of differences in M. More significant differences are observed for the more heavily mixed crystals: the 34.4 and 65.7 atom% ^{13}C results are no-

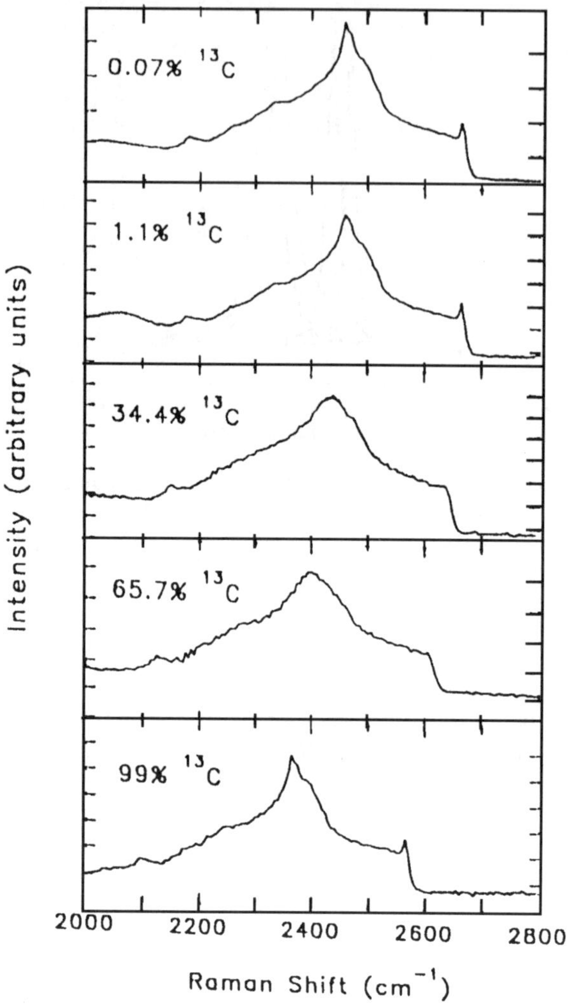

Fig. 2.15. Second-order Raman spectra for synthetic diamond with labeled compositions at room temperature. (After Hass et al. [97])

ticeably broader and do not exhibit the sharp peak near the high-frequency cutoff. As was shown above, this peak at the top of the second-order spectrum (2667 cm^{-1} for 1.1 atom% ^{13}C) has been the subject of intense controversy. Chrenko [98] also examined the second-order spectra of his samples and claims that he was able to see this peak at all compositions except 68% ^{13}C. His measurements may have been of somewhat higher resolution than in the paper of Hass et al., but it is clear that even in his 89 atom% ^{13}C spectrum (which are the only raw data presented), some broadening of this peak has occurred (for details, see [97]).

The IR absorption in mixed crystals can change in two ways, depending on the concentration: one-mode and two-mode (see, for example, the review by Elliott et al. [71]). In one-mode behavior, the spectrum always exhibits a single band whose maximum gradually shifts from one extreme position to another. The two-mode behavior corresponds to the presence of two bands in the spectrum, which are characteristic for each of the components of a mixed crystal. As the concentration of components changes, these bands shift, and their intensities undergo strong redistribution. In principle, the same system can exhibit different types of behavior at the opposite ends (Barker and Sievers [8]). This classification is only qualitative, and it is seldom realized in its pure form (Chang and Mitra [70]). The appearance of the localized mode in the limit of the isolated defect is considered the most important necessary condition for the two-mode behavior of phonons [and also electrons (Ipatova [85])]. In the review by Elliott et al. [71], a simple quantitative criterion was suggested for determining the type of behavior of the IR absorption in a crystal of the NaCl type (see also Ipatova [85]). Because the square of the frequency of the TO (Γ) phonon is inversely proportional to the reduced mass of the unit cell M, the shift caused by the defect is

$$\Delta = \omega_{\mathrm{TO}}^2 (1 - \bar{M}/\bar{M}'). \tag{2.119}$$

This shift is compared in the paper of Elliott and others with the width of the phonon optical zone. This width in the parabolic dispersion approximation, neglecting acoustic branches, is

$$W = \omega_{\mathrm{TO}}^2 \frac{\varepsilon_0 - \varepsilon_\infty}{\varepsilon_0 + \varepsilon_\infty}. \tag{2.120}$$

The localized or gap mode appears, provided $|\Delta| > \frac{W}{2}$. However, as noted by Elliott et al. [71], for the two peaks to be retained up to a concentration of about 0.5, the stricter condition of $|\Delta| > W$ should be satisfied. The substitution of numerical values in (2.119) and (2.120) shows that the relation

$$|\Delta| > \frac{1}{2} W \tag{2.121}$$

for LiH (LiD) is always valid because $|\Delta| = 0.44\,\omega_{\mathrm{TO}}^2$ and $W = 0.58\,\omega_{\mathrm{TO}}^2$. This means that the localized mode should be observed at low concentrations. This conclusion agrees with the experimental data described above (Fig. 2.9). As for the second theoretical relation $|\Delta| > W$, as noted above, for LiH (LiD) crystals, the reverse relation $W > |\Delta|$ is always valid (see, e.g., [25]). We will consider this question in more detail after a discussion of Raman spectra of mixed $\mathrm{LiH}_x\mathrm{D}_{1-x}$ crystals at high isotopic concentrations. Figure 2.16 shows the second-order Raman spectra of mixed $\mathrm{LiH}_x\mathrm{D}_{1-x}$ crystals at room temperature [81]. Along with the properties of Raman spectra at high concentrations discussed in the review of Plekhanov [25], note also that as the hydrogen concentration further increases ($x > 0.15$), the intensity

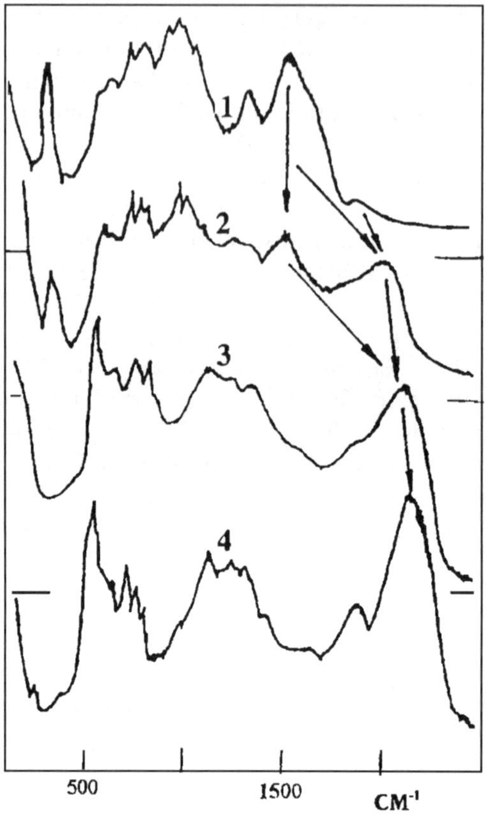

Fig. 2.16. Second-order Raman spectra of mixed LiH_xD_{1-x} crystals excited at $\lambda = 448.0\,nm$ at room temperature: spectra 1, 2, 3, and 4 for $x = 0.0$, 0.42, 0.76, and 1.0, respectively. The arrows show the bands corresponding to LO(Γ) phonons. (After [27])

of the 2LO (Γ) phonon peak in a LiD crystal decreases, whereas the intensity of the highest frequency peak in mixed LiH_xD_{1-x} crystals increases. The latter peak is related to the renormalized LO (Γ) modes in a mixed crystal. Thus, comparison of Raman spectra 1 and 2 in Fig. 2.16 shows that in the concentration range of $0.1 < x < 0.45$, the Raman spectrum exhibits LO (Γ) phonon peaks of a pure LiD and mixed LiH_xD_{1-x} crystal. A further increase in $x > 0.45$ is accompanied by two effects observed in the Raman spectra of mixed crystals. The first effect is manifested in a substantial rearrangement of the acousto-optical part of the spectrum (spectra 1–3 in Fig. 2.16), and the second consists of a further blue shift of the highest frequency LO (Γ) phonon peak. This peak shifts to the position of peak 12 in the spectrum of a pure LiH crystal [23]. This is most clearly seen by spectra 2 and 4 in

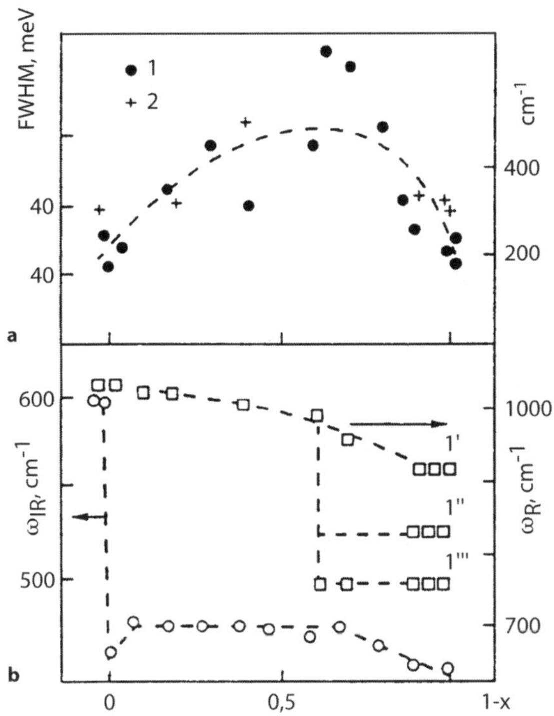

Fig. 2.17. Dependence of **(a)** the half-width *a* **(b)** frequency of optical phonons on the isotope concentration in the second-order Raman spectra of mixed $LiH_{1-x}D_x$ crystals. **(a)** (1) and (2) λ_{exc} = 253.7 and 488.0 nm, respectively. **(b)** (1) (Plekhanov [25]) and (2) data on IR absorption spectra for TO(Γ) phonons in thin LiH_xD_{1-x} films (Montgomery and Hardy [100]). (After [27])

Fig. 2.16. Note that the resonance increase in the intensity of the highest frequency peak is observed in all mixed crystals for $x > 0.15$ [27].

The dependence of the frequency of the highest frequency peak in Raman spectra of pure and mixed crystals on the concentration of isotopes is presented in Fig. 2.17 (curve 1). Figure 5.1 also shows the concentration dependence of the frequency of the TO (Γ) phonon measured from IR absorption spectra of thin LiH_xD_{1-x} films (Montgomery and Hardy [100]) (curve 2). One can see that both these dependences are nonlinear. Curve 2 was already discussed before in the analysis of the concentration dependence of the frequency and intensity of the localized mode. Note here that the concentration dependence in a broad range has not yet been self-consistently described by Tripathi and Behera [101]. It should be emphasized once more that the theory is in good agreement with experimental data in the region of low concentrations of isotopes but is not the case for high concentrations. One can see from Fig. 2.17 that for concentrations of $x < 0.45$, the LO (Γ) exhibit the two-mode behavior. This contradicts the prediction of the CPA (Elliott et al. [71] and Elliott and Leath [102]), according to which the width W of optical vibration bands should be smaller than the frequency shift (Δ) of the transverse optical phonon. However, calculations according to (3.30) and (3.31) show that in LiH(LiD) crystals, the reverse inequality takes place, i.e., $W > |\Delta|$. According to Plekhanov [81], this discrepancy between experimental results and the theory based on CPA ([102]) is explained mainly by the strong potential of scattering of phonons, caused by a large change in mass upon substitution of deuterium for hydrogen. This is also confirmed by the observation of the localized mode in these systems (see above).

2.3.4 Isotopically Induced Disorder Effects in Vibrational Spectra

As described in preceding sections, isotopically enriched samples of Ge show a frequency inversely proportional to the square root of the average mass (see Fig 2.13) and a linewidth inversely proportional to the mass (Fig. 2.18), in agreement with the harmonic approximation and Fermi's golden rule, respectively. However, in natural Ge and $^{70/76}$Ge alloy isotopes, with a Gaussian disorder, there is an additional shift of the phonon frequency and a broadening of the linewidth. This additional shift is $0.34 \pm 0.04\,\text{cm}^{-1}$ in natural Ge and $1.06 \pm 0.4\,\text{cm}^{-1}$ in the $^{70/76}$Ge alloy sample, which has nearly the maximum isotopic disorder possible with natural isotopes. Recent single-site CPA calculations (Fuchs et al. [89] and Cardona et al. [92]), based on the density of states obtained within a harmonic approximation, predict that the disorder-induced frequency shift of the Γ-point optical phonon amounts to

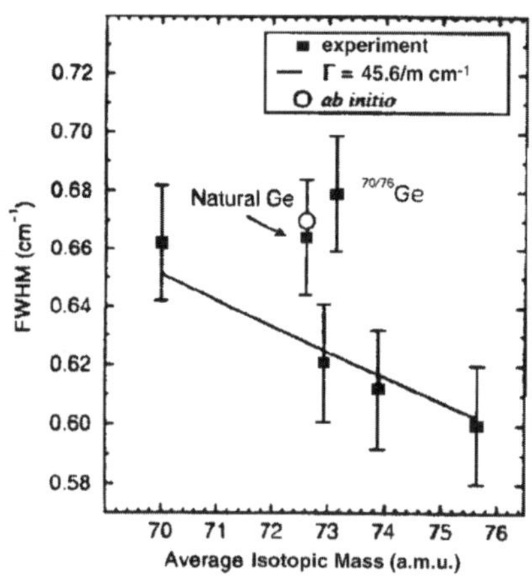

Fig. 2.18. Intrinsic phonon linewidth of isotopic Ge vs average mass, measured at 10 K, with an excitation of 6471 Å. The ab initio linewidth reported by Debernardi and co-workers (Debernardi et al. [92]) for isotopically pure Ge with $M = 72.6$ is indicated by a circle. (After Zhang et al. [93])

$(0.4 \pm 0.1)\,\mathrm{cm}^{-1}$ for natural Ge and $(1.2 \pm 0.1)\,\mathrm{cm}^{-1}$ for the alloy sample, depending on the dynamic lattice model used to obtain the phonon density of states. A simple estimate from second-order perturbation theory can also provide a reasonable prediction of these disorder-induced frequency shifts (Menendez et al. [103]). The disorder-induced shift is given by

$$\Delta\omega = g_2 \frac{\omega^2}{12} \int_0^\infty \frac{1}{\omega^2 - \omega_i^2} N_\mathrm{d}(\omega_i)\,\mathrm{d}\omega_i \tag{2.122}$$

where the phonon density of states (Nelin and Nilsson [104]) is normalized through $\int_0^\infty N_\mathrm{d}(\omega_i)\,\mathrm{d}\omega_i = 6$. The mass-fluctuation parameter g_2 is given by

$$g_2 = \sum_i x_i \left(1 - \frac{M_i}{\bar{M}}\right)^2, \tag{2.123}$$

where x_i is the fraction of isotope i, M_i its mass, and $\bar{M}$ is the average mass. This equation yields $g_2 = 5.87 \cdot 10^{-4}$ for natural Ge and $g_2 = 1.53 \cdot 10^{-3}$ for the alloy, respectively. Using (2.122), Zhang et al. [93] obtain a disorder-induced frequency shift of $0.41\,\mathrm{cm}^{-1}$ for the natural sample and $1.07\,\mathrm{cm}^{-1}$ for the $^{70/76}$Ge alloy sample, respectively.

An additional broadening [$(0.03 \pm 0.03)\,\mathrm{cm}^{-1}$ for natural Ge and $(0.06 \pm 0.03)\,\mathrm{cm}^{-1}$ for the alloy sample] is clearly seen (see Fig. 2.18) in the two isotopically mixed samples. Their linewidths lie significantly above those expected from the inverse-average-mass rule. This can be attributed to a combination of disorder-induced scattering and anharmonic decay (see also Fuchs et al. [89]). Including the anharmonic broadening in the phonon density of states $N_\mathrm{d}(\omega)$ yields a nonvanishing scattering probability on the mass defects. One can thus estimate the mass disorder broadening of the optical phonon with the expression (Cardona et al. [106])

$$\Gamma_\mathrm{isotope} = g_2 \frac{\pi\omega^2}{12} N_\mathrm{d}(\omega)\,. \tag{2.124}$$

This calculation gives an additional broadening of $0.017\,\mathrm{cm}^{-1}$ for natural Ge and $0.046\,\mathrm{cm}^{-1}$ for the $^{70}\mathrm{Ge}_{0.5}\,^{76}\mathrm{Ge}_{0.5}$ alloy, in agreement with the measured data. From these results, it can be concluded that the maximal shift possesses LO(Γ) phonons but has broadening acoustic and optical phonons with $\boldsymbol{q} \neq 0$ (see, however, Gobel et al. [105]).

Figure 2.19 compares the composition dependence of the Raman frequency of diamond in the VCA (virtual crystal approximation) and CPA according to Hass et al. [97] and Spitzer et al. [107]. The Raman data in Fig 2.19a are in excellent agreement with those of Chrenko [98] and Hanzawa et al. [94]. Both sets of data exhibit a pronounced bowing (nonlinearity) relative to the VCA that is described very well by CPA. Hass et al. concluded that the bowing is a direct consequence of scattering due to isotopic disorder. Similar nonlinearities are observed in many other properties of alloy

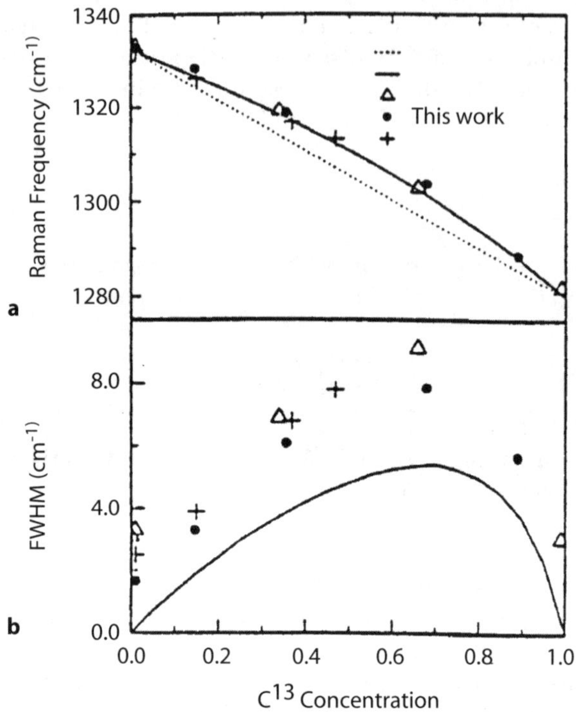

Fig. 2.19. Isotopic composition dependence of the **(a)** diamond Raman frequency and **(b)** linewidth (FWHM). CPA results for pure and adjusted Tubino et al. models and VCA results in **(a)** shown as *dashed*, *solid*, and *dotted lines*, respectively. Experimental data as indicated. Chrenko's data are taken from Chrenko [98]. (After Hass et al. [97])

systems (e.g., the band gaps of semiconductor alloys (Efros and Raikh [108] and isotopically-mixed crystals (Plekhanov [109]). The deviation from linearity is approximately 5 cm^{-1} near the middle of the composition range. This is much larger than the experimental uncertainties (about the size of the data points) and should certainly be considered if the Raman frequency is to be used as a measure of isotopic composition (Hass et al. [110]).

The measured Raman linewidths (Fig. 2.19b) are larger near the center of the composition range than near the end points. The variation is not symmetrical in x and $(1-x)$ and (as in LiH, see Fig. 2.17) the maximum width occurs at approximately 70 atom% ^{13}C. The CPA curves represent intrinsic contributions to the Raman linewidth due to the disorder-induced broadening of the zone center optic mode. The observed widths, according Hass et al. [97], contain additional contributions due to instrumental resolution ($\sim 1.8\,\mathrm{cm}^{-1}$) and anharmonic decay (Fuchs et al. [89] and Wang et al. [111]).

The anharmonic broadening of the Raman line calculated for diamond by Wang et al. [111] is of the order of $1\,\mathrm{cm}^{-1}$ at 300 K. Contributions other than disorder thus account well for the observed widths near $x = 0$ and 1. Assuming that such contributions are reasonably constant across the entire composition range, we see that both CPA calculations account very well for the qualitative trend in the data, including the peak near $x = 0.7$. The pure valence force-field model (Tubino et al. [56]) underestimates the magnitude of the variation, however, by about a factor of 2 (for details, see Hass et al. [97] and Spitzer et al. [107]).

Detailed calculations of the self-energy [(2.122) and (2.124)] and the first-order Raman lineshape were performed by Spitzer et al. [107]. They obtained qualitative agreement with experimental results. Comparing the Raman lineshape of Ge and C, note the large isotopic broadening for diamond, contrary to the small broadening observed for Ge (compare Figs. 2.19 and 2.13; 2.18). The reason lies in the fact that $\boldsymbol{k} = 0$ is not the highest point of the phonon dispersion relation for diamond (Schwoerer-Bohning et al. [112]). This maximum lies somewhat off $\boldsymbol{k} = 0$, resulting in a nonvanishing density of states at ω_0, considerably larger than that found from the relation $N_\mathrm{d} \sim \mathrm{Re}\left(\omega_0 - \omega + i\frac{\Delta\omega_0}{2}\right)^{\frac{1}{2}}$ (Cardona [113]). This density of states is strongly asymmetrical about ω_0, a fact that yields an asymmetrical phonon lineshape

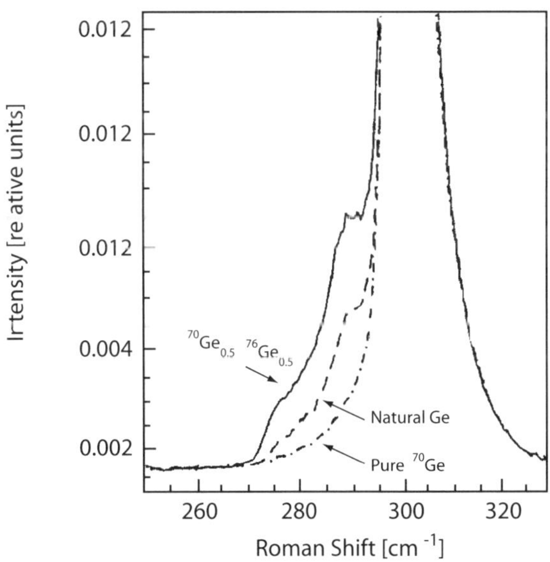

Fig. 2.20. Vertically expanded first-order Raman line of Ge with different amount of isotopic disorder, exhibiting disorder-induced scattering. (After Fuchs et al. [90])

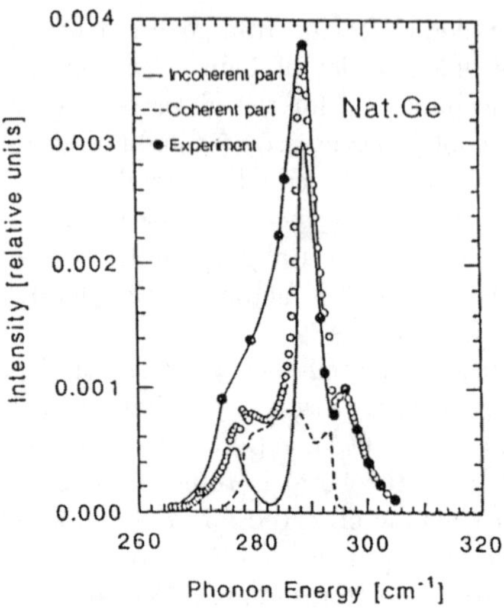

Fig. 2.21. Coherent and incoherent components of the disorder-induced Raman scattering of natural Ge calculated with the CPA. The sum of these two components (*open dots*) is compared with the experimental results (*filled dots*). The agreement is excellent for both the lineshape and the absolute intensity. (After Cardona [106])

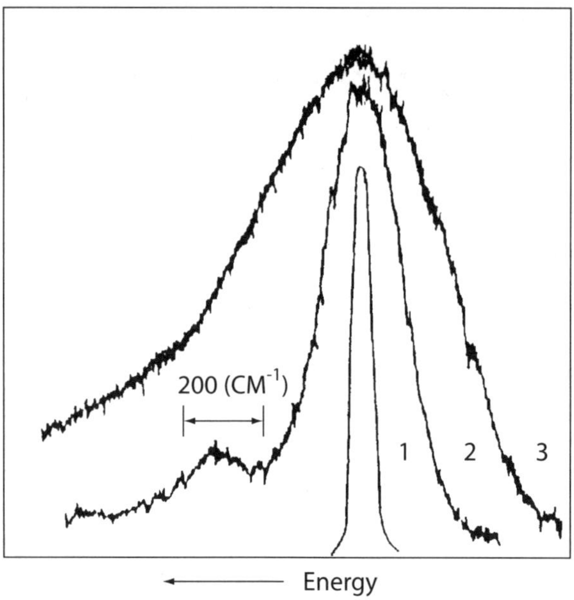

Fig. 2.22. Lineshape of excited light (1), 2LO(Γ) line scattering at 4.2 K in LiH (2) and mixed LiH_xD_{1-x} (3) crystals excited at $\lambda = 253.7$ nm. (After Plekhanov [27])

(Spitzer et al. [107]). This asymmetry also results in lopsided dependence of the linewidth versus concentration (Figs. 2.17; 2.19), which disagrees with the symmetrical dependence expected from the proportionality to g [(2.124)].

As already mentioned, because of the isotopic disorder, not only $\boldsymbol{k} = 0$ phonons should contribute to the first-order Raman spectrum but also all others, with maximum contributions for ω in the regions where the density of states has a maximum, especially for the TO phonon branches. The experimental results for Ge are shown in Fig. 2.20, where the vertical scale has been normalized to 1 at the Raman peak. The disorder-induced contribution is rather weak (less than 0.4 % of the peak) but clearly identifiable since it increases from natural Ge to $^{70}Ge_{0.5}\,^{76}Ge_{0.5}$ and is absent for ^{70}Ge (Fuchs et al. [114]). The two structures observed (at 275 and 290 cm^{-1}) correspond to maxima in the density of TO phonons. The disorder-induced contribution of Fig. 2.20 can be theoretically calculated using CPA techniques. Fuchs et al. found two different contributions: a "coherent" one due to structure in the imaginary part of the self-energy of the $\boldsymbol{k} = 0$ phonon, plus an "incoherent" part due to nonconservation of $\boldsymbol{k}$ (see also Etchegoin et al. [91]). The two contributions, together with their sum, are compared with the expiremental results in Fig. 2.21. The agreement is excellent (see, however, Gobel et al. [105]). This effect has also been observed for diamond (Spitzer et al. [107]), α-Sn (Wang et al. [115]), as well as for LiH_xD_{1-x} (Plekhanov and Altukhov [116], see also Fig. 2.22). The effects of isotopic disorder observed in Raman spectra of LiH_xD_{1-x} crystals are similar but have some important differences (see also [117] and references therein), in contrast to germanium and diamond, whose first-order Raman spectra exhibit one-mode behavior of LO phonons over the entire range of concentrations.

3 Thermal Properties

3.1 Dependence of the Thermal Conductivity on the Isotopic Composition

All solids can conduct heat – some better, others worse. In an isotropic solid, the spread of heat obeys Fourier's law, discovered in 1882

$$Q = -k\,\mathrm{grad}\,T = -k\left(\frac{\partial T}{\partial n}\right), \tag{3.1}$$

where Q is the surface density of the heat stream; it is a vector whose module is equal to the heat flow across the cross section perpendicular to Q; T is the temperature; $\frac{\partial T}{\partial n}$ is the gradient of temperature along a normal $\boldsymbol{n}$ to the isothermic surface; and k is heat conductivity. The minus sign to the right of the expression (3.1) is connected to the fact that heat flows in a direction opposite to the temperature gradient, from the hot to the cold side. In anisotropic solids, the tensor of second rank and its form depend on crystal symmetry.

In common solids, there are two main mechanisms of heat transfer: the transfer of the heat energy of free electrons and the transfer of heat energy by vibrations of crystal-forming particles. In metals, the effects of both mechanisms act simultaneously, whereas in insulators, heat energy is transferred by atomic (ionic) vibrations. The kinetic energy of vibrations is transferred from the heated strip to the cold one. The beam of the kinetic energy of atoms macroscopically looks like heat flow. This process is similar to that of the spread of elastic sound waves in solids. When explaining the phenomenon of heat conductivity, one is allowed to consider that the atoms have strong harmonic vibrations diffusing in the crystal lattice like a system of uninteracting elastic waves. Such waves would spread in a crystal freely without damping and, consequently, they would have an unlimited mean free path. Heat flow, even at small gradients of temperature, could exist indefinitely long before heat equilibrium occurred, and the heat conductivity in that case would be endless.

In real solids as it was shown in the experiments (Berman [121]) heat conductivity is finite. The finite value of heat conductivity is connected with the circumstance that in real crystals, the vibrations of crystal-forming particles are not purely harmonic. The reason for this circumstance is that the forces interacting between atoms do not depend linearly but depend on the displacement of atoms. Therefore, as will be shown below, heat transport by lattice vibrations in solids is governed by anharmonicities of lattice forces (which are responsible for thermal expansion), by various imperfections of the crystal lattice, and by external boundaries. The anharmonic character of vibrations is taken into account by the expansion of potential energy by the anharmonic term gx^3 (see above). Keeping the anharmonic term in the expansion of potential energy, we take into account the interaction between vibrational modes in a real situation. This interaction is usually described as the mutual scattering of phonons. In the first step, there are three phonon processes: the decay of one phonon into two others, sticking together of two phonons, etc. Mutual scattering of phonons is accompanied by the creation and annihilation of phonons (see also Ashcroft and Mermin [118]).

As shown by Peierls in 1929 (Peierls [119]), the probability of indicated transitions in three-phonons processes does not equal zero, if the next two conditions are fulfilled:

$$\hbar\omega_1 + \hbar\omega_2 = \hbar\omega_3, \tag{3.2}$$

$$\hbar\boldsymbol{k}_1 + \hbar\boldsymbol{k}_2 = \hbar\boldsymbol{k}_3 + \boldsymbol{G}. \tag{3.3}$$

Here $\boldsymbol{G} = 2\pi\boldsymbol{H}$, and $\boldsymbol{H} = \boldsymbol{h}\mathrm{a} + \boldsymbol{k}\mathrm{b} + \boldsymbol{l}\mathrm{c}$ is the vector of the reciprocal lattice. The expression (3.2) is the energy conservation law for the three phonon processes. The phonon with quasi-impulse $\boldsymbol{k}$ and frequency ω, commonly speaking, does not possess the same impulse as ordinary material particle. However, the value of $\hbar\boldsymbol{k}$, called the quasi-impulse (wave vector), in the common case is equivalent to a pulse. The expression (3.3) at $\boldsymbol{G} = 0$ corresponds to the wave vector conservation law. The interaction, at which the condition $\boldsymbol{G} = 0$ is fulfilled, is called a normal or N-process. This term originats from the analogy of the interaction of elementary particles (e.g., electrons) for which the laws of conservation of energy and momentum are fulfilled. In distinction from ordinary particles, the number of phonons not conserved in phonon interactions. Moreover, in phonon collisions, the wave vector can be conserved with precision equal to the vector of the reciprocal lattice. This means that the crystal lattice, in which the phonons move, also takes part in the collisions, taking away part of the momentum equal to $\boldsymbol{G} = 2\pi\boldsymbol{H}$.

Peierls called the interaction in which $\boldsymbol{G} \neq 0$ in expression (3.2), the process of throw over (of the transfer) or U-process, which is derived from the German word Umklapprozesse – the process of transfer. In the U-process, energy must be conserved as in a normal process. The difference between N- and U-processes is determined if $\boldsymbol{G}$ is larger or smaller then the value of wave vector of the third phonon ($\boldsymbol{k}_3$) in the collision of two phonons with wave vectors $\boldsymbol{k}_2$ and $\boldsymbol{k}_1$. If $\boldsymbol{k}_3 = \boldsymbol{k}_2 + \boldsymbol{k}_1 < \boldsymbol{G}$, e.g., $\boldsymbol{k}_3$ does not leave the border of the first Brillouin zone; in such a case, all three vectors have positive directions relating to k_x and for them the relations (3.2) and (3.3) at $\boldsymbol{G} = 0$ are justified. This picture is in accordance with the N-process. So far in this situation, $\boldsymbol{k}_3$ coincides with the direction, in which energy is effectively transferred by phonon modes with $\boldsymbol{k}_1$ and $\boldsymbol{k}_2$, so in this case, as was shown by Peierls, the N-processes by themselves do not lead to the reconstruction of the equilibrium distribution of phonons. The latter means that the completed transfer of energy can be conserved and in the absence of a temperature gradient, e.g., heat conductivity is infinitely large. However, after the U-process, heat energy is transferred in direction that does not coincide with the direction of group velocities in the modes of $\boldsymbol{k}_1$ and $\boldsymbol{k}_2$. Such essential changes in $\boldsymbol{k}$ always lead to the reconstruction of an equilibrium distribution of phonons and, consequently to a finite value of the conductivity.

In the light of the processes described, we analyze the dependence of heat conductivity on temperature. For this, we use the expression for the heat conductivity of phonons, obtained in the kinetic theory of gases (see also Ashcroft and Mermin [118]).

$$k_{\text{lattice}} = \frac{1}{3} C_{\text{v}} \langle v_{\text{sound}} \rangle \langle \lambda_{\text{ph}} \rangle = \frac{1}{3} C_{\text{v}} \langle v_{\text{sound}}^2 \rangle \tau, \tag{3.4}$$

where C_{v} is the heat capacity of a single volume of crystals connected with lattice vibrations; $\langle v_{\text{sound}} \rangle$ is the average velocity of phonons, approximately equal to the velocity of the sound in a crystal is weakly dependent on temperature; $\langle \lambda_{\text{ph}} \rangle$ is the average mean free path of phonons equal to the average distance which they pass between two sequential collisions; $\tau = \frac{\langle \lambda_{\text{ph}} \rangle}{\langle v_{\text{sound}} \rangle}$ is the effective time of the relaxation whose reverse meaning τ^{-1} corresponds to the frequency of phonon collisions.

In (3.4), the main values that determine the dependence of heat conductivity on temperature are C_{v} and $\langle \lambda_{\text{ph}} \rangle$. At high temperatures $T \gg \theta$, the specific heat approaches a limiting value which is determined by the Dulong and Petits law $3Nk_{\text{B}}$; it becomes independent of temperature. In this case the dependence of heat conductivity on temperature is determined mainly by the temperature variations of the mean free path of phonons. And at these temperatures, the number of phonons is rather large and their change with temperature linear;

$$\langle n(\boldsymbol{k}, s) \rangle = \frac{1}{\mathrm{e}^{\frac{\hbar\omega(\boldsymbol{k},s)}{k_{\text{B}}T}} - 1} \simeq \frac{k_{\text{B}}T}{\hbar\omega(\boldsymbol{k}, s)}. \tag{3.5}$$

The probability of creation of the processes to transfer heat increases with rising temperature, and it is expected that the frequency of collisions τ^{-1} rises proportionally to T. In this case, the change in the mean free path of a phonon is inversely proportional to temperature:

$$\langle \lambda_{\text{ph}} \rangle \sim \frac{1}{T}. \tag{3.6}$$

Then at $T \gg \theta$,

$$k_{\text{lattice}} \sim \frac{1}{T}. \tag{3.7}$$

For decreasing temperature ($T < \theta$), the average number of phonons able to take part in the processes of transfer, as follows from (3.5), decreases exponentially:

$$\langle n(\boldsymbol{k}, s) \rangle = \frac{1}{e^{\hbar\omega(\boldsymbol{k},s)/k_{\text{B}}T} - 1} \approx e^{-\frac{\theta}{T}}. \tag{3.8}$$

The probability of the transfer process is also diminished according to the exponential law, and this means that the mean free path (as the time of relaxation) of the phonon with decreasing temperature increases exponentially:

$$\langle \lambda_{\text{ph}} \rangle \sim e^{\frac{\theta}{T}}. \tag{3.9}$$

With decreasing temperature, the heat capacity decreases in accordance with Debye's law, as $\sim T^3$ (see also above), but the growth of heat conductivity happens predominantly because of the sharp exponentially increased term for $\langle \lambda_{\text{ph}} \rangle$. Then we obtain

$$k_{\text{lattice}} \sim T_{\text{lattice}}^3 \, e^{\frac{\theta}{T}}. \tag{3.10}$$

When temperature T approachs θ and the probability of the transfer process becomes small, $\langle \lambda_{\text{ph}} \rangle$ is comparable with the dimensions of the specimen and does not depend on temperature.

We should note that the defects (intrinsic and extrinsic) in the crystal lattice also influence the $\langle \lambda_{\text{ph}} \rangle$. This influence diminishes with the decrease, as in this case, the most significant is long wavelength phonons, whose length of waves at 1 K achieves the values $\sim$ 100 interatomic distances. Defects which have the size of the order of average interatomic distances do not influence such waves, but they scatter on the crystal's surface; therefore $\langle \lambda_{\text{ph}} \rangle$ is determined mainly by the size of the specimen.

The change in heat conductivity with temperature described above is proved by numerous experimental data (see, e.g., Klemens [120] and Berman [121], as well as below).

3.1.1 Theoretical Models

Before analyzing the experimental results of thermal conductivity of isotopically pure crystals, we briefly discuss the main theoretical models of thermal conductivity (Berman [121], Klemens [120], and Ziman [122]). Klemens obtained a scattering rate (similar to the familiar Raleigh scattering of photons)

$$\tau_{\mathrm{I}}^{-1} = A\omega^4, \tag{3.11}$$

$$A = \frac{gV}{4\pi v_{\mathrm{B}}^3} \tag{3.12}$$

with the constant A containing the mass variance $g = \frac{\sum c_i M_i^2 - (\sum c_i M_i)^2}{(\sum c_i M_i)^2}$, where c_i and M_i represent the concentration and the mass constituent isotopes, respectively. The mean free path obtained by Klemens is $L_i \sim gT^4$, where as mentioned above, g denotes the isotopic mass variance. In (3.11) and (3.12), V is the volume per atom, and the averaged sound velocity is v_{B}. Equation (3.11) corresponds to a Debye-like phonon density of states $D(\omega) \sim \omega^2$, which, it will be shown, suffices to account for global experimental results.

Boundary scattering at low temperatures leads to a T^3 dependence of k; the prefactor is determined by the geometric size of the sample and the details of the surface. The scattering rate can be written as

$$\tau_{\mathrm{B}}^{-1} = \frac{v_{\mathrm{B}}}{L_{\mathrm{E}}}, \tag{3.13}$$

where L_{E} represents an effective phonon mean free path, that includes effects resulting from sample size, geometry, aspect ratio, phonon focusing, specular (diffuse reflection at the surface), etc.

In the following we analyze the widely used scattering theory of $k(T)$ formulated by Callaway [123] and the modifications introduced by Holland [124].

Callaway's Model

This model assumes the following:

1. a Debye-like phonon spectrum, with no anisotropies or particular structures in the phonon density of states, i.e., no distinction of polarization (between longitudinal and transverse phonons);
2. one averaged sound velocity v_{B};
3. diffuse scattering at the surface of the sample [see (3.13)];
4. normal three-phonon process, including a relaxation rate $\tau_{\mathrm{N}}^{-1} = B_2\omega^2T^3$, which should be valid only for low-frequency longitudinal phonons;
5. three-phonon Umklapp processes have a relaxation rate like that of N processes $\tau_{\mathrm{U}}^{-1} = B_1\omega^2T^3$ (see also Klemens, 1959);
6. that all phonon scattering processes can be represented by relaxation times depending on frequency and temperature;
7. the additivity of reciprocal relaxation times for independent scattering processes.

Then the total thermal conductivity k can be written as (see also Berman [121]):

$$k = k_1 + k_2, \tag{3.14}$$

where k_1 and k_2 are defined by

$$k_1 = cT^3 \int_0^{\frac{\theta}{T}} \tau_{\mathrm{C}}(x) J(x) \mathrm{d}x, \tag{3.15}$$

$$k_2 = cT^3 \frac{\left[\int_0^{\frac{\theta}{T}} \frac{\tau_{\mathrm{C}}(x)}{\tau_{\mathrm{N}}(x)} J(x) \mathrm{d}x\right]^2}{\int_0^{\frac{\theta}{T}} \frac{\tau_{\mathrm{C}}(x)}{\tau_{\mathrm{N}}(x)\tau_{\mathrm{R}}(x)} J(x) \mathrm{d}x} = cT^3(\beta I) \tag{3.16}$$

with

$$\beta = \frac{\int_0^{\frac{\theta}{T}} \frac{\tau_{\mathrm{C}}(x)}{\tau_{\mathrm{N}}(x)} J(x) \mathrm{d}x}{\int_0^{\frac{\theta}{T}} \frac{\tau_{\mathrm{C}}(x)}{\tau_{\mathrm{N}}(x)\tau_{\mathrm{R}}(x)} J(x) \mathrm{d}x}; \quad I = \int_0^{\frac{\theta}{T}} \frac{\tau_{\mathrm{C}}(x)}{\tau_{\mathrm{N}}(x)} J(x) \mathrm{d}x \tag{3.17}$$

and

$$J(x) = \frac{x^4 \mathrm{e}^x}{(\mathrm{e}^x - 1)^2}; \quad \frac{1}{\tau_{\mathrm{C}}(x)} = \frac{1}{\tau_{\mathrm{N}}(x)} + \frac{1}{\tau_{\mathrm{R}}(x)};$$
$$x = \frac{\hbar\omega}{k_{\mathrm{B}}T}; \quad m = \frac{k_{\mathrm{B}}}{\hbar}; \quad c = \frac{k_{\mathrm{B}} m^3}{2\pi^2 v_{\mathrm{B}}}. \tag{3.18}$$

In (3.18) k_{B} is Boltzmann's constant, $\hbar$ is Planck's constant, and τ_{N} (τ_{R}) denotes the relaxation time of N processes (resistive processes). The corresponding combined relaxation rate τ_{C}^{-1} can be written as the sum of the normal, nonresistive rate (N) and resistive rate R of (3.18). In the Callaway formulation, in contrast to the earlier models of Klemens [120] and Ziman [122], all resistive scattering probabilities are taken as additive $\left(\frac{1}{\tau_{\mathrm{R}}}\right) = \sum_i \left(\frac{1}{\tau_i}\right)$ (here τ_i represents the isotopic- (τ_i), the boundary- (τ_{B}), and the Umklapp- (τ_{U})-scattering times), i.e., the corresponding scattering mechanisms are assumed to be independent.

The k_2 term is not only a correction term to k_1 (as sometimes stated in the literature (Callaway [123]) but is essential to counteract the effect of treating N processes in τ_{C} as if they were entirely resistive. Consequently, k_2 is a non-negligible part of Callaway's theory. The magnitude of k_2 is essentially controlled by the concentration of point defects. In the majority of cases of physical interest, resistive scattering dominates ($\tau_{\mathrm{N}} \gg \tau_{\mathrm{R}} \Longrightarrow \tau_{\mathrm{C}} \approx \tau_{\mathrm{R}} \Rightarrow k_2 \ll k_1$)

and only k_1 is important. Therefore, in the literature, very often only the k_1 term is included. However, when N processes become comparable to resistive processes ($\tau_N \approx \tau_R$), e.g., in very pure, defect-free (i.e., isotopically pure) samples, k_2 integrals contribute significantly to total thermal conductivity (see also Berman [121]). Thus, in isotopically pure samples, normal three-phonon scattering rather than Umklapp processes determine the phonon mean free path.

Holland's Model

In the next step, we consider briefly Holland's theory (Holland [124]), which extended the Callaway theory to include explicitly the thermal conductivity of both transverse and longitudinal phonons, under the assumption $k_2 = 0$.

1. Since the variation of phonon relaxation times with frequency and temperature strongly depend on the actual phonon branch and its dispersion, contributions to thermal conductivity by the two kinds of differently polarized phonons(transverse and longitudinal), are considered separately whereas normal processes are taken into account for the class of crystal at hand, as suggested by Herring [125].
2. A more realistic representation of the very dispersive transverse acoustic modes of Ge is used (for details, see Nelin and Nilsson [104]). It involves splitting the range of integration in two parts, a low- and a high-frequency range with different temperature and frequency dependences.

The four scattering mechanisms assumed for the analysis with Holland's model are chosen to have the following temperature and frequency dependence (see, also Table 3.1):

$$\tau_{\mathrm{I}}^{-1} = A\omega^4, \tag{3.19}$$

$$\tau_{\mathrm{B}}^{-1} = \frac{v_{\mathrm{B}}}{L_{\mathrm{E}}}, \tag{3.20}$$

$$\tau_{\mathrm{TO}}^{-1} = B_{\mathrm{T}}\omega T^4 \quad \text{for} \quad 0 \leq \omega \leq \omega_1, \tag{3.21}$$

$$\tau_{\mathrm{L}}^{-1} = B_{\mathrm{L}}\omega^2 T^3 \quad \text{for} \quad 0 \leq \omega \leq \omega_3, \tag{3.22}$$

Table 3.1. Parameters for Debye and Callaway models fit for diamond. The Callaway model takes into account phonon scattering in N processes

Sample	A (cm)	B (cm^3)	C (cm/k)	DK
Debye model				
Natural type 11[a]	0.1	$4.0 \cdot 10^{-25}$	$2.0 \cdot 10^{-12}$	550
Synthetic enriched ^{12}C	0.1	$0.6 \cdot 10^{-25}$	$2.0 \cdot 10^{-12}$	550
Callaway model				
Natural type 11[a]	0.055	$1.5 \cdot 10^{-26}$	$1.4 \cdot 10^{-12}$	730
Synthetic enriched ^{12}C	0.055	0	$1.4 \cdot 10^{-12}$	730

$$\tau_{\mathrm{TU}}^{-1}\frac{B_{\mathrm{TU}}\omega^2}{\sinh(x)} \quad \text{for} \quad \omega_1 \leq \omega \leq \omega_3, \tag{3.23}$$
$$\tau_{\mathrm{TU}}^{-1} = 0 \quad \text{for} \quad \omega \prec \omega_1,$$

where $x = \frac{\hbar\omega}{k_{\mathrm{B}}T}$ and $T(L)$ represent transverse (longitudinal) acoustic phonons. Holland's integral k_1 (Callaway's notation) has been separated into TA and LA contributions k_{T} and k_{L}. The term k_{T} splits into the contribution of N processes k_{TO} and that of U-processes k_{TU}:

$$k = k_{\mathrm{T}} + k_{\mathrm{L}} = k_{\mathrm{TO}} + k_{\mathrm{TU}} + k_{\mathrm{L}}, \tag{3.24}$$

with

$$k_{\mathrm{TO}} = \frac{2}{3}H_{\mathrm{TO}}T^3 \int_0^{\frac{\theta_1}{T}} \tau_{\mathrm{C}}^{\mathrm{TO}}(x)J(x)\mathrm{d}x, \tag{3.25}$$

$$k_{\mathrm{TU}} = \frac{2}{3}H_{\mathrm{TU}}T^3 \int_{\frac{\theta_1}{T}}^{\frac{\theta_2}{T}} \tau_{\mathrm{C}}^{\mathrm{TU}}(x)J(x)\mathrm{d}x, \tag{3.26}$$

$$k_{\mathrm{L}} = \frac{1}{3}H_{\mathrm{L}}T^3 \int_0^{\frac{\theta_3}{T}} \tau_{\mathrm{C}}^{\mathrm{L}}(x)J(x)\mathrm{d}x, \tag{3.27}$$

where

$$\tau_{\mathrm{C}}^{\mathrm{TO}}(x) = \left(\frac{v_{\mathrm{B}}}{L_{\mathrm{E}}} + Am^4x^4T^4 + B_{\mathrm{T}}mxT^5\right)^{-1},$$

$$\tau_{\mathrm{C}}^{\mathrm{TU}}(x) = \left(\frac{v_{\mathrm{B}}}{L_{\mathrm{E}}} + Am^4x^4T^4 + \frac{B_{\mathrm{TU}}m^2x^2T^2}{\sinh(x)}\right)^{-1},$$

$$\tau_{\mathrm{C}}^{\mathrm{L}}(x) = \left(\frac{v_{\mathrm{B}}}{L_{\mathrm{E}}} + Am^4x^4T^4 + B_{\mathrm{L}}m^2x^2T^5\right)^{-1} \quad \text{and}$$

$$H_i = \frac{k_{\mathrm{B}}m^3}{2\pi^2 v_i}, \quad m = \frac{k_{\mathrm{B}}}{\hbar}.$$

In each of the three integrals, the constant H_i contains the corresponding sound velocity v_i. We should emphasize that U processes are neglected in (85) because they should not contribute below $\omega_1(\theta_1 = 101\,\mathrm{K}$ for Ge). The terms for N processes were omitted from (3.26) since they should be relatively small above ω_1. Both assumptions have been checked to be quantitatively justified (Holland [124]). Nevertheless, the integral formulation makes each of the scattering mechanisms operative over a large temperature interval. Thus, the effect of varying one of the coefficients always induces modifications in the influence of the other coefficients on thermal conductivity.

3.1.2 Experimental Results

Diamond

Since the early work of Pomeranchuk in 1942 demonstrating the role of isotopes as phonon scatterers with a resulting influence on thermal conductivity, and the work performed by Geballe and Hull [126] it has been known that the maximum thermal conductivity k_{m} is strongly affected by isotopic composition (see also [127–129]). This fact received considerable attention in recent years for diamond: an $\simeq 1\%$ reduction of the ^{13}C content in natural diamond enhances k_{m} by 50% (see, e.g., Anthony et al. [130–132], Onn et al. [133], Olson et al. [134], and Graebner et al. [135]). As we can see from results obtained Geballe and Hull observed an increase in the k_{m} of an enriched ^{74}Ge sample (with 95.8% of ^{74}Ge) by a factor of 3 with respect to natural germanium (see also [47]).

It has long been known (see, e.g., Wilks and Wilks [136]) that diamond has extremely high thermal conductivity (see also Berman [121] and references therein). The thermal conductivity of high purity type-11^a diamond at room temperature is a five times larger that of copper, and is the highest value of any known material (Graebner et al. [135] and Graebner [137]). Diamond is thus very useful for heat dissipation, and a 50% enhancement of conductivity at room temperature would therefore be of considerable practical interest. The magnitude of the enhancement is surprising, since in similar systems like LiF (substitution of $^7\mathrm{Li} \longrightarrow {}^6\mathrm{Li}$ and $^4\mathrm{He} \longrightarrow {}^3\mathrm{He}$), the isotopic effect is only $1-2\%$ at the same temperature. On the experimental side, the enhancement in diamond has since been confirmed in different papers (Banholzer and Anthony [138], Onn et al. [133], Hass et al. [97], Olson et al. [134], and Graebner et al. [135]) using different experimental methods (see also Graebner [137]). The theoretical picture, however, remains rather cloudy, because the standard theories predicted only a few percent effect (see, e.g., Ziman [122]). At present, the thermal conductivity of diamond has been investigated in detail and in wide temperature region ($10-1200\,\mathrm{K}$) (see also [23]).

The thermal conductivities of two natural specimens are shown in Fig. 3.1 [134]. The measurements indicated here at and below room temperature agree very well with these previously reported values. The authors of the paper citated noted that there is a $\sim 10\%$ difference between the two samples, which is due to the different quality of these samples. All data are in close agreement.

One is confronted with a situation where small concentrations of isotopic impurities lead to an expectably large reduction in thermal conductivity. In an attempt to quantify the phonon scattering resulting from isotopic impurities in diamond, the experimental data (see Fig. 3.1) have been analyzed using the Debye model of thermal conductivity

$$\Lambda(T) = N k_{\mathrm{B}} \sum_{i=1}^{3} v_i \left[\frac{T}{\theta}\right]^3 \int_0^{\frac{\theta_{\mathrm{D}_i}}{T}} l(x) \frac{x^4\,\mathrm{e}^x}{(\mathrm{e}^x - 1)^2} \mathrm{d}x, \tag{3.28}$$

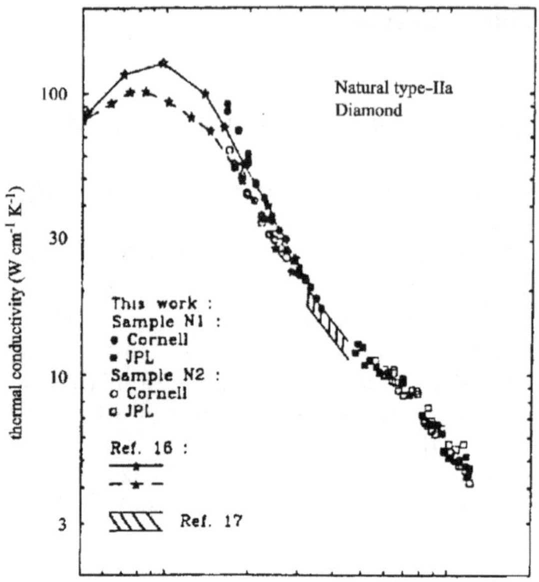

Fig. 3.1. Thermal conductivity of two type 11^a diamonds. Also shown the data by Berman (Berman [121]) as well as Burgmeister (Burgmeister, 1978) on type 11^a diamonds. Burgmeister measured 30 different samples, all of which lie within the range shown. (After Olson et al. [134])

where N is the number density of atoms (for diamond $N = 1.762 \cdot 10^{23}\,\mathrm{cm}^{-3}$) and the sum over i denotes a sum over the one longitudinal and the two transverse phonon modes, v_i is the sound velocity for that mode (for diamond (Novikov [139]) $v_l = 1.75 \cdot 10^6\,\mathrm{cm/s}$ and $v_\mathrm{t} = 1.28 \cdot 10^6\,\mathrm{cm/s}$). θ_{D_i} is the effective Debye temperature for mode i, given by

$$\theta_{\mathrm{D}_i} = 2.997 \cdot 10^{-11} v_i N^{\frac{1}{3}} sK. \tag{3.29}$$

$l(x)$ is the phonon mean free path, and

$$x = \frac{\hbar v_i}{k_\mathrm{B} T \lambda}, \tag{3.30}$$

where λ is the phonon wavelength. As before [see (3.18)], the assumption is made that the resistive scattering rates add, so that

$$l(x) = \left(l_\mathrm{boun}^{-1} + l_\mathrm{def}^{-1} + l_\mathrm{umkl}^{-1}\right)^{-1} + \frac{\lambda}{2}. \tag{3.31}$$

The terms l_boun, l_def, and l_umkl are the phonon mean free path associated with the sample boundaries, point defects, and Umklapp processes, respectively. The term $\frac{\lambda}{2}$, half the phonon wavelength, is included to avoid the

Table 3.2. Comparison of the effect of isotopic scattering on the thermal conductivity of natural Ge, Si, and diamond at 300 K. The % increase in k is the increase in the thermal conductivity of the nearly isotopically pure sample compared to the natural sample. The isotope is defined as $\Gamma = \sum_i f_i \left(\frac{\Delta M_i}{M}\right)^2$. (After Asen-Palmer et al. [141] and Capinski et al. [145])

	% increase in k	θ	$\Gamma \cdot 10^{-4}$
Ge	30	376	5.80
Si	60	658	2.01
Diamond	50	1860	0.76

nonphysical case where the mean free path becomes short compared to the phonon wavelength.

The Debye model is sometimes referred to as the Klemens–Callaway model, but Olson et al. [134] felt that the Debye model is more appropriate. Klemens made the assumption that the scattering rates add and ignored the N processes, whereas Callaway took N processes into account. At the same time, we note that when Callaway applied his theory to Ge, he ignored N processes (see above) and simply used the Debye model. The data of Berman et al. [121] and one set of high-temperature data (sample N1 in the paper of Olson et al. [134]) have been fitted, using phonon mean free paths of the form

$$l_{\text{boun}}^{-1} = \frac{1}{A}, \tag{3.32}$$

$$l_{\text{def}}^{-1} = \frac{B}{\lambda^4}, \tag{3.33}$$

$$l_{\text{umkl}}^{-1} = \text{C}\frac{T}{\lambda^2}\text{e}^{-\frac{D}{T}}. \tag{3.34}$$

Formulae (3.33) and (3.34) were used by Klemens [120] and Peierls [119]. A good fit (see Fig. 2.9) is found for the following parameters: $A = 0.1$ cm, $B = 4.0 \cdot 10^{-25}$ cm^3, $C = 2.0 \cdot 10^{-12}$ cm/K, and $D = 550$ K. Those parameters are listed in Table 3.2 and are close to those published (see also Hass et al. [07], Onn et al. [133], and Graebner et al. [137]). In the paper of Olson et al., a reasonable fit was found for $B = 0.6 \cdot 10^{-25}$ cm^3 (see Table 3.1). Thus, it was necessary to reduce B by $3.4 \cdot 10^{-25}$; this is the Raleigh term caused by isotopic scattering. Klemens calculated the Raleigh term taking into account not only the mass difference in the impurities but also the difference in volume occupied. It was found that

$$l_{\text{def}}^{-1} = \frac{\Omega_0}{4\pi}\delta\left[\frac{\Delta M}{M} + 2\gamma\alpha\right]^2\left[\frac{2\pi}{\lambda}\right]^4, \tag{3.35}$$

where Ω_0 is the volume per atom equal to $\frac{1}{N} = 5.68 \cdot 10^{-24}$ cm^3, δ is the atomic fraction of the impurity (assumed $\ll 1$), $\frac{\Delta M}{M}$ is the fractional mass difference for the impurity, γ is the Gruneisen parameter equal to 1.1 for diamond (Novikov [139]), and α is the fractional volume difference of the impurity. For 1.1% ^{13}C in ^{12}C, where $\frac{\Delta M}{M} = \frac{1}{12}$ and $\alpha = 0.0005$ (Olson et al. [134]), l_{def} calculated from (3.35) is $5.38 \cdot 10^{-26}\,\text{cm}^3$. This is 6.3 times smaller than the scattering rate obtained experimentally. Moreover, the same conclusion was obtained by Onn et al. [133], who found that the calculated scattering rate is 5.4 times smaller than the measured scattering rate. The main reason for this observation is that in the Debye model, nonresistive three-phonon N processes are ignored (see also Hass et al. [97]). These scattering processes do not themselves lead to degradation of heat flow but instead restore the phonon distribution to a displaced Planck distribution compatible with the heat flowing. Suppose there were a defect that strongly scattered phonons at a particular frequency f_0. In the Debye model, phonons with frequency f_0 would not contribute to thermal conductivity, and phonons of other frequencies would be unaffected. N processes would act to channel other phonon modes into that frequency, effectively reducing the thermal conductivity contribution from the other modes as well. N-process scattering rates in diamond are largely unknown, so a quantitative computation is not possible. In the paper of Olson et al. [134], it was assumed that the N-process scattering rate is much larger than the resistive scattering rate; then both the Callaway model and the Ziman variational method led to same result (Berman [140]):

$$\Lambda(T) = N k_{\text{B}} \sum_{i=1}^{3} v_i \left[\frac{T}{\theta_{\text{D}_i}}\right]^3 \frac{\left[\displaystyle\int_0^{\frac{\theta_{\text{D}_i}}{T}} x^4\, \text{e}^x \left(\text{e}^x - 1\right)^{-2}\right]^4}{\displaystyle\int_0^{\frac{\theta_{\text{D}_i}}{T}} l^{-1}(x) x} \text{e}^x \left(\text{e}^x - 1\right)^{-2} \text{d}x, \tag{3.36}$$

where $l^{-1}(x)$ is given by the inverse of (3.31). The data of natural type 11^a diamond have been fitted using the phonon mean free paths, as given in (3.32)–(3.34). A good fit has been found for the following parameters: $A = 0.055\,\text{cm}$, $B = 1.5 \cdot 10^{-26}\,\text{cm}^3$, $C = 1.4 \cdot 10^{-12}\,\text{cm/K}$, and $D = 730\,\text{K}$. These values are listed in Table 3.1 (Callaway's model). Not surprisingly, these parameters are somewhat different from those found using the Debye model. The Rayleigh scattering term was $l_{\text{def}}^{-1} = 0$ in this case. Thus $1.5 \cdot 10^{-26} \lambda^{-4}\,\text{cm}^3$ had to removed. This corresponds to 3.6 times less scattering than predicted by (3.35). In this case, it was assumped that the N processes dominate and enhance isotopic scattering by a factor of 23 over the case where N processes are ignored. The same conclusion was reached by Hass et al. [97]. So, the observed isotopic effect can be explained by existing theory by taking into account the effect of N processes, but the assumption that the N processes dominate leads to an incorrect temperature dependence of thermal conductivity.

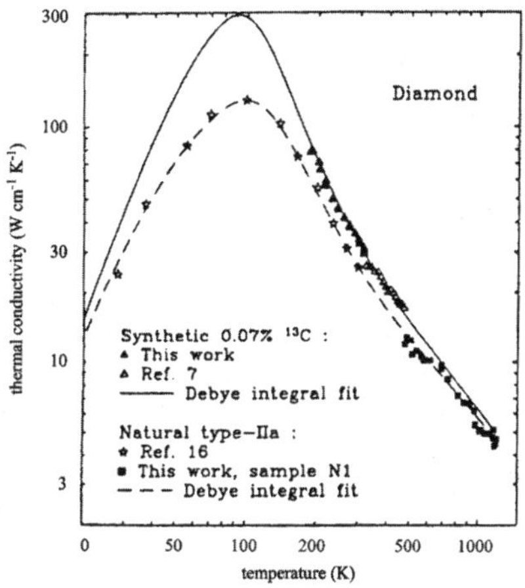

Fig. 3.2. Thermal conductivity of natural type 11^{a}diamond and synthetic diamond with 0.07% ^{13}C isotope concentration. The *dashed line* is the fit calculated from the Debye model. The *solid line* is the fit for the nearly isotopically pure diamond, where the point defect scattering (Raleigh term) was the only parameter changed (after Olson et al. [134])

Germanium

At present, there are few papers devoted to investigating the dependence of thermal conductivity on the isotopic compositions of germanium and silicon crystals. In view of the high purity ($|N_d - N_a| < 10^{14}\,cm^{-3}$) and the perfection of the samples used, next, we will follow in the main the experimental results of the paper of Asen-Palmer et al. [141] (see also [128]). In this paper, only four scattering mechanism were considered:

1. Normal (N) three-phonon scattering;
2. three-phonon Umklapp (U) processes;
3. boundary (B) scattering;
4. isotopic (I) mass fluctuations (point defects).

The parameters for the two latter mechanisms are fixed by theory and are therefore not adjustable. Dislocation scattering was also considered.

The thermal conductivity versus temperature measured for various isotopic compositions are displayed as log–log plots in Figs. 3.3–3.4. The data for the isotopically purest sample ^{70}Ge (99.99%) are shown together with the results for the less pure ^{70}Ge (86%), natural Ge, and the most isotopically disordered sample containing 43% of ^{70}Ge and 48% of ^{76}Ge. The maximum $k(T)$ amounts to 10.5 kW/mK near 16.5 K for ^{70}Ge (99.99%), which is the highest value of k measured for Ge, higher than the thermal conductivity maximum of sapphire (6 kW/mK near 35), and comparable to that of silver (11 kW/vK near 8 K, Asen-Palmer et al. [141]). The isotopically most

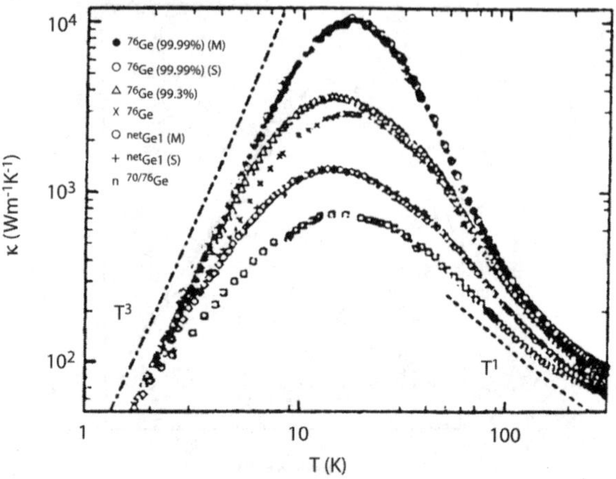

Fig. 3.3. Thermal conductivity versus temperature of five Ge samples with different isotopic compositions: ^{70}Ge (99.99%), ^{70}Ge (96.3%), ^{76}Ge (86%), ^{nat}Ge1, and $^{70/76}$Ge. Two samples, ^{70}Ge (99.99%) and ^{nat}Ge1, have been measured with two different experimental setups in Stuttgart (S) and in Moscow (M). The *dot-dashed line* represents simply a T^3 law, expected for pure boundary scattering, whereas the *dashed line* shows a $\frac{1}{T}$ dependence expected for phonons at high temperatures. (After Asen-Palmer et al. [141])

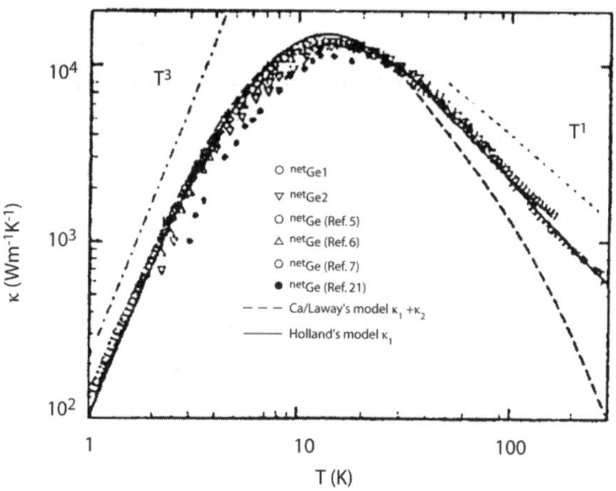

Fig. 3.4. Thermal conductivity of various ⟨100⟩ oriented natural Ge bars. The *dashed line* represents the thermal conductivity calculated with the full Callaway's model ($k_1 + k_2$), where the parameters $B_1 = B_2 = 2.6 \cdot 10^{-23}\mathrm{s}/k^3$ were used. The *continuous line* represents the heat conductivity calculated with the model of Holland (only k_1) using the single set of parameters for all samples: $B_{\mathrm{T}} = 1.5 \cdot 10^{-11}\ 1/k^4$, $B_{\mathrm{TU}} = 4.5 \cdot 10^{-18}\,\mathrm{s}$, $B_{\mathrm{L}} = 9.0 \cdot 10^{-24}\,\mathrm{s}/k^3$, $L_{\mathrm{E}} = 3.8\,\mathrm{mm}$, and $g = 58.7 \cdot 10^{-5}$ for ^{nat}Ge. (After Asen-Palmer et al. [141])

disordered sample shows, as expected, the lowest thermal conductivity for undoped Ge (0.75 kW/mK near 15.4 K). In this sample, isotopic scattering is dominant, in contrast to pure ^{70}Ge (99.99%), where it becomes negligible. As pointed out by Asen-Palmer et al., except for sample ^{70}Ge (95.6%), all samples had similar geometry (within 10% equal cross-sectional dimensions and identically prepared surfaces). They were cut with a diamond saw and the surfaces lapped with a 20 pm diamond powder slurry (for details, see [141]). The overall features of the $k(T)$ curves displayed in Figs. 3.3 and 3.4 are those found for defect-free insulators: The T^3 behavior at sufficiently low temperatures (see also Zhernov [128] and references therein), are due to boundary scattering, and a maximum results from normal and Umklapp phonon processes which lead to a $\frac{1}{T}$ dependence above 100 K. Comparing the results depicted in Fig. 3.3, one can see that the maximum thermal conductivity k_{m} of $^{70/76}$Ge is 14 times smaller than that of ^{70}Ge (91.91%). The value of k_{m} for natural Ge is increased by a factor of ~ 8 in the ^{70}Ge (99.99%) sample. The increase in k, however, is only 30% at 300 K. We should add also: (1.) The maximum of $k(T)$ shifts slightly to higher temperatures with increasing isotopic purity in accordance with theory (Jackson and Walker [142]). (2.) The strong influence of isotopic disorder on k is clearly displayed over the entire temperature range in which $k(T)$ was measured. Point defect scattering from isolated atoms of different isotopes or different elements with very similar force constants is one of the rare cases which can be calculated analytically without adjustable parameters for phonons.

Using the Callaway model in its original form, i.e., keeping both k_1 and k_2 [see (3.15) and (3.16)] it was adjusted for two free parameters (B_1, B_2) in the combination of the four scattering mechanisms considered, with the scattering rates for isotopic and boundary scattering fixed at the values given in (3.11)–(3.13) ($v = 22.6 \cdot 10^{-30}\,\mathrm{m}^3$, $v_{\mathrm{B}} = 3500\,\mathrm{m/s}$, and $0.01 \leq g \leq 58.7 \cdot 10^{-5}$, and $3.6 \leq L_{\mathrm{E}} \leq 4.8$). In this manner, an acceptable representation of all the data was achieved but only below about 30 K (see, however, [143]). Moreover, the adjustable coefficients obtained were not the same for the various samples (for detail, see Asen-Palmer et al. [141]). As a typical example, the fit to the data for natural Ge with Callaway's theory is shown in Fig. 3.4 (*dashed line*). The convexity of the calculated thermal conductivity above the maximum, describing a steeper decrease of $k(T)$ with increasing temperature than found experimentally, cannot be removed by changing the parameters. The reason can be traced to an underestimation of the U processes in that model. Instead of using an exponential function for the Umklapp scattering probability, as proposed in the literature (Peierls [119] and Klemens [120]), the N processes, as well as the U processes, are represented by the same temperature and frequency dependences, $B_{1/2}\omega^2 T^3$. The prefactors B_1 and B_2 are thus indistinguishable in Callaway's theory. This analysis points out that in isotopically pure samples, normal three-phonon scattering rather than Umklapp processes determine the phonon mean free path.

Further, we should briefly analyze the experimental results with the Holland theory. It was Holland, who extended the Callaway theory to include explicitly the thermal conductivity by both transverse and longitudinal phonons, under the assumption $k_2 = 0$. The four scattering mechanisms assumed for the analysis with Holland's model were chosen by Asen-Palmer et al. [141] to have the temperature and frequency dependence described by (3.19)–(3.23). In this model, the three freely adjustable coefficients – B_T, B_L, and B_{TU} (equal to $1.0 \cdot 10^{-11}\,1/k^4$, $6.9 \cdot 10^{-24}\,\mathrm{s}/k^3$, and $5.0 \cdot 10^{18}\,\mathrm{s}$) – have been obtained by linear regression. The isotopic and boundary scattering rates were fixed by (3.11)–(3.13), the corresponding mass variance g, the effective mean free path L_E, and the sound velocity v_B respectively. With this model, it has been possible to obtain a good representation of the thermal conductivity of all samples studied in the paper of Asen-Palmer et al. Using a unique set of parameters in the temperature range $2 - 200\,\mathrm{K}$, the agreement between experimental data and fitted curves is rather good ($\pm 5\%$), as exemplified by the *solid line* in Fig. 3.4 for natural Ge.

3.1.3 High Thermal Conductivity Silicon

As is well known natural silicon contains three isotopes, ^{28}Si (92%), ^{29}Si (5%), and ^{30}Si (3%). By removing the minority isotopes, silicon-28 crystal would have a more perfect crystal lattice with less heat and electromagnetic noise and higher thermal conductivity to dissipate the heat that is generated more efficiently. The mechanisms for this improvement are reduced efficacy of electron–phonon and phonon–phonon interactions (for details, see [23]).

It has been proven with several semiconductor materials that removal of minority isotopes results in substantially higher thermal conductivity. As pointed out before, the removal of only 1% carbon-13 to produce isotopically pure carbon-12 diamond resulted in a 50% improvement in room temperature thermal conductivity and over a 10-fold improvement at low temperatures (see preceding sections). Germanium has five stable isotopes with abundances of $7 - 36\%$ (see above). By concentrating one of the isotopes to 99.9%, the thermal conductivity of isotopically pure germanium-70 was increased 20% at room temperature and 1000% at 15 K [141].

The results of thermal conductivity measurements by Ruf et al. [146] are shown in Fig. 3.5. In the umklapp range, where k has been measured before, the data from SI 284 (*filled circles*), obtained with the steady-state heat-flow technique, essentially confirm the earlier results [144] (*filled triangles*) except for the value at 100 K, where the optical pump-probe method becomes less reliable. The results for the reference sample SIN 1 (*open circles*), measured only for $T > 90\,\mathrm{K}$, agree very well with the "recommended" thermal conductivity curve for natural silicon ("plus" symbols) from [159], which was extracted from several measurements. The agreement with the results from [144] (*open triangles*) is also good, given their somewhat larger uncertainty. The cited author found a value of $k = 237(8)\,\mathrm{Wm^{-1}K^{-1}}$ for SI 284 at 300 K. The reference

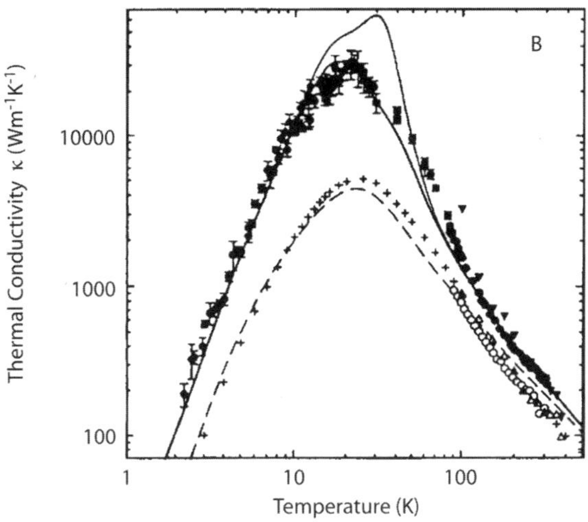

Fig. 3.5. Thermal conductivity of the highly isotopically enriched ^{28}Si sample SI 284 (*filled circles*) and the natural Si, reference SI N1 (*open circles*). The *filled* and *open triangles* are other measurements for highly isotopically enriched ^{28}Si and natural Si, respectively (from [159]; "plus" symbols denote the "standard" curve for natural Si (from [159]). The *thin solid* and *dashed lines* are the theoretical results of [144] for ^{28}Si and natural Si, respectively. The *thick solid line* has been calculated with the same theory using the actual mass variance $g_2 = 2.33 \times 10^{-6}$ of sample SI 284. (After Ruf et al. [146])

value for natural silicon (SIN 1 and [159]) is 150 $\mathrm{Wm^{-1}K^{-1}}$. This represents an enhancement of almost 60% and 40% greater at 100°C (chip operating temperature). At liquid nitrogen temperature (LNT) $T = 77$ K, the thermal conductivity values, according to Ruf et al., are $k(\mathrm{SI\,284}) = 3300\,\mathrm{Wm^{-1}K^{-1}}$ and $k(\mathrm{nat.\ Si}) = 1400\,\mathrm{Wm^{-1}K^{-1}}$, yielding enhancement by a factor of about 2.4. Around 20 K, k reaches a maximum of about 30,000 (5000) $\mathrm{Wm^{-1}K^{-1}}$ and is thus six times larger than in natural silicon (5140 $\mathrm{Wm^{-1}K^{-1}}$ [159]). A number of applications of this high k can be envisaged (see also [27] and below).

The calculations (see Fig. 3.5), based on Ref. [141], describe, according to [146], the data quite well for both SI 284 and natural silicon. As expected, the maximum k is reduced compared to the prediction for an ideally pure sample, even for a rather small g_2 value of SI 284. For both samples, a slight "overshoot" of the experimental data compared to theory occurs around 50–80 K. The cited authors attribute the dispersion of silicon to the anomalously flat transverse acoustic phonon, which has a large impact on freeze-out of the umklapp process in this temperature range. For temperatures below the thermal conductivity maximum, $k(T)$ approaches the boundary-scattering region where the T^3 behavior due to the temperature dependence of the phonon specific heat dominates. In this regime, k is very sensitive to geometric and surface effects, and strong variations, depending on sample orientation, may occur [141]. This is manifested by the difference between the above L_E values and the sample diameter with which they should be compared. These effects need to be investigated in more detail. In the highly isotopically enriched sample, the T^3 behavior of k is observed for temperatures up to !0 K, in contrast to that of natural silicon. Note that the low-temperature data of Ruf et al. are in reasonable agreement with the calculations of Omini and Sparavigna [129]. These authors, on the other hand, fail to describe the thermal conductivity enhancement in the umklapp region. Further interesting effects occur in the behavior of thermal conductivity at high temperatures. Ruf et al. found an almost constant enhancement in k of about 60% between 100 and 300 K and estimate that this effect persists up to at least 400 K. This observation also contradicts the predictions of a recent calculation for thermal conductivity enhancement in isotopically pure silicon [145]. More clearly than in [141], the Rufs data unambiguously show a linear dependence of $\frac{1}{k}$ on temperature above 80 K for both samples [$k\,(\mathrm{SIN}\,1)^{-1} - k(\mathrm{SI}\,284)^{-1} = 0.987\,T \cdot 10^{-5}\,\mathrm{Wm^{-1}K^{-1}}$, T in K] (see also [47]).

Improvement in the thermal conductivity of silicon is important because as feature size continuously decreases, current density increases, and more heat is generated per unit volume, causing device operating temperatures to rise. The integrated circuit industry (AMD and other companies) is moving toward lower operating voltages and sophisticated mechanical means to minimize this heat buildup, but greater heat dissipation on the microscale is becoming essential to improve device performance and ability.

Other, more subtle benefits may also be available with isotopically pure silicon. As device dimensions decrease, the gate oxide thickness in CMOS devices also must decrease. As the transistor size approaches 0.1 micron, the gate oxide thickness will be about 2 nanometers (0.002 microns). This is equivalent to 5 atomic layers of silicon dioxide. The gate oxide integrity or breakdown voltage of oxides of this thickness is a key stumbling block to smaller devices. The more perfect crystal structure of isotopically pure silicon and silicon dioxide may lead to improved gate oxide integrity and fewer surface traps at the silicon–silicon dioxide interface. It has already been shown that the use of deuterium, an isotope of hydrogen (see also above), instead of normal hydrogen, during the processing of CMOS (complementary metal oxide semiconductor) devices has led to reduced "hot electron" damage and 50 to 100 times longer lifetimes before oxide wear-out (for details, see Lyding et al. [147]).

Isonics Corporation (Colorado, USA), a world leader in isotopically engineer material, has taken the concept of isotopically pure silicon and made it a reality. The first product to be offered for sale is an epitaxial silicon wafer, which is the starting wafer for all advanced microprocessors. This wafer has a thin layer of isotopically pure silicon-28 grown on the surface of prime quality, natural silicon wafer. Epitaxial wafers made to custom specifications are currently available in a range of sizes to integrated circuit manufacturers. Many of the advantages of isotopically pure silicon can be obtained with this type of wafer, while keeping costs down.

Thus, the analysis of the thermal conductivity of Ge and Si (as well as C) samples with several isotopic compositions, using a modified Callaway–Holland formalism, works well below 200 K. To conclude this section, we should mention that at present, there are few unique samples where much detailed investigation of the influence of isotopic effects on the thermal conductivity of the crystals of diamond and Ge (Si) (see also Table 3.2) was done. In this connection, we must add that analogous investigation of the isotopic effects on the thermal conductivity in LiH_xD_{1-x}crystals will be very intriguing taking into account the presence of local vibrations at low concentrations and two-mode behavior of LO phonons at large concentrations [25].

3.2 Lattice Constant Dependence on Temperature and Isotopic Composition

The lattice parameter at any given temperature is determined by three different contributions. First and most obvious is the size of the atomic radii and the nature of the chemical bonding between them, which are most important in determining interatomic spacings and crystal structure. Second is the effect of temperature on the distance between atoms which normally produces a volume expansion with increasing temperature. Finally, there is the effect of the zero-point displacement, which is a purely quantum effect that has no classical analogue. This last contribution results from the fact that the lowest energy state of the system, zero-point energy, generally corresponds in an anharmonic potential to an atomic displacement somewhat larger than that associated with the potential minimum.

It is well-known that the phonon–phonon interaction (anharmonicity) results, at low T, in a dependence of the lattice constant on isotopic mass (see, e.g., Reissland [5]). The effect can be understood as related to "zero-temperature thermal expansion" produced by the quantum motion of atoms at $T = 0$. At high temperatures the lattice constant expands linearly with T. If we extrapolate this linear dependence to $T = 0$, we obtain the unrenormalized lattice constant, i.e., the lattice constant which we would obtain if the atoms were at rest. The difference between this extrapolated lattice constant and that measured for $T \to 0$ represents Δa_0 $(T = 0)$ which should vary with M like $M^{-\frac{1}{2}}$. The dependence of a_0 on M results in a dependence of energies of interband van Hove singularities on M since these gaps depend on a_0.

Since the zero-point displacement is usually a small contribution to the lattice parameter at 0 K, its contribution is often ignored, particularly since its magnitude is difficult to determine experimentally. An important exception (see below), however, are crystals that are isotopic variants of lithium hydride. Since they are chemically identical, the contribution to the lattice parameters due to the atomic radii and chemical bonding may be taken as constant in all of them. Consequently, the differences in lattice parameters at various temperatures may be attributed solely to differences in the thermal expansion and the zero-point displacement. As it happens, these differences are relatively large in these crystals mostly because of the large relative differences in the atomic masses of the three isotopic forms of hydrogen and the relatively large changes which these produce in the anion–cation reduced masses (see also review by Plekhanov [25]).

The first paper devoted to the calculation of a change in the molecular volume upon isotopic substitution was that of London [148], which has now become classic. He started with an expression for free energy using the Einstein free energy function to reach the expression

$$\frac{M}{V}\frac{\mathrm{d}V}{\mathrm{d}M} = \frac{\gamma\beta}{V}\left[U - E_0 - TC_\mathrm{v}\right]\left[\frac{\mathrm{d}\ln\nu}{\mathrm{d}\ln M}\right], \tag{3.37}$$

where, V is molecular volume, M is the atomic mass, γ is the Gruneisen constant $\gamma = \frac{\alpha V}{\beta C_\mathrm{v}}$, α is the volume expansion coefficient, β is compressibility, C_v is the molar specific heat, E_0 is the potential energy, ν is the phonon frequency, and U is the total energy.

For monoatomic solids,

$$\frac{\mathrm{d}\ln\nu}{\mathrm{d}\ln M} = -\frac{1}{2}, \tag{3.38}$$

and $\frac{\mathrm{d}V}{\mathrm{d}M}$ can be expressed through a Debye function with characteristic temperature θ_D. At high temperatures, one can express the Debye function by a power series in $\left(\frac{\theta_\mathrm{D}}{T}\right)^2$. Then (3.37) becomes (London [148])

$$\frac{M}{V}\frac{\mathrm{d}V}{\mathrm{d}M} = -\frac{1}{20}\alpha T\left(\frac{\theta_\mathrm{D}}{T}\right)\left[1 + \frac{11}{420}\left(\frac{\theta_\mathrm{D}}{T}\right)^2 + \cdots\right], \tag{3.39}$$

and at absolute zero,

$$\frac{M}{V}\frac{\mathrm{d}V}{\mathrm{d}M} = -\frac{9}{16}\frac{\gamma\beta}{V}R\theta_\mathrm{D}. \tag{3.39$'$}$$

Here R is the gas constant.

For a diatomic cubic crystal with atomic masses M and m (see also Born and Huang [1]),

$$\nu^2 \sim \left(\frac{1}{M} + \frac{1}{m}\right). \tag{3.40}$$

Therefore,

$$\frac{\mathrm{d}\ln\nu}{\mathrm{d}\ln M} = -\frac{1}{2}\frac{1}{\left(1+\dfrac{M}{m}\right)}, \tag{3.41}$$

which leads to

$$\frac{M}{V}\frac{\mathrm{d}V}{\mathrm{d}M} = -\frac{1}{20}\alpha T\left(\frac{\theta_{\mathrm{D}}}{T}\right)^2\frac{1}{1+\dfrac{M}{m}}. \tag{3.42}$$

As pointed out by London, to obtain more accurate results, one should have detailed knowledge of the frequency spectrum since ν does not depend on M in a simple fashion. The values predicted by London's analysis are in reasonable agreement with experimental findings (for details, see London [148, 149]).

Lithium Hydride

Closely related to molar volume is thermal expansion (see, e.g., Kogan [150]). Reference [23, Table VIII] gives values of lattice constants and thermal expansion coefficients for isotopic LiH. The data indicate that the isotopic effect, e.g., lighter isotopes having larger lattice constants, is reduced at higher temperatures. Heavier isotopes have larger thermal expansion coefficients. Similar findings were also reported by other workers (for details, see, e.g., [23]). These findings are expected since Debye theory predicts a larger heat capacity [118] for heavier isotopes, and then from the Gruneisen relation, $\alpha = \frac{\gamma C_{\mathrm{v}}}{3\beta V}$, one can obtain $\alpha_{\mathrm{heavy}} > \alpha_{\mathrm{light}}$.

After the classical Londons paper [148], later in Anderson et al. [151], a simple empirical expression that related changes in the reduced mass μ of the unit cell and in the lattice constant a upon isotopic substitution was suggested:

$$a = A\mu + B, \tag{3.43}$$

where A and B are constants, which are, however, dependent on temperature. This relation can be readily obtained by taking into account a linear temperature dependence of the lattice constant, which is typical for high temperatures. The nonlinear temperature dependence of the lattice constant of LiH and LiD crystals (Smith and Leider [152] and Anderson et al. [151]) observed in experiments (Fig. 3.6) can be described by the second-degree polynomial

$$\frac{\Delta a}{a} = \frac{\sqrt{\mu_{\mathrm{LiD}}} - \sqrt{\mu_{\mathrm{LiH}}}}{\sqrt{\mu_{\mathrm{LiH}}\mu_{\mathrm{LiD}}}}\left[A + B(T-T_0) + C(T-T_0)^2\right], \tag{3.44}$$

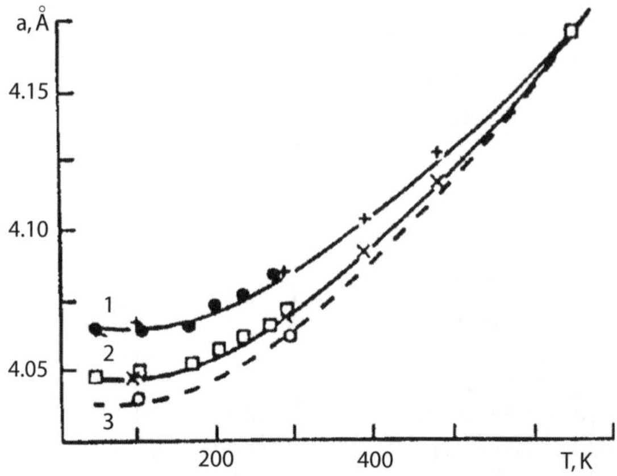

Fig. 3.6. Temperature dependence of the lattice constant of (1) LiH, (2) LiD, and (3) LiT crystals. Experimental data are taken from Smith and Leider [152] and Anderson et al. [151]. The *solid line* is theoretically calculated. (After Plekhanov [25])

where $\Delta a = a_{\mathrm{LiH}} - a_{\mathrm{LiD}}$, $\frac{I}{\mu_{\mathrm{LiH}}} = \frac{1}{M_{\mathrm{H}}} + \frac{1}{M_{\mathrm{Li}}}$, $T = 25°\mathrm{C}$, and $\theta = 1080\,\mathrm{K}$ is the Debye temperature of a LiH crystal. The values of constants A, B, and C, determined by the method of least squares, are presented in Table 3.3. For comparison, the values of these constants, calculated in a similar way for diamond, silicon, and germanium (Pavone and Baroni [153]]), are also given. The value of temperature $T = 810\,\mathrm{K}$, at which the lattice constant is the same for LiH and LiD crystals, was found from theoretical calculations. The lattice constant is equal to 4.165 Å. This means that the temperature dependence $a(T)$ in LiD crystals (a heavy isotope), which have a smaller Debye temperature, is stronger than that in LiH crystals. This general conclusion is valid for a broad class of compounds, from an ionic LiH crystal to a covalent germanium crystal. It follows from theoretical calculations (Fig. 3.6) that for $T > 810\,\mathrm{K}$, the lattice constant of LiD crystals is larger than that of LiH crystals. This agrees qualitatively with the results of microscopic calculations of the temperature dependence of a change in the unit cell upon isotopic substitution, according to which $a_{\mathrm{LiH}} = a_{\mathrm{LiD}}$ at $900 - 950\,\mathrm{K}$ (Shpilrain et al. [154]). The change in the lattice constant upon isotopic substitution is determined mainly by the anharmonicity of vibrations, which results in the dependence of the distance between atoms on the vibration amplitude, i.e., on the mean vibrational energy. It is well known that the vibrational energy depends not only on temperature, but on the isotopic composition as well. Therefore, to take into account the thermal expansion of a crystal lattice, one should consider the effect of anharmonic terms in the expression for potential energy of pair interaction between atoms at temperature T.

Table 3.3. Values of coefficients of polynomial (3.44) describing the temperature dependence of the lattice constant upon isotopic substitution of a mass. (After Plekhanov [25]

Substance	$A \cdot 10^3$	$B \cdot 10^6$	$C \cdot 10^9$
LiH	55.40	−55.54	−102.80
C	−5.48	3.55	8.21
Si	−1.60	3.94	−6.90
Ge	−0.72	2.27	−6.40

Germanium

Among semiconducting crystals, the Ge crystal was the first for which the dependence of the lattice constant on an isotopic effect was investigated both theoretically and experimentally (Buschert et al. [155], Pavone and Baroni [153], and Noya et al. [156]). In the very first paper (Buschert et al. [155]), this dependence was experimentally studied in a perfect crystal of natural isotopic composition (average $M = 72.59$), and a second crystal was isotopically enriched, containing 95.8% of ^{74}Ge (average $M = 73.93$). Using the following equation (analogous to (3.44)) for the relative changes in the lattice constant a with isotopic mass at low temperature ($e^{-\frac{\hbar\omega}{k_B T}} \gg 1$)

$$\frac{\Delta a}{a} = -\frac{C}{a^3}\frac{\Delta M}{M}\left(\gamma_0 \hbar\omega_0 + \frac{3}{4}\gamma_a k_B \theta\right), \tag{3.45}$$

where $\gamma_0 = 1.12$ and $\gamma_a = 0.40$ are the Gruneisen parameters for optical and acoustical phonon modes in Ge, $\theta = 374\,\mathrm{K}$ is the Debye temperature, and $\hbar\omega_0 = 37.3\,\mathrm{meV}$. Buschert et al. evaluated (3.45) for a 95.8% enriched ^{74}Ge crystal in comparison with a natural crystal. Equation (3.45) predicts $12 \cdot 10^{-6}$ and $6 \cdot 10^{-6}$ reduction in a for $T = 0$ and $T = 300\,\mathrm{K}$, respectively. Buschert et al. experimentally found reductions of 14.9 and 6.3 ppm at 77 K and $T = 300\,\mathrm{K}$, respectively. The agreement between calculated and measured values is very good considering the uncertainties of the Gruneisen parameter values used in the theory. After this paper (Buschert et al. [155]), two theoretical papers (Pavone and Baroni [153] and Noya et al. [156]) were published where the isotopic effect and its temperature dependence were studied. In the paper of Pavone and Baroni [153], the dependence of the lattice constants of C, Si, and Ge upon their isotopic purity were calculated using first principles by treating nuclear vibrations by density-functional perturbation theory. The main results of this paper are depicted in Fig. 3.7. The values of the constants A, B, and C [see (3.44] fitted to theoretical data for the three materials studied in the paper of Pavone and Baroni as well as for LiH [25] are reported in Table 3.3. Noya et al. [156] studied the dependence of the lattice parameters upon the isotope mass for five isotopically pure Ge crystals by quantum path-integral Monte Carlo simulations. The interatomic

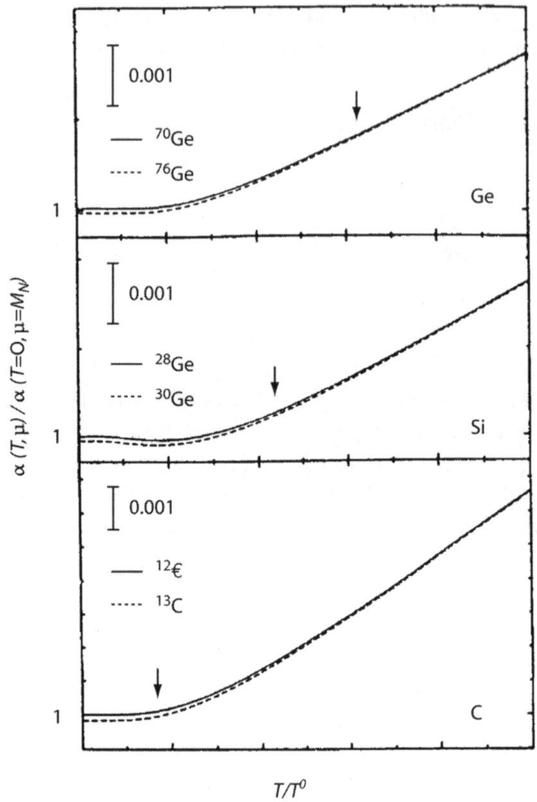

Fig. 3.7. Dependence of the equilibrium lattice constants of C, Si, and Ge on temperature for different isotopic masses. The temperature is given in units of T^* [T^* (C) = 1941 K, T^* (Si) = 744 K, T^* (Ge) = 440 K]. The arrows indicate room temperature (25°C). The lattice constants are in units of zero-temperature lattice constants at the natural isotopic compositions ($a_{\mathrm{C}} = 6.71$, $a_{\mathrm{Si}} = 10.23$, and $a_{\mathrm{Ge}} = 10.61$ a.u.). Note the different units in the three panels, which are indicated by vertical bars. (After Pavone et al. [153])

interactions in the solid were described by an empirical of the Stillinger–Weber type. At 50 K, the isotopic effect leads to an increase of $2.3 \cdot 10^{-4}$ Å in the lattice parameter of ^{70}Ge with respect to ^{76}Ge. Comparison of the simulation results with available experimental data for ^{74}Ge (see also [155]) shows that the model employed provides a realistic description of this anharmonic effect. It was shown by Noya et al. that the calculated fractional change in the lattice parameter of ^{74}Ge with respect to a crystal whose atoms have the average mass of natural Ge amounts to $\frac{\Delta a}{a} = -9.2 \cdot 10^{-6}$ at $T = 0$ K (see also Fig. 3.8), that is not badly in agreement with the results of Buschert et al.

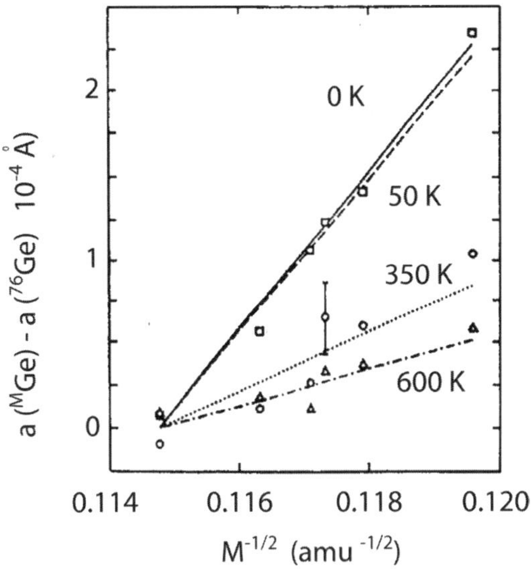

Fig. 3.8. Difference between the lattice parameter of stable germanium isotopes and that of ^{76}Ge versus the inverse square of the isotope mass for different temperatures. Symbols are the results from quantum simulations: *squares* at 50 K, *circles* at 350 K, and *triangles* at 600 K. Lines are the results from quantum quasi-harmonic approximation: *continuous line* at 0 K, *dashed* at 50 K, *dotted* at 350 K, and *dash-dotted line* at 600 K. (After [156])

Diamond

The isotopic dependence of the lattice constant of diamond has also attracted interest in connection with the above thermal properties (see above section of thermal conductivity). Banholzer et al. [131] reported the lattice constants of natural and ^{13}C diamond by X-ray diffraction using powder samples and single crystals. Holloway et al. [40,157] examined the influence of the isotopic ratio on the lattice constant of mixed crystals of $^{13}C_x^{12}C_{1-x}$ by single-crystal X-ray diffractometry. According to their results, the lattice constant for, as in case of LiH, decreased linearly with the ^{13}C content according to the expression

$$a(x) = 3.56714 - 5.4 \cdot 10^{-4} x. \tag{3.46}$$

The fractional difference $\frac{\Delta a}{a}$ between both end compositions is $-1.5 \cdot 10^{-4}$. The lattice constants of the five samples of mixed diamond with different isotopic compositions were studied in the paper of Yamanaka et al. [158]. In that paper, it was shown that the standard deviations of the lattice constant were in the range of $5-9 \cdot 10^{-5}$ Å. The lattice constant varies (see also Fig. 3.8) with the isotopic ratio, and it can be expressed in quadratic form as

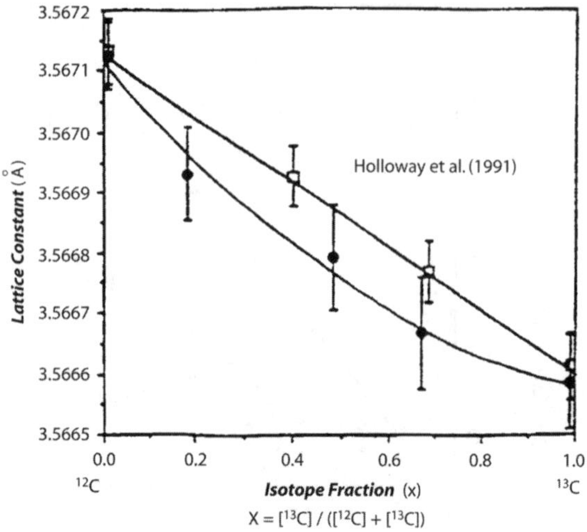

Fig. 3.9. Isotope dependence of the lattice constant of diamond. The *curved line* shows the quadratic (3.47), which fits to *solid circles* obtained by Yamanaka et al. The *straight line* fits to *open squares* obtained by Holloway et al. [40]. (After Yamanaka et al. [158])

$$a(x) = 3.56712 - 9.0 \cdot 10^{-4}x + 3.7 \cdot 10^{-4}x^2, \tag{3.47}$$

where $x = \left[\frac{^{13}\mathrm{C}}{^{12}\mathrm{C}+^{13}\mathrm{C}}\right]$. This expression contrasts (Fig. 3.9) with the linear relation reported by Holloway et al. [157]. A linear relation between the lattice constant and isotopic ratio would be somewhat puzzling because the compressibility and the Gruneisen parameter are not the same for different isotopes (see also [25]).

In concluding this section, we should stress the premier role of the anharmonic effect in the dependence of a on the isotopic effect. As pointed out by Vogelgesang et al. [33], the concentration-dependent lattice parameter incorporating zero-point motion in combination with anharmonicity, can be deduced from (2.99)

$$a(x) = a_{12} - \frac{\hbar g_1}{(6k_1^3 M_{12})}\left[1 - \left(\frac{M_{12}}{M_x}\right)^{\frac{1}{2}}\right], \tag{3.48}$$

with $M_x = (1-x)\,M_{12} + xM_{13}$. A comparison of (3.48) with the data of Holloway et al. for $a(x)$ and $k_1 = 3Ba = 4.76 \cdot 10^5\,\mathrm{dyn/cm}$ for natural diamond yields $g_1 = (4.5 \pm 0.4) \cdot 10^{14}\,\mathrm{erg/cm^3}$. Yamanaka's et al. results analyzed in the same manner yield $g_1 = (4.7 \pm 0.4) \cdot 10^{14}\,\mathrm{erg/cm^3}$, very close to Holloway's data. Lattice constants for the isotopic dependence of compound semiconductors can be found in monograph [23].

4 Isotopic Renormalization of the Electronic Excitation Energy Spectrum

4.1 Exciton States

It has been more than six decades since the introduction of the quasi-particle exciton by Frenkel [161], and the extreme fertility of this idea has been demonstrated most powerfully. According to Frenkel, the exciton is an electron excitation of one of the atoms (ions) of a crystal lattice because of the translation at symmetry that moves through the crystal in an electrically neutral formation. Since Frenkel, the concept of an exciton has been developed in the papers of Peierls [162]] and Slater and Schokley [163]. Problems concerning light absorption by the solid state have been considered somewhat differently (Wannier [164] and Mott [165]). According to the Wannier–Mott results, the exciton is the state of an electron and hole bonded by the Coulomb force. The electron and hole the in exciton state are spatially separated, and their charges are screened. In the Frenkel papers, the excitations localized on the lattice site were described thus; after the Wannier–Mott papers, the excitons became divided into Frenkel (small radius) excitons (for details, see Davydov [166]) and Wannier–Mott (large radius) excitons [167]. However, a description of the basic difference between these two models is absent (Davydov [166], Knox [167], and Agranovich and Ginsburg [168]). The experimental discovery (see, e.g., Gross [169] of the Wannier–Mott exciton in the hydrogen-like absorption spectrum of semiconducting crystals was the basis for a new subject – exciton physics (see also [109]). The influence of external perturbation (electrical and magnetic fields, one-axial and hydrostatic deformation) on the optical spectra of Wannier–Mott excitons (see, e.g., Gross [169]) and their energetic characteristics has been demonstrated repeatedly. These investigations permitted high-accuracy measurements of exciton binding energy and also of their translational mass, values of effective masses of the electron and hole, their g-factors, etc. Moreover, the detailed account of photon–exciton interaction has led to the concept of polaritons (Pekar [170]). From the time of the experimental discovery of the Wannier–Mott exciton, the problem of the interaction of excitons and the crystal lattice has persisted for more than four decades (Haken [171]).

As noted above, the theory of excitons with a large radius has been reviewed by several authors (Knox [167], Dexter and Knox [172], Elliott [173], and Dimmock [174]). All of these discussions are exceptionally lucid, and each emphasizes somewhat different aspect of the problem. In view of Dimmock's review of the various approximate exciton wave functions and techniques for evaluating optical transition matrix elements, only a very brief discussion of the results of immediate use will be given here. The description of the exciton wave function in the effective mass approximation (see also Dresselhaus [175]) differs from the impurity function only in that we must consider both the electron and hole on equal footing. The slowly varying envelope function for the exciton depends on the coordinates of both particles. It is given as the solution of (compare (46) of Dimmock [174])

$$\left[-\frac{\hbar^2}{2m_\mathrm{e}}\nabla_e^2 - \frac{\hbar^2}{2m_\mathrm{h}}\nabla_\mathrm{h}^2 - \frac{e^2}{k\left|\boldsymbol{r}_\mathrm{e}-\boldsymbol{r}_\mathrm{h}\right|} - \varepsilon\right]\Phi\left(\boldsymbol{r}_\mathrm{e},\boldsymbol{r}_\mathrm{h}\right) = 0. \tag{4.1}$$

The total exciton wave function is

$$\Psi_x = \Phi\left(\boldsymbol{r}_\mathrm{e},\boldsymbol{r}_\mathrm{h}\right) u_{\mathrm{c,o}}\left(\boldsymbol{r}_\mathrm{e}\right) u_{\mathrm{v,o}}\left(\boldsymbol{r}_\mathrm{h}\right). \tag{4.2}$$

Making the usual center-of-mass transformation

$$\boldsymbol{r} = \boldsymbol{r}_\mathrm{e} - \boldsymbol{r}_\mathrm{h}, \quad \boldsymbol{R} = \frac{\left(m_\mathrm{e}\boldsymbol{r}_\mathrm{e} + m_\mathrm{h}\boldsymbol{r}_\mathrm{h}\right)}{M}, \tag{4.3}$$

where

$$M = m_\mathrm{e} + m_\mathrm{h}, \tag{4.4}$$

the equation separates into translational and Coulomb interaction terms. Upon substitution of

$$\Phi\left(\boldsymbol{r}_\mathrm{e},\boldsymbol{r}_\mathrm{h}\right) \mathrm{e}^{\mathrm{i}\boldsymbol{KR}} F\left(\boldsymbol{r}\right), \tag{4.5}$$

we obtain

$$\left[-\hbar^2\nabla^2/2M - e^2/kr - \varepsilon_\mathrm{n}\right] F\left(\boldsymbol{r}\right) = 0 \tag{4.6}$$

with

$$\varepsilon = \varepsilon_\mathrm{n} + \hbar^2K^2/2M. \tag{4.7}$$

Equation (2.6) is identical to the corresponding equation for the impurity (Dimmock [174]) problem. In the hydrogenic approximation,

$$F_\mathrm{n}\left(\boldsymbol{r}\right) = \left(\pi a_x^3\right)^{-\frac{1}{2}} \mathrm{e}^{-\frac{r}{a_x}}. \tag{4.8}$$

Still following the analogy with the impurity problem, the exciton envelope function is conventionally expanded into a Fourier series

$$\mid\Phi\rangle = \sum_{k_\mathrm{e}k_\mathrm{h}} \mid \boldsymbol{k}_\mathrm{e}\boldsymbol{k}_\mathrm{h}\rangle\langle\boldsymbol{k}_\mathrm{e}\boldsymbol{k}_\mathrm{h} \mid \Phi\rangle = \sum_{k_\mathrm{e}k_\mathrm{h}} \mid \boldsymbol{k}_\mathrm{e}\boldsymbol{k}_\mathrm{h}\rangle a(\boldsymbol{k}_\mathrm{e}\boldsymbol{k}_\mathrm{h}), \tag{4.9}$$

where

$$| \boldsymbol{k}_{\mathrm{e}}\boldsymbol{k}_{\mathrm{h}}\rangle = \mathrm{e}^{\mathrm{i}\boldsymbol{k}_{\mathrm{e}}\boldsymbol{r}_{\mathrm{e}}}\mathrm{e}^{\mathrm{i}\boldsymbol{k}_{\mathrm{h}}\boldsymbol{r}_{\mathrm{h}}}$$

and

$$a\left(\boldsymbol{k}_{\mathrm{e}}\boldsymbol{k}_{\mathrm{h}}\right) = \langle \boldsymbol{k}_{\mathrm{e}}\boldsymbol{k}_{\mathrm{h}} \mid \Phi\rangle = \iint \mathrm{e}^{-\mathrm{i}\boldsymbol{k}_{\mathrm{e}}\boldsymbol{r}_{\mathrm{e}}}\mathrm{e}^{-\mathrm{i}\boldsymbol{k}_{\mathrm{h}}\boldsymbol{r}_{\mathrm{h}}}\Phi\left(\boldsymbol{r}_{\mathrm{e}}\boldsymbol{r}_{\mathrm{h}}\right)\mathrm{d}\boldsymbol{r}_{\mathrm{e}}\,\mathrm{d}\boldsymbol{r}_{\mathrm{h}}. \tag{4.10}$$

Expressing $\boldsymbol{r}_{\mathrm{e}}$ and $\boldsymbol{r}_{\mathrm{h}}$ in center-of-mass coordinates $\boldsymbol{r}$ and $\boldsymbol{R}$, $\boldsymbol{r}_{\mathrm{e}} = \boldsymbol{R} + \left(\frac{m_{\mathrm{h}}}{M}\right)\boldsymbol{r}$, $\boldsymbol{r}_{\mathrm{h}} = \boldsymbol{R} - \left(\frac{m_{\mathrm{e}}}{M}\right)\boldsymbol{r}$,

$$a\left(\boldsymbol{k}_{\mathrm{e}}\boldsymbol{k}_{\mathrm{h}}\right) = \int \mathrm{e}^{\mathrm{i}(\boldsymbol{K}-\boldsymbol{k}_{\mathrm{e}}-\boldsymbol{k}_{\mathrm{h}})}\mathrm{d}\boldsymbol{R}\int \mathrm{e}^{-\boldsymbol{k}\boldsymbol{r}}F\left(\boldsymbol{r}\right)\mathrm{d}\boldsymbol{r}, \tag{4.11}$$

where

$$\boldsymbol{k} = \left(\frac{m_{\mathrm{h}}}{M}\right)\boldsymbol{k}_{\mathrm{e}} - \left(\frac{m_{\mathrm{e}}}{M}\right)\boldsymbol{k}_{\mathrm{h}}.$$

The first integral is the delta function $\delta_{\boldsymbol{K},\boldsymbol{k}_{\mathrm{e}}+\boldsymbol{k}_{\mathrm{h}}}$, and the second has been evaluated for the impurity problem (see, e.g., Dimmock [174])

$$\begin{aligned} a\left(\boldsymbol{k}_{\mathrm{e}}\boldsymbol{k}_{\mathrm{h}}\right) &= \int \mathrm{e}^{-\mathrm{i}\boldsymbol{k}\boldsymbol{r}}F\left(\boldsymbol{r}\right)\mathrm{d}r\,\delta_{\boldsymbol{K},\boldsymbol{k}_{\mathrm{e}}+\boldsymbol{k}_{\mathrm{h}}} \\ &= \left\{\frac{8\pi^{\frac{1}{2}}a_x^{\frac{3}{2}}}{\left[1+\left(ka_x\right)^2\right]^2}\right\}\delta_{\boldsymbol{K},\boldsymbol{k}_{\mathrm{e}}+\boldsymbol{k}_{\mathrm{h}}}. \end{aligned} \tag{4.12}$$

In most cases, we are interested in excitons that participate in optical transitions and thus conserve momentum with a photon. Therefore, $\boldsymbol{k}$ is usually quite small. Letting $\boldsymbol{K} = 0$, $\boldsymbol{k}_{\mathrm{e}} = -\boldsymbol{k}_{\mathrm{h}}$ reduces $\boldsymbol{k}$ to $\boldsymbol{k}_{\mathrm{e}}$, and

$$\Phi\left(\boldsymbol{r},\boldsymbol{R}\right) = \mathrm{e}^{\mathrm{i}\boldsymbol{K}\boldsymbol{R}}F\left(\boldsymbol{r}\right)F\left(\boldsymbol{r}\right), \tag{4.13}$$

hence

$$\begin{aligned} | \Phi\rangle - | F\left(\boldsymbol{r}\right)\rangle &= \sum_{k_{\mathrm{e}}k_{\mathrm{h}}} | \boldsymbol{k}_{\mathrm{e}}\boldsymbol{k}_{\mathrm{h}}\langle a(\boldsymbol{k}_{\mathrm{e}}\boldsymbol{k}_{\mathrm{h}})\delta_{0,\boldsymbol{k}_{\mathrm{e}}+\boldsymbol{k}_{\mathrm{h}}} \\ &= \sum_{k_{\mathrm{e}}k_{\mathrm{h}}} \mathrm{e}^{\mathrm{i}(\boldsymbol{k}_{\mathrm{e}}+\boldsymbol{k}_{\mathrm{h}})\boldsymbol{r}}a(\boldsymbol{k}_{\mathrm{e}}\boldsymbol{k}_{\mathrm{h}})\delta_{\boldsymbol{k}_{\mathrm{e}}-\boldsymbol{k}_{\mathrm{h}}} \\ &= \sum_{k} \mathrm{e}^{\mathrm{i}\boldsymbol{k}\boldsymbol{r}}a\left(\boldsymbol{k}\right) = \sum_{k} | \boldsymbol{k}\rangle\langle \boldsymbol{k} \mid F\rangle. \end{aligned} \tag{4.14}$$

We will also need to evaluate the expansion of the wave function $F\left(\boldsymbol{r}\right)$ at $\boldsymbol{r} = 0$. The exponential $\mathrm{e}^{\mathrm{i}\boldsymbol{k}\boldsymbol{r}}$ is unity at $\boldsymbol{r} = 0$, and $a\left(\boldsymbol{k}\right)$ is independent of $\boldsymbol{r}$; hence,

$$| F\left(0\right)\rangle = \sum_{k} a\left(\boldsymbol{k}\right) \equiv \sum_{k}\langle \boldsymbol{k} \mid F\rangle. \tag{4.15}$$

Finally, the total exciton wave function can be written in a number of different forms:

$$\Psi_x(\boldsymbol{r}_\mathrm{e}\boldsymbol{r}_\mathrm{h}) \equiv |\,x\rangle = \Phi(\boldsymbol{r}_\mathrm{e}\boldsymbol{r}_\mathrm{h})\,u_{\mathrm{c,o}}(\boldsymbol{r}_\mathrm{e})\,u_{\mathrm{v,o}}(\boldsymbol{r}_\mathrm{h}) = u_{\mathrm{c,o}(\boldsymbol{r}_\mathrm{e})}u_{\mathrm{v,o}}\sum_k \mathrm{e}^{\mathrm{i}\boldsymbol{k}\boldsymbol{r}}a(\boldsymbol{k})$$

$$= \sum_{k_\mathrm{e},k_\mathrm{h}} |\,c,\boldsymbol{k}_\mathrm{e};v,\boldsymbol{k}_\mathrm{h}\rangle a(\boldsymbol{k}_\mathrm{e}\boldsymbol{k}_\mathrm{h})\,\delta_{k_{\mathrm{e}'},-k_\mathrm{h}}, \tag{4.16a}$$

where

$$|\,c,\boldsymbol{k}_\mathrm{e};v,\boldsymbol{k}_\mathrm{h}\rangle = \mathrm{e}^{\mathrm{i}\boldsymbol{k}_\mathrm{e}\boldsymbol{r}_\mathrm{e}}u_{\mathrm{c,o}}(\boldsymbol{r}_\mathrm{e})\,\mathrm{e}^{\mathrm{i}\boldsymbol{k}_\mathrm{h}\boldsymbol{r}_\mathrm{h}}u_{\mathrm{v,o}}(\boldsymbol{r}_\mathrm{h})\,. \tag{4.16b}$$

Now we come to a rather different consideration. The exciton state $|\,x\rangle$ is an excited state of the crystal arising from the removal of an electron from the valence band and leaving behind a hole. The ground state of the system is the absence of an exciton, say $|\,0\rangle$. Further, the exciton state is not properly described by assuming that the product of one electron functions because of the Coulomb interaction between the electron and hole (note $|\,x\rangle$ cannot be written as a product function). Thus the familiar Hartree–Fock approximation is not adequate, and recourse to many-body techniques becomes necessary.

Rather than delve into details of treating the many-body problem, the reader is referred to the standard texts (e.g., Kittel [176] and Dexter and Knox [172]). Often, the evaluation of the optical matrix element is suppressed with little comment. However, it is not a trivial problem either conceptually or in its execution, it is straightforward though in the proper formalism. On the other hand, the final result is similar to the impurity matrix element, and we will lean on this analogy in preference to a detailed treatment. The momentum matrix element between the vacuum state $|\,0\rangle$ and the creation of an exciton $|\,x\rangle$ is

$$\langle x\,|\,\hat{\varepsilon}\cdot\boldsymbol{p}\,|\,0\rangle = \sum_k a(\boldsymbol{k}) < u_{\mathrm{v,o}}(\boldsymbol{r})\,|\hat{\varepsilon}\cdot\boldsymbol{p}|\,u_{\mathrm{c,o}}(\boldsymbol{r}) = \sum_k a(\boldsymbol{k})\,p_{\mathrm{cv}} = F(0)\,p_{\mathrm{cv}}. \tag{4.17}$$

It might be instructive to compare the form of the exciton matrix element with the analogous band impurity matrix element. Consider the hydrogenic 1s state

$$\langle x\,|p_{\hat{\varepsilon}}|\,0\rangle^2 = \left|\sum a(\boldsymbol{k})\right|^2 |p_{\mathrm{cv}}|^2 = |F(0)|^2\,|p_{\mathrm{cv}}|^2 \tag{4.18}$$

and

$$|\langle I|\,p_{\hat{\varepsilon}}\,|\,n,\boldsymbol{k}\rangle\,|^2\,|a(\boldsymbol{k})|^2\,|p_{\mathrm{cv}}|^2\,, \tag{4.19}$$

where $|\,a(\boldsymbol{k})\,|^2$ is given by (4.12). The difference in the form of the exciton and impurity matrix elements arises from differences in energy and momentum selection rules. Band impurity transitions involve one localized state (described by a sum over all $\boldsymbol{k}$ states) and a band state where $\boldsymbol{k}$ is a good quantum number. Thus, the band Bloch state $\mid n, \boldsymbol{k}\rangle$ couples only the $a(\boldsymbol{k})$-Fourier coefficient eliminating the sum over $\boldsymbol{k}$. In the exciton, both the electron and hole must be simultaneously Fourier transformed. Since both particles are localized, $\boldsymbol{k}$ is not a good quantum number for either particle; the sum over $\boldsymbol{k}$ remains though the double sum over $\boldsymbol{k}_{\mathrm{e}}$ and $\boldsymbol{k}_{\mathrm{h}}$ is reduced to a single sum by the requirement $\boldsymbol{k}_{\mathrm{e}} = \boldsymbol{k}_{\mathrm{h}}$.

Since the exciton is an electron–hole complex, each excited state, namely, the exciton itself, must be associated with a ground state, the absence of an exciton, i.e., the excited state and the ground state are mutually exclusive. Therefore, the emission probability is proportional to the number of excitons, N_x (number of electron–hole pairs) rather than the product of the number of electrons and holes np. Accordingly, the spontaneous emission rate is

$$\begin{aligned} R_{\mathrm{sp}}(\hbar\omega) &= \left(\frac{2\pi}{\hbar}\right)\sum_{E_x}\left|\langle x\left|\Re\right|0\rangle\right|^2 G(\hbar\omega)\, N_x(E_x)\,\delta(\hbar\omega - E_x) \\ &= n(\hbar\omega)\left(4e^2\hbar\omega/m^2\hbar^2c^3\right)\left|p_{\mathrm{cv}}\right|^2 \\ &\quad\times\sum_{E_x}\left|F(0)\right|^2 N_x(E_x)\,\delta(\hbar\omega - E_x)\,\delta_{\boldsymbol{K},\boldsymbol{k}'}, \end{aligned} \tag{4.20}$$

where $\boldsymbol{k}$ is the photon wave vector and $\hbar\boldsymbol{K}$ is the momentum of the exciton with kinetic energy

$$\varepsilon_{\mathrm{kin}}(\boldsymbol{K}) = \hbar^2\boldsymbol{K}^2/2M, \qquad M = m_{\mathrm{e}} + m_{\mathrm{h}}. \tag{4.21}$$

Assuming that photon momentum is negligible, only zero kinetic energy excitons $\boldsymbol{K} = 0$ can recombine.

Below, we briefly treat the general solution of the hydrogenic equation to emphasize the commonality of the different states. The radial solution of the hydrogenic equation can be written in a unifying way for both positive and negative energies as (e.g., Bethe and Salpeter [177] and Elliott [178])

$$R_{n,l}(r) = N\rho^l e^{-\frac{\rho}{2}} F\left(l + 1 - \eta/2l + \frac{2}{\rho}\right), \tag{4.22}$$

where η is an energy parameter,

$$\eta = \left(-\frac{E_{\mathrm{B}}}{\varepsilon}\right), E_{\mathrm{B}} = \hbar^2/2\mu a_x^2, \tag{4.23}$$

and ε is the energy eigenvalue in (4.15). Two cases are distinguished, the negative energy discrete states and the positive energy continuum states (see also Fig. 4.1). The parameters ρ, N, and η and the s-state solutions $\left|F(0)\right|^2$ at $\boldsymbol{r} = 0$ are quoted for two cases.

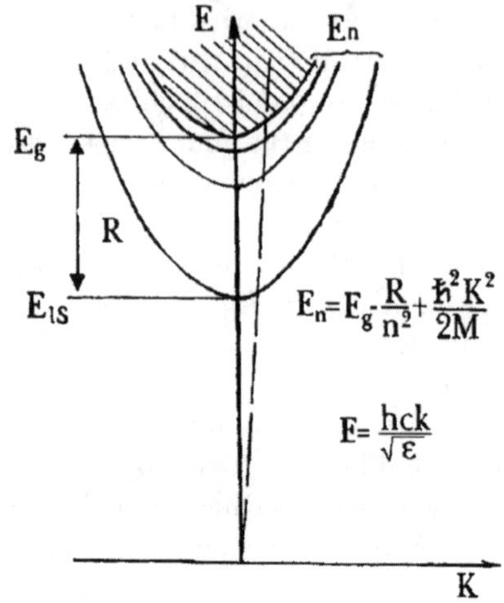

Fig. 4.1. Discrete and continuous (*hatched area*) Wannier–Mott exciton energy spectrum taking into account its kinetic energy. The *broken line* connects to the dispersion of light in the medium

(a) Discrete states ($\varepsilon < 0$). Here, η is the principal quantum number n,

$$\varepsilon_{\mathrm{n}} = -\frac{E_{\mathrm{B}}}{n^2}, \tag{4.24a}$$

$$\rho = \frac{2r}{na_x}, \tag{4.24b}$$

$$N = \left\{ \frac{\left(\frac{2}{na_x}\right)}{(2l+1)!} \right\} \left[\frac{(n+l)!}{(n-l-1)!2n} \right], \tag{4.24c}$$

and

$$|F(0)|^2 = \left(\pi a_x^3 n^3\right)^{-1}. \tag{4.24d}$$

Only s-states are nonzero at $r = 0$.

(b) Continuum states ($\varepsilon > 0$). For the energy ε to become positive, η must go over to an imaginary number, $\eta = i\gamma = \frac{i}{ka_x}$, then

$$\varepsilon(k) = -\frac{E_{\mathrm{B}}}{(i\gamma)^2} = E_{\mathrm{B}}(ka_x)^2 = \hbar^2 k^2/2\mu, \tag{4.25a}$$

$$\rho = -2ikr, \tag{4.25b}$$

$$N = \left(\frac{i^l}{(2^l+1)!}\right) |\Gamma(l+1-i\gamma)| \, e^{\pi\gamma/2}, \tag{4.25c}$$

and

$$|F(0)|^2 = \frac{z\,e^z}{\sinh z}, \quad z = \pi\gamma. \tag{4.25d}$$

From (4.25a)

$$\varepsilon(k) = \frac{E_B}{\gamma^2} = \hbar\omega - \varepsilon_g; \quad \gamma = \left[\frac{E_B}{(\hbar\omega - \varepsilon_g)}\right]^{\frac{1}{2}} \tag{4.25e}$$

Note that k (or γ) is nothing more than a quantum number; although the energy $\varepsilon(k) = \frac{\hbar^2 k^2}{2\mu}$ has the familiar appearance of kinetic energy, it actually contains both potential and kinetic energy contributions of the relative motions of the electron and hole (but not the translational energy of the exciton as whole). Substituting the results in (4.20) and noting that

$$E_x = \varepsilon_g - E_B/n^2, \tag{4.26}$$

we see that the emission from the discrete states occurs below the band gap energy at

$$\hbar\omega = \varepsilon_g - E_B/n^2 \quad \text{for } n = 1, 2, 3, \ldots \tag{4.27}$$

We should stress that the center of mass can move through a crystal by diffusional or drift processes, just like individual electronic particles. However, this exciton migration does not of itself produce electrical conductivity, since an exciton contains a pair of charges of opposite sign. The exciton binding energy $E_B = \frac{\mu e^4}{2\hbar^2\varepsilon^2 n^2}$ does not depend on the nuclear mass (see also [109]). From the last formula (see also Fig. 4.1) the isotopic effect in the frozen crystal lattice we come to the natural conclusion that is absent on the levels of the Wannier–Mott exciton. This crude estimate does not take into account the exciton–lattice interaction. The coupling of the exciton–phonon interaction depends on the phonon's frequency and thus on the mass of the vibrational atoms (ions), in other words, on the isotope's mass. This dependence of the exciton binding energy in LiH (LiD) crystals was observed in Plekhanov et al. [179], where the reflectance spectra at low temperature were measured for the first time (see also [180]). As will be shown below, the discovery of the dependence of E_B on the isotopic composition of a crystal lattice revealed the unique possibility of the experimental reconstruction of the value of the coupling of Coulomb and Fröhlich interactions and also control of the effectivity of the exciton–phonon interaction.

4.2 Exciton–Phonon Interaction

The foundations of the theory of exciton–phonon interaction were laid in the 1950s. The interaction between an electron and a nonpolar optical phonon in a crystal can be described simply in terms of a deformation potential (see, e.g., Bir and Picus [181]). The net effect of the lattice displacement on the electron is assumed to be a small shift in the electronic energy band of the crystal. The constant of proportionality between this energy shift and the lattice displacement is defined as the deformation potential. The exciton–phonon (deformational potential) Hamiltonian can, therefore, be expressed as (Ansel'm and Firsov [182])

$$H_{\mathrm{EP}} = \sqrt{\frac{\hbar}{2\mu N\omega_0}} \frac{(W_{\mathrm{e}} - W_{\mathrm{h}})}{a} a^{+}_{\boldsymbol{k}++\boldsymbol{q}} a_{\boldsymbol{k}} \left[b_{\boldsymbol{q}} + b^{+}_{-\boldsymbol{q}} \right], \tag{4.28}$$

where $W_{\mathrm{e}}, W_{\mathrm{h}}$ denote the deformation potentials of the electron and hole, respectively; $a^{+}_{\boldsymbol{k}+\boldsymbol{q}}$, $a_{\boldsymbol{k}}$ are the creation and annihilation operators of an exciton with wave vector $\boldsymbol{k}$; $b^{+}_{-\boldsymbol{q}}, b_{\boldsymbol{q}}$ are the creation and annihilation operators of an optical phonon with momentum $\hbar\boldsymbol{q}$; μ is the reduced mass of atoms in the unit cell; N is the number of unit cells in the crystal; a is the lattice constant of the crystal; and $\hbar\omega_0$ is the energy of the optical phonon. In addition to a pioneering study (Ansel'm and Firsov [182]), the interaction of excitons with longitudinal optical phonons was considered by many authors [167,171,183–185]. In ionic crystals, there are two main mechanisms for the interaction of excitons with lattice vibrations. One mechanism, the short-range deformation interaction, is caused by modulation of the wave function of excitons by longitudinal acoustic vibrations. The magnitude of this interaction is characterized by the deformation potential (see above). The deformation interaction strongly affects the energy spectrum and dynamics of excitons of relatively small radius (e.g., the ground state of excitons in AHC (alkalic halide crystals) and crystals of inert gases [167]). As the radius of the exciton increases, this interaction becomes less important, since the wave vector of actual phonons is $q \propto r_{\mathrm{ex}}^{-1}$ [186] where r_{ex} is the exciton radius and the number of such phonons is proportional to q^3.

The second mechanism – the polarization or Fröhlich interaction [187] – is caused by the coulombic interaction of the charge carriers forming the exciton with the macroscopic field created by longitudinal optical oscillations (see, e.g., Pekar [170] and Firsov [184]). If the exciton radius is much greater than the lattice constant, then the exciton–phonon interaction may be regarded as the sum of independent interactions of electrons and holes with phonons (see Klochikhin [188]). The interaction operator of charge e and mass m (m_{e} or m_{h}), neglecting the dispersion of the latter, is (Fröhlich [187])

$$H_{\mathrm{el}} = \sum_{q} W_{\boldsymbol{q}} \exp(\mathrm{i}\boldsymbol{q}\boldsymbol{r})(b^{+}_{-\boldsymbol{q}} + b_{\boldsymbol{q}}), \tag{4.29}$$

where

$$W_{\boldsymbol{q}} = \frac{\hbar\omega_{\mathrm{LO}}}{|\boldsymbol{q}|} \left[\frac{4\pi q^2_{\mathrm{e,h}}}{V} \right]^{1/2}, \tag{4.30}$$

where V is the volume of the system, and $\boldsymbol{r}$ are the coordinates of the particles. In this expression, we introduce the main parameters that determine the interaction of the electron (hole) with optical vibrations: the polaron "radius"

$$r^*_{\mathrm{e,h}} = \left(\frac{\hbar}{2} m_{\mathrm{e,h}} \omega_{\mathrm{LO}}\right)^{1/2} \tag{4.31}$$

and the dimensionless Fröhlich constant of interaction

$$g^2_{\mathrm{e,h}} = \frac{1}{\alpha}\left(\frac{1}{\varepsilon_\infty} - \frac{1}{\varepsilon_0}\right) \frac{e^2}{\hbar \omega_{\mathrm{LO}} r^*_{\mathrm{e,h}}} . \tag{4.32}$$

The first of these quantities characterizes the size of the polarization region of the lattice by the extra charge, and the second describes the strength of the electron–phonon interaction (see also Knox [167] and Haken [171])).

As follows from (4.29), the interaction operator $W_{\boldsymbol{q}}$ does not depend on the mass of a quasi-particle and is the same for electrons and holes. Accordingly, the interaction Hamiltonian of Wannier–Mott excitons with optical phonons has the same form (4.19); the only difference is that $\exp(\mathrm{i}\boldsymbol{qr})$ is replaced with $(\exp \mathrm{i}\boldsymbol{q}_{\mathrm{e}}\boldsymbol{r}) - \exp(\mathrm{i}\boldsymbol{q}_{\mathrm{h}}\boldsymbol{r})$, where $\boldsymbol{r}_{\mathrm{e,h}}$ are the coordinates of the electron (hole). In the center-of-mass system, the interaction operator becomes (Ansel'm and Firsov [182])

$$H_{\mathrm{EXL}} = \sum_q W_{\boldsymbol{q}} \left[\exp(\mathrm{i}\boldsymbol{q}_{\mathrm{e}}\boldsymbol{r}) - \exp(\mathrm{i}\boldsymbol{q}_{\mathrm{h}}\boldsymbol{r})\right] \left(b^+_{-\boldsymbol{q}} + b_{\boldsymbol{q}}\right) . \tag{4.33}$$

Replacing $m_{\mathrm{e,h}}$ by the reduced mass μ by analogy with (4.29), we can define the characteristic size of the polarization region $r^*_{\mathrm{ex}} = \left[\left(r^*_{\mathrm{e}}\right)^2 + \left(r^*_{\mathrm{h}}\right)^2\right]^{1/2}$ and the interaction constant g^2_{ex}. Making use of the characteristics of the Wannier–Mott exciton (r_{ex}, E_{b}), one can express the latter as

$$g^2_{\mathrm{ex}} = \left(\frac{E_{\mathrm{b}}}{\hbar\omega_{\mathrm{LO}}}\right)^{1/2} \frac{\varepsilon_0}{\varepsilon_\infty} - 1 = \frac{r^*_{\mathrm{ex}}}{r_{\mathrm{ex}}} \left(\frac{\varepsilon_0}{\varepsilon_\infty} - 1\right) , \tag{4.34}$$

where $r_{\mathrm{ex}} = a_0 \left(\varepsilon m_{0/\mu}\right)$ and $a_0 = \hbar^2/m_0 e^2 = 0.53\,\text{Å}$ is the Bohr radius of a hydrogen atom.

The scattering of excitons by LO phonons is determined by the magnitude and the wave-vector dependence of the matrix element

$$H^{\lambda_1\lambda_2}_{\mathrm{EXL}} = \langle \Psi_{\lambda_1} \left| H_{\mathrm{EXL}} \right| \Psi_{\lambda_2} \rangle , \tag{4.35}$$

where Ψ_{λ_1} and Ψ_{λ_2} are the wave functions of the initial and final states of the exciton with wave vectors $\boldsymbol{k}_1$ and $\boldsymbol{k}_2 = \boldsymbol{k}_1 + \boldsymbol{q}$. The properties of the matrix element of exciton–phonon scattering, as first noted in Bulyanitza [189], depend crucially on the parities of the initial λ_1 and final λ_2 states. If the parity is the same (scattering occurs within the same band, like 1 s–1 s or

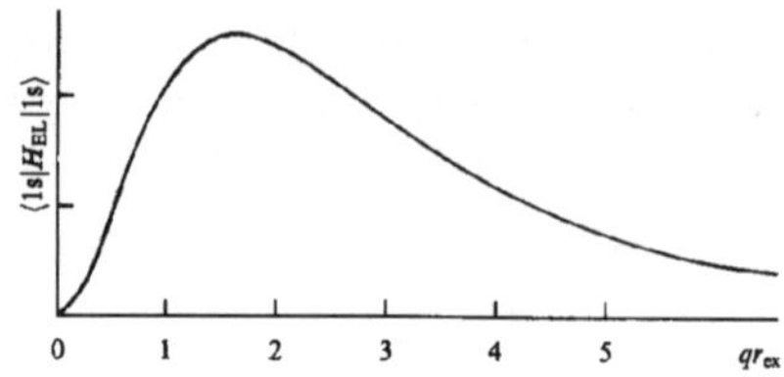

Fig. 4.2. Dependence of the matrix element of intraband scattering on qr_{ex} for the mass ratio $\frac{m_{\text{h}}}{m_{\text{c}}} = 3.5$ (Plekhanov [78])

2 s–2 s, etc., or in the case of interband scattering 1 s–2 s, 1 s–3 s, etc.), this mechanism of exciton–phonon scattering is forbidden, because $H_{\text{EXL}}^{\lambda_1\lambda_2} \to 0$ when $q \to 0$. When excitons are scattered in a ground state band (1 s–1 s), as shown in the paper of Ansel'm and Firsov, the matrix element has the form

$$H_{\text{EXL}}^{\lambda_1\lambda_2} \propto \left(\frac{g_{\text{ex}}\omega_{\text{LO}}}{r_{\text{ex}}}|q|\right)\left\{\left[1+\left(\frac{q_{\text{e}}r_{\text{ex}}}{2}\right)^2\right]^{-2} - \left[1+\left(\frac{q_{\text{h}}r_{\text{ex}}}{2}\right)^2\right]^{-2}\right\}. \quad (4.36)$$

When q is small (it is the small values that are of special importance; see below), the matrix element is proportional to

$$H_{\text{EXL}}^{\lambda_1\lambda_2} \propto \left(\frac{m_{\text{h}} - m_{\text{e}}}{m_{\text{h}} + m_{\text{e}}}\right) qr_{\text{ex}}. \quad (4.37)$$

As follows from (2.43), the Fröhlich mechanism of intraband scattering is absolutely forbidden when the effective masses are equal because the centers of the distribution of masses and charges coincide at $m_{\text{e}} = m_{\text{h}}$, so the polarization interactions of electrons and holes cancel out completely.

In the general case, the matrix element (4.36) arrives at the maximum near $qr_{\text{ex}} \sim 1$, and then falls off rapidly (Fig. 4.2). Similar behavior is displayed by the matrix elements of the scattering processes of the same symmetry (is – js, ip – jp, etc.). In such cases, the matrix element attains its maximum at the value of the inverse radius (r_{n}) of the corresponding exciton state, that is,

$$q_{\max} \sim \frac{1}{nr_{\text{n}}}, \quad r_{\text{n}} = nr_1. \quad (4.38)$$

According to the results of the paper by Bulyanitza, when scattering occurs between bands of different symmetry (is – kp), when $q \to 0$, the matrix element tends to a constant (nonzero) value, and such processes are therefore allowed. As in our first case, however, the matrix element $H_{\text{EXL}}^{\lambda_1\lambda_2}$ falls off rapidly as $q \succ r_{\text{ex}}^{-1}$ increases. The dependence of matrix elements on q for some cases of allowed and forbidden scattering for CdS is discussed in Permogorov [185].

The behavior of matrix elements (4.36) is definitive for the structure and properties of luminescence spectra of free excitons and Raman scattering in the resonance region. The existing theory of exciton–phonon interaction describes the case of an isotropic band with a parabolic dispersion of exciton states. Its successful application to LiH (with high anisotropy of the valence band, see Fig. 2.6), once again testifies to the fact that the dispersion of the exciton band is parabolic (Plekhanov and Altukhov [191]) in the range of low kinetic energies of excitons. Let us add that, according to Permogorov [185], the exciton band exhibits parabolic dispersion even for such anisotropic crystals as CdS, which is supported by the successful application of the theory of exciton–LO–phonon interaction to the analysis of luminescence spectra and RRLS (resonant Raman light scattering) of free excitons in CdS crystals (see also Kröger [190]).

Very recently, Zheng and Matsuura [192] published a paper devoted to electron–phonon interaction in mixed crystals. Their results are briefly discussed. The total Hamiltonian of an electron–LO–phonon interaction system in a mixed crystal can be written as the sum of a free-electron Hamiltonian, the two-mode LO–phonon Hamiltonian, and the electron–phonon interaction Hamiltonian; it is given by

$$\begin{aligned} H = {} & \frac{\hbar^2}{2m}\nabla^2 + \sum_{j=a,b} \hbar\omega_{\mathrm{L}j}\hat{a}^{+}_{\boldsymbol{k}j}\hat{a}_{\boldsymbol{k}j} \\ & + \sum_{j=a,b}\sum_{\boldsymbol{k}} i\sqrt{\frac{e^2\hbar\omega_{\mathrm{L}j}}{2V\bar{\varepsilon}_j\varepsilon_{\mathrm{o}}k^2}}\left(\hat{a}^{+}_{-\boldsymbol{k}j} + \hat{a}_{\boldsymbol{k}j}\right)\mathrm{e}^{\mathrm{i}\boldsymbol{kr}}, \end{aligned} \tag{4.39}$$

where m is the conduction band mass of the electron. This problem is analogous to that of the polaron in polyatomic crystals (Matsuura [193]). They considered the first terms in (4.39) as the unperturbed Hamiltonian H_0 and the last term as a perturbation H_{I}. The energy of the system is calculated by the perturbation method. Let $|\,\boldsymbol{q},0\rangle$ denote the zero-order wave function. The corresponding zero-order energy is $E^0(\boldsymbol{q}) = \hbar^2 q^2/2m$. The first-order correction is obviously zero, that is, $\langle\boldsymbol{q},0\,|H_{\mathrm{I}}|\,\boldsymbol{q},0\rangle = 0$. The second-order correction can be written as

$$E''(\boldsymbol{q}) = \sum_{j=a,b}\sum_{\boldsymbol{k}} \frac{\left|\langle\boldsymbol{q}-\boldsymbol{k},l_{j\boldsymbol{k}}\,|H_{\mathrm{I}}|\,\boldsymbol{q},0\rangle\right|^2}{E^0(\boldsymbol{q}) - E^0(\boldsymbol{q}-\boldsymbol{k}) - \hbar\omega_{\mathrm{L}j}} \tag{4.40}$$

with

$$\left|\langle\boldsymbol{q}-\boldsymbol{k},l_{j\boldsymbol{k}}\,|H_{\mathrm{I}}|\,\boldsymbol{q},0\rangle\right|^2 = \frac{e^2\hbar\omega_{\mathrm{L}j}}{2V\bar{\varepsilon}_j\varepsilon_{\mathrm{o}}k^2}. \tag{4.41}$$

Following the same procedure as Wang and Liang [194], Zheng and Matsuura replaced the sum over $\boldsymbol{k}$ by an integral in $\boldsymbol{k}$ space; then they obtain

$$E(\boldsymbol{q}) = \frac{\hbar^2 q^2}{2m} + \sum_{j=a,b}\frac{e^2\hbar\omega_{\mathrm{L}j}}{8\pi^2\bar{\varepsilon}_j\varepsilon_{\mathrm{o}}}\int_{o}^{\infty}\frac{\mathrm{d}k\,\sin\theta\,\mathrm{d}\theta}{(\hbar^2/2m)\left(2kq\cos\theta - k^2\right) - \hbar\omega_{\mathrm{L}j}}. \tag{4.42}$$

We restrict ourselves to the energy range $\hbar^2 q^2/2m < \hbar\omega_{\mathrm{Lj}}$ and expand the integrand in (4.42) for small q. The polaron energy is given by

$$E(\boldsymbol{q}) \approx \frac{\hbar^2 q^2}{2m} - \sum_{j=a,b} \alpha_j \left[\hbar\omega_{\mathrm{Lj}} + \frac{\hbar^2 q^2}{12m}\right]$$
$$= -\sum_{j=a,b} \alpha_j \hbar\omega_{\mathrm{Lj}} + \frac{\hbar^2 q^2}{2m^*}, \qquad (4.43)$$

with

$$\alpha_j = \frac{e^2}{8\pi^2 \bar{\varepsilon}_j \varepsilon_{\mathrm{o}}} \left[\frac{2m}{\hbar^3 \omega_{\mathrm{Lj}}}\right], \quad j = a, b, \qquad (4.44)$$

$$m^* = m\left[1 - \frac{\alpha_{\mathrm{a}} + \alpha_{\mathrm{b}}}{6}\right]^{-1}, \qquad (4.45)$$

where α_{a} and α_{b} are the Fröhlich electron–phonon coupling constants of two LO–phonon modes in the mixed crystal, respectively; and m^* is the polaron mass. The Fröhlich coupling constants α_{a} and α_{b} and the total Fröhlich coupling constant $(\alpha_{\mathrm{a}} + \alpha_{\mathrm{b}})$ as functions of the composition x are shown in Fig. 4.3. The polaron energy shift $\alpha_{\mathrm{a}}\hbar\omega_{\mathrm{La}}$ and $\alpha_{\mathrm{b}}\hbar\omega_{\mathrm{Lb}}$ and the total polaron energy shift $(\alpha_{\mathrm{a}}\hbar\omega_{\mathrm{La}} + \alpha_{\mathrm{b}}\hbar\omega_{\mathrm{Lb}})$ as functions of the composition x are plotted in Fig. 4.4.

It is seen that although the concentration dependences of α_{a} and α_{b} display strong nonlinearity, the total Fröhlich coupling constant $(\alpha_{\mathrm{a}} + \alpha_{\mathrm{b}})$ is almost a straight line in the figures. The polaron energy shifts of $\alpha_{\mathrm{a}}\hbar\omega_{\mathrm{La}}$ and $\alpha_{\mathrm{b}}\hbar\omega_{\mathrm{Lb}}$ are also nonlinear, but the total polaron energy shift has approximately a linear relation with the concentration x. These theoretical results strongly support the linear interpolation approximation used by many

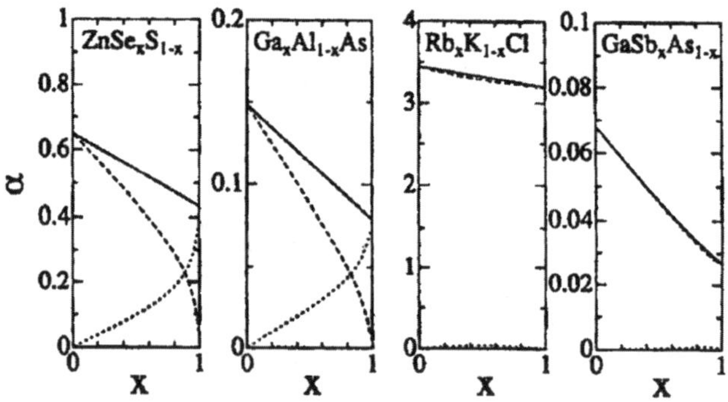

Fig. 4.3. Concentration dependence of Fröhlich constants α_{a} (*dotted line*), α_{b} (*dashed line*), and $\alpha_{\mathrm{a}} + \alpha_{\mathrm{b}}$ (*full line*) of several mixed crystals. (After [192])

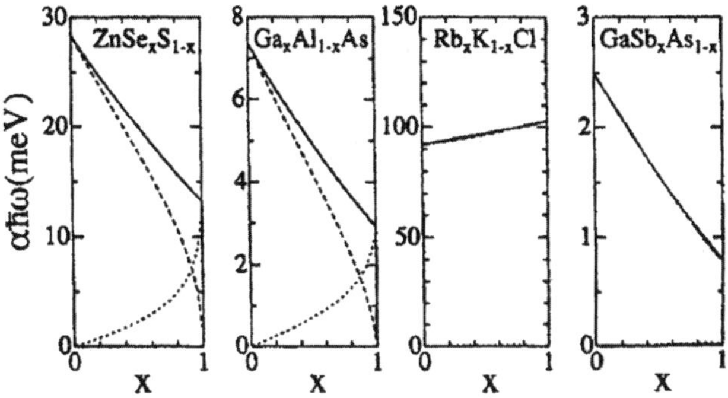

Fig. 4.4. Concentration dependence of the polaron energy shift $\alpha_a\hbar\omega_{La}$ (*dotted line*), $\alpha_b\hbar\omega_{Lb}$ (*dashed line*), and the total polaron energy shift $\alpha_a\hbar\omega_{La} + \alpha_b\hbar\omega_{Lb}$ (*full line*) of several mixed crystals. (After [192])

experimenters to estimate the Fröhlich constants and the polaron energy shifts of weak-coupling mixed crystals (see, e.g., Nicholas et al. [195] and Adachi [196]). The results of Zheng and Matsuura are very different from the theoretical results of Wang and Liang [194], where the expressions of the displacements of the two LO–phonon modes and the electron–phonon interaction Hamiltonian are not entirely correct.

The sentence "the total Fröhlich coupling constant $(\alpha_a + \alpha_b)$ is almost exactly a linear function of the concentration x" implies that $\alpha_a(x)+\alpha_b(x) \simeq x\alpha_1 + (1-x)\alpha_2$, where α_1 and α_2 are the Fröhlich coupling constants of the two end members of the mixed crystal, respectively. Theoretical calculations are too complex to see this result transparently. In the end, we should note that the theoretical results of Zheng and Matsuura are correct for a weak-polar mixed crystal $(\alpha_a + \alpha_b) \leq 1$.

4.3 Giant Isotopic Effect in the Energy Spectrum of Wannier–Mott Exciton in LiH Crystals

The results of papers [197–203] have shown that LiH has a direct band-gap $E_g \sim 5.0\,\text{eV}$ (for details, see also [78,109]). At room temperature and ambient pressure, lithium hydride forms a NaCl type cubic lattice, i.e., it consists of two centered cubic lattices shifted relative to each other by half the lattice constant (see, e.g., Calder et al. [205]). Compounds with such a crystalline structure belong to the O_h^5 space group. According to the type of chemical bond, LiH is generally assigned to 1 – V11 ionic compounds [an analogue of alkali halide crystals (AHC)]. However, more detailed investigation shows that its ionic character lies in the range 0.8 to 1 (see, e.g., [205]). A comparison of ionic radii $r_{Li+} = 0.68\,\text{Å}$ and $r_{H-} = 1.5\,\text{Å}$ with lattice constant $\ell = 2.042\,\text{Å}$ shows that the lithium and hydrogen atoms in LiH strongly overlap (see also Pauling [206]). More detailed information on LiH and its physicochemical parameters are presented in Shpilrain et al. [154]. Samples of LiH were grown from the melt using modified Bridgman–Stockbarger techniques, which have been described many times elsewhere (see, e.g., [207]). To improve the hydrogen (deuterium) stoichiometric composition, the bulk crystals were additionally annealed in a hydrogen or deuterium atmosphere of 2–5 atm at a temperature of $500 - 550°$C.

Taking into account the high reactivity of the freshly cleaved surface of lithium hydride crystals in air, it was necessary to develop a method of crystal cleavage that would allow us to obtain a pure surface and also retain it for several hours (the time required for the experiment). The well-known technique of crystal cleavage directly in a helium bath of an optical cryostat in liquid or superfluid helium proved suitable for this purpose (Plekhanov et al. [208]). This technique allows one to obtain samples with a pure surface. We found no changes in the shape of reflection, luminescence [204], and resonant Raman scattering (RRS) spectra [116] of samples cleaved this way during $10-16$ hours of the experiment. Using the cleavage facility with three degrees of freedom and a degree of rotation of up to 90° enabled us to carry out our experiments. As a rule, the investigated samples of LiH crystals are cleaved from bulk quality crystals. X-ray diffraction studies show (Zimmerman [57]) that mixed $\mathrm{LiH}_x\mathrm{D}_{1-x}$ crystals form a continuous series of solid solutions and behave like virtual crystals with the crystal lattice constant changing according to Vegard's law. The experimental setup for measuring low-temperature ($2 - 300$ K) reflection and luminescence (RRS) spectra was described elsewhere (see, e.g., [109, 116, 210]). The results described in this book were obtained for pure crystals surface cleaved directly in liquid helium, as mentioned above. Samples characterized by a weak topographic dependence of exciton reflection and luminescence spectra were selected for study (see also [78,204]). Figure 4.5 shows the specular reflection spectrum of a LiH crystal with a pure surface at 4.2 K [211]. One can see that the spectrum exhibits a fine structure. The long-wavelength reflection maximum, which shows the singularity from the short-wavelength side at 4.960 eV (Plekhanov [212]), corresponds to the energy $E_{1\mathrm{s}} = 4.950$ eV. Analogous structures of this maximum have been observed frequently in reflection spectra of crystals with pure surfaces (Kink et al. [213]). In the short-wavelength region after the dispersion dip, a second maximum at $E_{2\mathrm{s}} = 4.982$ eV is observed. The maximum of this peak shifts to the red with increasing temperature, its intensity decreases and cannot be detected above 115 K (Fig. 4.6). This suggests that a decrease in the $E_{2\mathrm{s}}$ peak intensity results from ionization of the states related to this peak. As in Segall and Marple [214], the ionization energy of this state can be roughly estimated assuming that $\Delta E_{2\mathrm{s}} = k_{\mathrm{B}}T$ (where k_{B} is Boltzmann's constant). This estimate gives $\Delta E_{2\mathrm{s}} = 10$ meV. Note that the redshift of this state with increasing temperature differs from the shift of the long-wavelength reflection band (see below). High values of the reflection co-

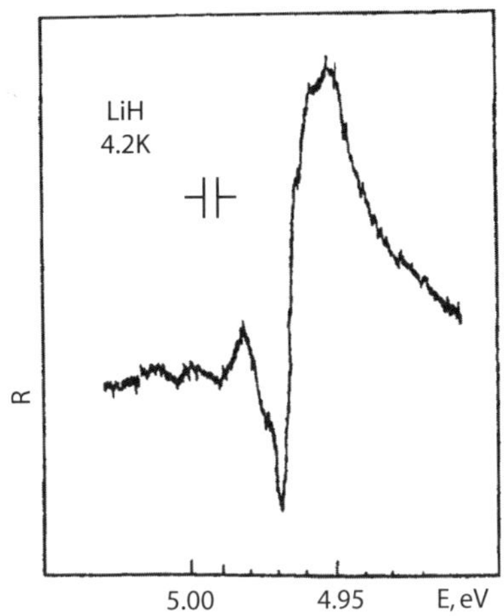

Fig. 4.5. Mirror reflection spectrum of a LiH crystal cleaved in liquid helium. Angle of incidence of 45°. (After [78])

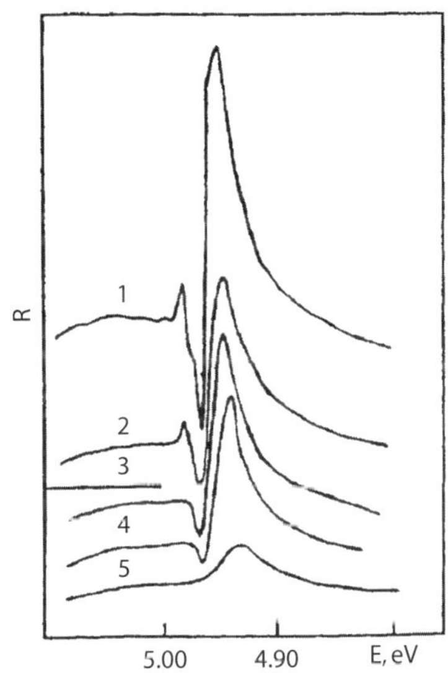

Fig. 4.6. Reflection spectrum of a LiH crystal measured at 1: 2 K; 2: 71 K; 3: 138 K; 4: 220 K; 5: 288 K. (After [78])

efficient (0.7 – 0.9) for the first long-wavelength peak and its comparatively small half-width ($\Delta E \leq 25\,\text{meV}$) along with a sharp temperature dependence (see below) suggest that E_{1s} and E_{2s} peaks are related to the first two terms of the hydrogen-like series of the Wannier–Mott exciton. According to Plekhanov et al. [179, 215], the electronic transition is direct and occurs at

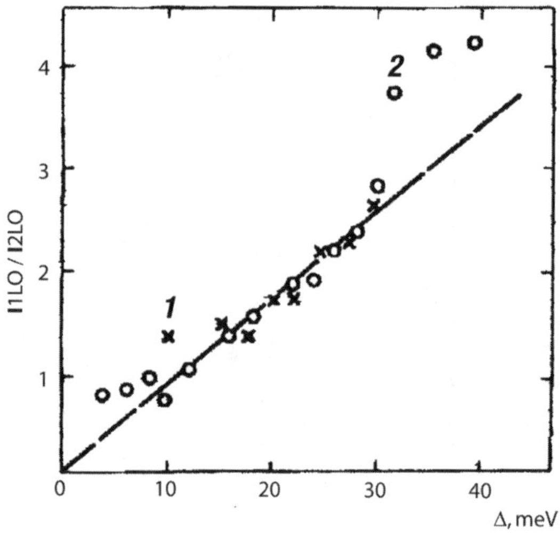

Fig. 4.7. The dependence of the ratio $I_{1\mathrm{LO}}/I_{2\mathrm{LO}}$ on the exciton kinetic energy ($1 - T = 4.4\,\mathrm{K}$; $2 - 78\,\mathrm{K}$). (After [78])

the point X at the Brillouin zone boundary [216–218] where the distance between points X_1 and X_4 is a minimum (for details, see [204]). The assumption that the exciton spectrum of LiH crystals is hydrogen-like (especially in the parabolic region of the exciton band for energies $\leq 40\,\mathrm{meV}$ (Fig. 4.7) [212]) seems natural for the following reasons:

1. The ratio of intensities observed for the first two peaks in the reflection spectrum is close to the n^{-3} dependence, according to the theory [178] for direct allowed transitions, where n is the principal quantum number.
2. The distance Δ_{12} between these peaks is substantially smaller than the LO–phonon energy ($\hbar\omega_{\mathrm{LO}} = 140\,\mathrm{meV}$ [25, 51]). For LiH, this ratio $\Delta_{12}/\hbar\omega_{\mathrm{LO}} = 0.3$ (see, e.g., [109]). For comparison, note that, for excitons in CdS, where the large-radius exciton model is applied, this ratio is 0.74 (Segall and Mahan [219]).
3. The applicability of the Wannier–Mott model to excitons in LiH is supported by the fact that a similar structure consisting, as a rule, of two lines was observed in reflection spectra of many semiconducting crystals [185, 219, 220].

According to the above, taking into account the energies of the long-wavelengths peaks $E_{1\mathrm{s}}$ and $E_{2\mathrm{s}}$, the following values for the exciton Rydberg constant in LiH are obtained: $E_{\mathrm{b}} = (4/3)\Delta_{12\mathrm{s}} = 42\,\mathrm{meV}$ (where $\Delta_{12\mathrm{s}} = E_{2\mathrm{s}} - E_{1\mathrm{s}}$ and $\Delta E_{2\mathrm{s}} = 11\,\mathrm{meV}$). Here, $\Delta E_{2\mathrm{s}}$ is the binding energy of the first excited exciton state. The $\Delta E_{2\mathrm{s}}$ value found from the hydrogen-like expression for the exciton binding energy correlates well with the above estimates of the thermal ionization energy for the $n = 2\mathrm{s}$ exciton state. The exciton binding energy in these crystals was also determined from the temperature quenching of luminescence of free excitons. By assuming that it results from ionization of the exciton ground state, the binding energy in LiH obtained was $40 \pm 3\,\mathrm{meV}$ (see also [109]).

Isotopic substitution affects only the wave function of phonons; therefore, the energy values of electron levels in the Schrödinger equation ought to have remained the same. This, however, is not so, since isotopic substitution modifies not only the phonon spectrum, but also the constant of electron–phonon interaction. For this reason, the energy values of pure electron transition in molecules of hydride and deuteride are different [222]. This effect is even more prominent [23] when we are dealing with a solid. Intercomparison of absorption spectra for thin films of LiH and LiD at room temperature revealed that the long wave maximum (as we know now, the exciton peak [179]) moves 64.5 meV toward the shorter wavelengths when H is replaced with D. As will be shown below, this effect becomes even more pronounced at low temperatures.

The mirror reflection spectra of mixed and pure LiD crystals cleaved in liquid helium are presented in Fig. 4.8. For comparison, on the same diagram we have also plotted the reflection spectrum of LiH crystals with clean surfaces. All spectra have been measured with the same apparatus under the same conditions. As the deuterium concentration increases, the long-wave maximum broadens and shifts toward shorter wavelengths. As can clearly be seen in Fig. 4.8, all spectra exhibit a similar long-wave structure. This circum-

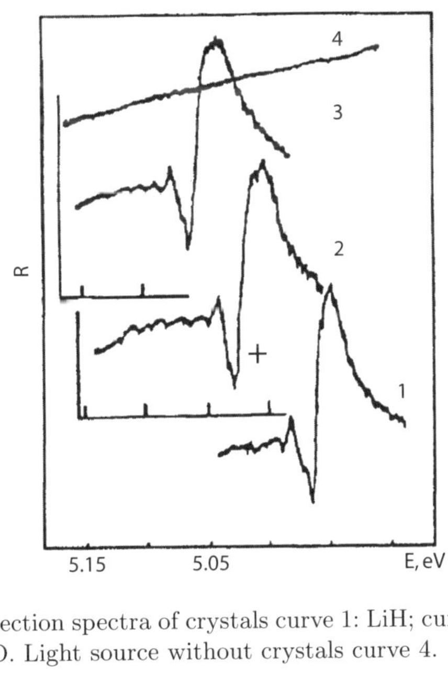

Fig. 4.8. Mirror reflection spectra of crystals curve 1: LiH; curve 2: LiH_xD_{1-x}; and curve 3 at 4.2 K: LiD. Light source without crystals curve 4. (After [78])

Table 4.1. Values of the energy of maxima in exciton reflection spectra of pure and mixed crystals at 2 K and energies of exciton binding E_b band-to-band transitions E_g. (After [23])

Energy, meV	LiH	$LiH_{0.82}D_{0.18}$	$LiH_{0.40}D_{0.60}$	LiD	Li^6H (78 K)
E_{1s}	4950	4967	5003	5043	4939
E_{2s}	4982	5001	5039	5082	4970
E_b	42	45	48	52	41
E_g	4992	5012	5051	5095	4980

stance allows us to attribute this structure to the excitation of the ground (1s) and the first excited (2s) exciton states. The energy values of exciton maxima for pure and mixed crystals at 2 K are presented in Table 4.1. The binding energies of excitons E_B, calculated by the hydrogen-like formula, and the energies of interband transitions E_g are also given in Table 4.1.

The ionization energy, found from temperature quenching of the peak of the reflection spectrum of the 2s state in LiD, is 12 meV. This value agrees fairly well with the value of ΔE_{2s} calculated by the hydrogen-like formula. Moreover, $E_B = 52$ meV for LiD agrees well with the energy of activation for thermal quenching of free-exciton luminescence in these crystals [212].

Going back to Fig. 4.8, it is hard to miss the growth of Δ_{12}, [211], which in the hydrogen-like model, causes an increase of the exciton Rydberg with the replacement of isotopes (see below). When hydrogen is completely replaced with deuterium, the exciton Rydberg (in the Wannier–Mott model) increases by 20% from 40 to 50 meV, whereas E_g exhibits a 2% increase, and at 2–4.2 K is $\Delta E_g = 103$ meV. This quantity depends on temperature, and at room temperature is 73 meV, which agrees well enough with $\Delta E_g = 64.5$ meV as found in the paper by Kapustinsky et al. [198]. The continuous change of the exciton Rydberg was observed earlier in crystals of the solid solutions A_3B_5 [222–224] and A_2B_5 [225–228]. Isotopic substitution of the light isotope (S^{32}) by the heavy one (S^{34}) in CdS crystals [227] reduces the exciton Rydberg (see also [228]); this was attributed to the tentative contribution from the adjacent electron bands, which, however, are not present in LiH (for more details, see [23]).The single-mode nature of the exciton reflection spectra of mixed crystals LiH_xD_{1-x} agrees qualitatively with the results obtained with the virtual crystal model (see, e.g., [71, 229]) and is at the same time its extreme realization, since the difference between ionization potentials ($\Delta\zeta$) for this compound is zero. According to the virtual crystal model, $\Delta\zeta = 0$ implies that $\Delta E_g = 0$, which is in contradiction with the experimental results for LiH_xD_{1-x} crystals. By now, the change in E_g caused by isotopic substitution has been observed for many broad-gap and narrow-gap semiconductor compounds (see, e.g., [230]).

4.4 Nonlinear Dependence of Band-Gap Energy on the Isotopic Effect

Numerous studies of reflection (absorption) spectra of mixed compounds A_1B_7, A_2B_6, and A_3B_5, carried out during the past five decades, point to the existence of large-radius excitons in these materials. The widths of exciton bands, it was shown, depend on the composition of the solid solution (see also [231]). The random relief of the potential is caused by disordering of the crystal lattice. According to modern views, if the kinetic energy of an exciton exceeds the potential energy of localization, the exciton will not "sense" the random relief of the potential. If the kinetic energy of the exciton is small, then the exciton will be localized by the potential created by fluctuations in the composition of the crystal lattice. Since the localization energy is different at different points of the sample, the exciton transition energy will also be different, which leads to broadening of the exciton line. The localized and delocalized states are separated by the mobility threshold (Belitz and Kirkpatrick [80]). From the above, it becomes clear that exciton states in solid solutions will be observed when and only when the exciton is not ionized by the field of such a fluctuation and broadening of the exciton lines is much less than the binding energy E_b of excitons in such states. In the localized exciton states, the relative motions of excitons and holes are localized and also the translational motion of the exciton as a whole. It ought to be added that quasi-momentum is no longer a good quantum number for excitons with low kinetic energy, and therefore excitons with different energies may be generated through the absorption of light. The density of states in the exciton band in the low-energy range is much different from the density of states of free excitons [78]. In a paper of Efros and Raikh [108], the method of optimal fluctuation (for more details, see Lifshitz [11]) was used to find expressions of the absorption coefficient $k(E)$ in the region of the long-wave wing of the exciton line, as well as the density of the localized states $\rho(E)$. Both functions fall off toward the depth of the forbidden zone as

$$K\left(E\right) \propto \rho\left(E\right) \propto \exp\left(\frac{E}{E_0}\right)^{1/2} . \tag{4.46}$$

In chemically mixed semiconductor crystals, especially those of the A_2B_6 group (Klochikhin et al. [233]), smearing of the conduction and valence band edges occurs because of fluctuations in the electron potential; the band is smeared more, the heavier the carrier. Numerous comparisons of the results on the broadening of exciton lines by large-scale compositional fluctuations (whose geometric size is commensurate with the wavelength of the exciton) with the proposed theoretical model point to good qualitative and quantitative agreement, especially for A_2B_6 compounds.

As follows from Fig. 4.8, excitons in LiH_xD_{1-x} crystals display a unimodal character, which facilitates the interpretation of their concentration

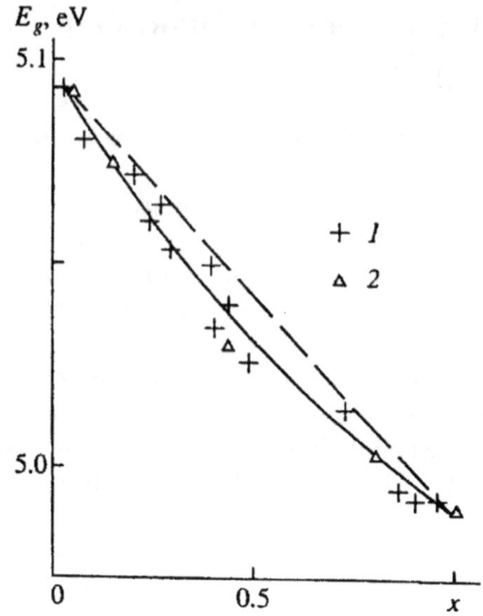

Fig. 4.9. Energy of band-to-band transitions E_g as a function of isotope concentration in mixed LiH_xD_{1-x} crystals at 2 K. (1) linear dependence of E_g on x in virtual crystal model; (2) calculation according to (2.28); points derived from reflection spectra are indicated by crosses and from luminescence spectra by triangles. (After Plekhanov [109])

dependence. Figure 4.9 shows the concentration dependence of the energy of interband transitions E_g. Each value of E_g was found by adding together the energy of the long-wave band in the reflection spectrum and the binding energy of the exciton. The latter was found from the hydrogen-like formula using the experimental values of the energies levels of 1s and 2s exciton states. We see that 100% replacement of hydrogen with deuterium changes E_g by $\Delta E_g = 103$ meV at $T = 2$ K (for details, see also [78]). This constitutes 2% of the energy of the electron transition, which is two orders of magnitude greater than the value corresponding to the isotopic replacement of atomic hydrogen with deuterium reported earlier [109].

The nonlinear concentration dependence of E_g can be sufficiently well approximated with a second order polynomial

$$E_g(x) = E_b + (E_a - E_b - b)x - bx^2, \tag{4.47}$$

where E_a, E_b are the values of E_g for LiD and LiH, respectively, and b is the curvature parameter equal to 0.046 eV. This result generally agrees with the published data (see also Elliott and Ipatova [234] and references therein). For comparison, let us indicate that in isotopic substitution in germanium, the energy E_g depends linearly on the isotopic concentration for both direct (E_0, $E_0 + \Delta_0$, $E_1 + \Delta_1$) and indirect electron transitions [235]. Unfortunately, today there is no information on the form of the function $E_g \propto f(x)$ for isotopic substitution in C, ZnO, CdS, CuCl, Cu_2O, GaAs,GaN, Si crystals, etc., although, as noted above, the values of E_g have been measured for isotopically pure crystals.

4.5 Renormalization of Binding Energy of Wannier–Mott Excitons by Isotopic Effect

In the original work [179], where it was found that the exciton binding energy E_b depends on the isotopic composition, this change in E_b was attributed to the exciton–phonon interaction (originally with LO phonons). The preferential interaction of excitons with LO phonons in LiH (LiD) crystals was later repeatedly demonstrated in luminescence spectra [212] and resonant Raman light scattering [116, 117], which consist of a phononless line (in the former case) and its LO repetitions. The effects of the Fröhlich mechanism of exciton–phonon interaction on the energy spectrum of Wannier–Mott excitons has been considered over and over again (see, e.g., [182, 185, 190, 214]). Today, we know that the main consequences of electron and hole interaction in excitons with polarization vibrations are the static screening of the lattice charges (introducing ε_0) and the change in the effective masses of the particles. Both effects of electron–(hole)–phonon interaction can easily be taken into account and lead to a change in the exciton Rydberg E_b. These corrections do not destroy the hydrogen-like structure of the exciton spectrum. At the same time, the non-Coulomb corrections of the electron–hole Hamiltonian modify the hydrogen-like structure and remove, for example, degeneration of levels with respect to orbital and magnetic quantum numbers (see, e.g., [109]). The very fact, however, that the problem of renormalization of the energy spectra of Wannier–Mott excitons does not admit an exact solution, even in the limiting cases, often gives rise to a situation in which there is no agreement among the results obtained by different authors. Starting with the classical works of Haken [171], all papers may be divided into two broad classes depending on how they deal with Coulomb interaction: between "bare" electrons and holes or between electrons and holes in the polaron state. In other words, first the interaction of band electrons and holes with LO optical phonons is taken into account, and then the Coulomb interaction between electrons and holes clad in "polarization coats" is considered. As will be shown below, the study of exciton–phonon interaction in crystals with the isotopic effect provides entirely new information and also allows us to reconstruct experimentally the values of Fröhlich and Coulomb interaction constants. From Fig. 4.10, we see that when hydrogen is completely replaced with deuterium, the binding energy of the exciton increases 20% from 42 to 52 meV [179]. It is easy to see that in a virtual crystal model, the binding energy of the exciton in LiT crystals [156]) must be equal to 57 meV (see Fig. 4.10). Hence, it follows that

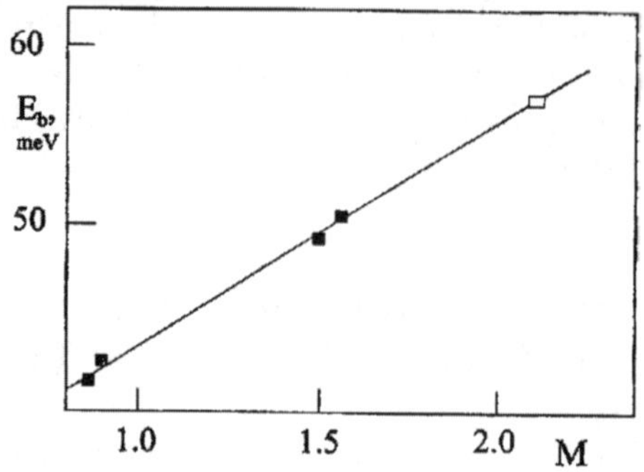

Fig. 4.10. Binding energy of Wannier–Mott excitons as a function of reduced mass of ions based on values of reduced mass of ions for ^{6}LiH, ^{7}LiH, ^{6}LiD, ^{7}LiD, and LiT. (After [78])

in a linear approximation, the isotopic dependence of the binding energy of Wannier–Mott excitons may be expressed as

$$E_\mathrm{b} = E_\mathrm{b}(0)\,(1+\gamma)\,, \tag{4.48}$$

where $E_\mathrm{b}(0)$ is the purely coulombic binding energy of the exciton (in the frozen lattice), which in our case is equal to 31.5 meV, and the angular coefficient is $\beta = 12.18\,\mathrm{meV/M}$, where M is the reduced mass of lithium and hydrogen (deuterium, tritium) ions; $\gamma = \beta M / E_\mathrm{b}\,(0)$ (see also [211, 215]). From the standard equation for the Coulomb binding energy of the exciton,

$$E_\mathrm{b} = \frac{e^4 \mu}{\hbar^2 \varepsilon_\infty^2}\,, \tag{4.49}$$

we get the dimensionless constant of Coulomb interaction:

$$\eta^2 = \frac{E_\mathrm{b}\,(0)}{\hbar\omega_\mathrm{LO}} = 0.47. \tag{4.50}$$

Comparing the value of $\eta^2 = 0.47$ and the constant of Fröhlich exciton–phonon interaction $g^2 = 0.33$ [191], we see that they are close enough. This implies that both the Fröhlich and the Coulomb interactions between electrons (holes) and LO phonons in excitons must be treated with equal attention, as has already been emphasized in Klochikhin's paper [188]. That paper deals from the start with "bare" electrons and holes, and all renormalizations are calculated in a two-particle configuration. Such an approach enables us to avoid the considerable difficulty which arises when polarons [236] are used as start-up particles. This difficulty is primarily associated with the fact that the momentum of each particle is conserved when the particles are treated separately, whereas the center-of-mass momentum is conserved when a pair moves as a whole. As demonstrated in Klochikhin's paper [188], this approach also makes it possible to calculate higher order corrections of the exciton–phonon interaction. It was also shown that the use of the pole parts of polaron Green functions in place of complete expressions in [236] and Mahanti and Varma [237] leads to a situation where the corrections of the order of $\eta^2 g^2$ and g^4 to the potential energy are lost because the corrections to the vertex parts and Green functions cancel out. The quantity lost is of the same order (g^2) as the correction to the residue but has the opposite sign (for more details, see [236, 237]). The approach developed in Klochikhin's paper [188] allowed calculating corrections of the order $\eta^2 g^2$ and g^4; the latter is comprised of the correction to the Fröhlich vertex and the correction to the Green functions in the exciton–phonon loop. It is important that the latter have opposite sign and cancel out exactly in the limit $E_\mathrm{b} \ll \hbar\omega_\mathrm{LO}$. As a result, because of the potential nature of the start-up Coulomb interaction, the correction to the Coulomb vertex of the order $\eta^2 g^2$ does not vanish. As a result, the following expression was obtained in Klochikhin's paper for the binding energy E_b of Wannier–Mott excitons when $E_\mathrm{b} \ll \hbar\omega_\mathrm{LO}$ (the spectrum of excitons remains hydrogen-like):

$$E_\mathrm{b} = \hbar\omega_\mathrm{LO}\left[\frac{\eta^2 - g^2 + \eta^2 g^2 (c + v)}{2}\right]^2, \tag{4.51}$$

where $c,\, v = \left(\frac{m_\mathrm{c,v}}{\mu}\right)^{1/2}$ and m_c, m_v are the electron and hole masses. Now, E_b depends explicitly on g^2 (the Fröhlich constant of exciton–phonon interaction) and hence, depends on the isotopic composition of the lattice, whereas the standard expression for the binding energy $E_\mathrm{b} = \hbar\omega_\mathrm{LO}\left(\eta^2 - g^2\right) = \frac{e^4\mu}{2\varepsilon_0^2\hbar^2}$, which describes the exciton spectrum of many semiconductors accurately enough, exhibits no dependence on the isotopic effect. In (4.51), the exciton spectrum remains hydrogen-like. When higher order corrections are taken into account, (4.51) becomes

$$E_\mathrm{b} = \frac{e^4\mu}{2\varepsilon_0\hbar^2}\left[1 + g^2\varepsilon_0\frac{m_\mathrm{c} + m_\mathrm{v}}{\varepsilon_\infty} + g^4\frac{\varepsilon_0}{\varepsilon_\infty}\left(\zeta_1 + \zeta_2\frac{1 - \varepsilon_\infty}{\varepsilon_0}\right)(m_\mathrm{c} + m_\mathrm{v})\right]. \tag{4.52}$$

The order-of-magnitude evaluation of the coefficients ζ_1, ζ_2 gives $\zeta_1 \approx 0.15$ and $\zeta_2 \approx 0.02$; when $g^2(m_\mathrm{c} + m_\mathrm{v}) \ll 3.3$, the correction of the order $\eta^2 g^4$ is much less than the term of the order $\eta^2 g^2$ [188]. So from (4.51), we see that the correction to the purely Coulombic binding energy is important primarily because the values of η^2 and g^2 are close to each other. Setting $\frac{m_\mathrm{v}}{m_\mathrm{c}} = 3.5$ and $\frac{g^2}{\eta^2} = 1 - \varepsilon_0/\varepsilon_\infty$, and $\left(\frac{\varepsilon_\infty}{\varepsilon_0}\right) = \left(\frac{\omega_\mathrm{TO}}{\omega_\mathrm{LO}}\right) = \frac{1}{3.5}$, in paper [180], it was found that $E_\mathrm{b(theor)} = 48$ and 42 meV for LiD and LiH, respectively. Comparing these results with the experimental values (see Table 4.1), we observe good agreement between theory and experiment. Hence follows the natural conclusion that the isotopic dependence of exciton binding energy is due primarily to the Fröhlich interaction mechanism between excitons and phonons.

4.6 Nonlinear Dependence of Binding Energy on Isotopic Concentration

A principal matter for further theoretical development is the question concerning the effect of crystal lattice disordering on the binding energy E_b of Wannier–Mott excitons. This problem has been treated theoretically in the papers of Elliott and co-workers (Kanehisa and Elliott [238] and Schwabe and Elliott [239]), where they study the effect of weak disordering on E_b (the disordering energy is comparable with E_b). The binding energy indicated in the papers was calculated under the coherent potential approximation by solving the Bethe–Salpeter equation as applied to the problem of Wannier–Mott excitons in a disordered medium. One of the principal results of this paper (Kanehisa and Elliott [238]) is the nonlinear dependence of E_b on concentration. As a consequence, the binding energy E_b at half-and-half concentrations is less than the value derived from the virtual crystal model. The exciton binding energy is reduced because energy E_g is less, owing to the fluctuation smearing of the edges of the conduction and valence bands. This conclusion is in qualitative agreement (although not in quantitative agreement, the discrepancy is about an order of magnitude (see also Kanehisa and Elliott [238]) with the experimental results for the mixed crystal $GaAs_{1-x}P_x$ with $x = 0.37$, where the reflection spectra exhibited two exciton maxima (see also Fig. 4.8) used for finding the value of E_b (see Nelson et al. [223] and references therein). Let us add that the pivotal feature of the model of Elliott and co-workers is the short-range nature of the Coulomb potential (for more details, see [211]).

Before the comparison of our experimental results with the theory developed by Elliott and Kanehisa, it would be prudent to review briefly the main properties of their theoretical model. According to [238], this model considers an exciton with a direct gap of a semiconductor alloy. Such a system consists of an electron (particle 1) in the conduction band (c) with mass m_c and a hole (particle 2) in the valence band (v) with mass m_v. The problem of the exciton in disordered systems is to solve the Hamiltonian

$$H = \boldsymbol{p}^2/2m_c + \boldsymbol{p}^2/2m_v + u(\boldsymbol{r}_1 - \boldsymbol{r}_2) + V_c(\boldsymbol{r}_1) + V_v(\boldsymbol{r}_2), \tag{4.53}$$

with both the Coulomb interaction u and the potential V_v due to disorder ($\nu = c, v$). Reference [213] neglected disorder-induced interband mixing. As is well known, in place of the electron–hole coordinates, ($\boldsymbol{r}_1$, $\boldsymbol{p}_1$) and ($\boldsymbol{r}_2$, $\boldsymbol{p}_2$), one may introduce the center-of-mass and relative coordinates, ($\boldsymbol{R}$, $\boldsymbol{P}$) and ($\boldsymbol{r}$, $\boldsymbol{p}$) to rewrite (4.53) as

$$H = \boldsymbol{p}^2/2m_r + u(\boldsymbol{r}) + \boldsymbol{P}^2/2M + V_c(\boldsymbol{R} + m_v\boldsymbol{r}/M) + V_v(\boldsymbol{R} - m_c\boldsymbol{r}/M), \tag{4.54}$$

where m_r and M are the reduced and total masses, respectively. Because of the random potential, the translational and relative degrees of freedom cannot be decoupled. This is difficult when considering the two-body problem in a disordered system [158, 215]. However, when the exciton state in question is well separated from other states so that the energy spacing is much larger than the translational width and disorder, one can forget about the relative motion and just apply any single-particle alloy theory (see, e.g., [215] and references therein) solely to their translational motion. For each exciton state, the translational part of the Hamiltonian in this case is

$$H_{\mathrm{t}} = \boldsymbol{P}^2/2M + \bar{V}_{\mathrm{c}}(\boldsymbol{R}) + \bar{V}_{\mathrm{v}}(\bar{R}). \tag{4.55}$$

Here, $\bar{V}_{\mathrm{c}}$ and $\bar{V}_{\mathrm{v}}$ are averages of V_{c} and V_{v}. This approach is similar to the Born–Oppenheimer adiabatic approximation. Such situations hold in some mixed alkali halide crystals and probably II–VI crystals. On the contrary, when exciton binding energy is comparable to disorder energy, the adiabatic approximation breaks down, and it is essential to take into account the effect of disorder on both the translational and relative motions. This is the case with the Wannier–Mott exciton in III–V alloys, for which the Elliott and Kanehisa model was developed. In this case, the solution task is to start from the independent electron and hole by neglecting u in (4.55) and then to take into consideration the Coulomb interaction between the average electron and average hole. A further simplified approach adopted in the literature [240] in solving the Bethe–Salpeter equation [177] is to suppose a free-electron-like a one-particle Green's function with a built-in width to allow for the random potential due to disorder. In the theoretical model cited, the average (or "virtual crystal") gap is given by

$$E_{\mathrm{g}}^{vc}(x) = E_0 + (\delta_{\mathrm{c}} - \delta_{\mathrm{v}})(x - 1/2), \tag{4.56}$$

where E_0 is the average gap and δ_{c}, δ_{v} are the values of the fluctuation broadening of the conduction and valence bands, respectively. Reference [238] also assumed the Hubbard density of states for both the conduction and valence bands with width W_{c} and W_{v}, respectively, as well as similar dispersion in both bands. With this assumption, the exciton binding energy has been calculated according to the CPA model (see also [71]). It should be added here that the key feature of the model developed in [213] is the short-range nature of the Coulomb potential.

The data from Table 4.1 and other published sources [78, 204] were used for plotting energy E_{b} as a function of isotopic concentration x in Fig. 4.11. The values of binding energy E_{b} were calculated using the hydrogen-like formula (see below) with the energies of exciton levels of 1s and 2s states from reflection spectra (see Fig. 4.8). The theoretical description of the binding energy of Wannier–Mott excitons as a function of x was based on the polynomial derived by Elliott at al. [238]:

$$E_{\mathrm{b}} = E_{\mathrm{b}}^{\mathrm{crys}} - E_{\mathrm{bow}} \left[\frac{1 - W}{2U_0} \right] - E_{\mathrm{eff}}, \tag{4.57}$$

$$E_{\mathrm{eff}} = x(1 - x) \frac{\delta_{\mathrm{c}} \delta_{\mathrm{v}}}{W}, \tag{4.58}$$

$$E_{\mathrm{b}}^{\mathrm{crys}} = U_0 + \frac{W}{2U_0} - W. \tag{4.59}$$

where $W = W_{\mathrm{c}} + W_{\mathrm{v}}$, and W_{c} and W_{v} are the widths of the conduction band and the valence band which are equal to 21 eV [241] and 6 eV [242, 243], respectively. Here, E_{bow} is the curvature parameter found from the function $E_{\mathrm{g}} \propto f(x)$ and δ_{c} and δ_{v} are the magnitudes of the fluctuation smearing of the valence band and the conduction band edges, $\delta_{\mathrm{c}} = 0.103\,\mathrm{eV}$ and $\delta_{\mathrm{v}} = -0.331\,\mathrm{eV}$. As follows from Fig. 4.11, these values of the parameters give a good enough description of the nonlinear dependence of the binding energy of Wannier–Mott excitons in a disordered medium. This agreement between theory and experiment once again proves the inherent consistency of the model proposed by Kanehisa and Elliott, since the isotopic substitution affects the short-range part of the interaction potential.

In this way, the nonlinear dependence of the binding energy of Wannier–Mott excitons is caused by isotopic disordering of the crystal lattice. As seen from Fig. 4.11, the exciton binding energy decreasing (relative linear law (VCA) – see *dashed line* in Fig. 4.11) in the vicinity of the middle concentration points out the fluctuational broadening of the edge of the conduction and valence bands. In accordance with the theoretical model, the last reason gives rise to reduced E_{g} and thereby the shallowing of the exciton levels and, respectively, the reduction of E_{b} (see also [211, 215]).

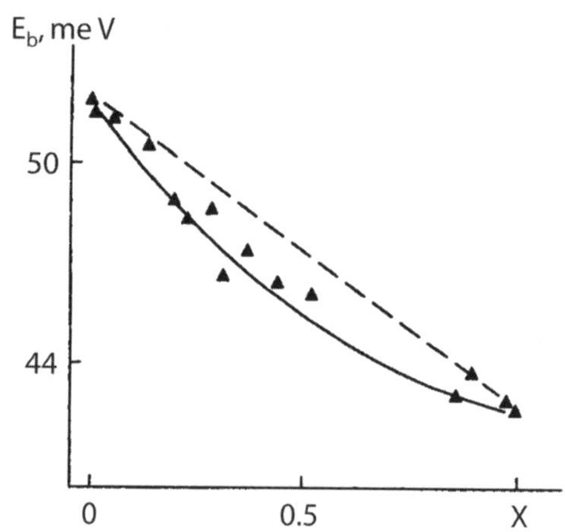

Fig. 4.11. Concentration dependence of binding energy of Wannier–Mott excitons. (1): approximation of virtual crystal model; (2) calculation according to (2.93); experimental points indicated by *triangles*. (After Plekhanov [109])

4.7 Isotopic Effect in the Luminescence Spectrum

Because of the low intensity of scattered light and thanks to the high resolution of modern spectroscopic instruments, the development of highly sensitive techniques of detecting of weak optical signals (photon counting mode, optical multichannel analyzers, optical linear arrays, and other specialized systems; (see, for example [179]), the luminescence method has become one of the most common techniques for studying excitons in dielectrics and semiconductors. The structure of the spectra of fundamental reflection (absorption) depends on the internal degrees of freedom of Wannier–Mott excitons, but the structure and shape of the luminescence spectrum are determined primarily by its external degrees of freedom. As is well-known, the latter are associated with the translational motion of large-radius excitons as a whole, with the translation mass $M = m_c + m_v$. The results on the luminescence of LiH and LiD crystals presented below were obtained from clean surfaces of crystals cleaved directly under liquid superfluid helium in the cell of an optical cryostat [208]. The effects of surface habitus on optical spectra (including luminescence spectra) of excitons in hygroscopic LiH and LiD crystals were briefly described in early papers [52, 208]. The luminescence of LiH crystals was first observed in 1959 by Gavrilov [177].

As demonstrated earlier (see, e.g., [78]), most low-energy electron excitations in LiH (LiD) crystals are large-radius excitons. Exciton luminescence is observed when LiH (LiD) crystals are excited in the midst of fundamental absorption. The spectrum of exciton photoluminescence of crystals of lithium hydride cleaved in liquid helium consists of a narrow (in the best crystals, its half-width is the $\Delta E \leq 10\,\mathrm{meV}$) phononless emission line and its broader phonon repetitions, which arise due to radiative annihilation of excitons with the production of one to five longitudinal (LO) phonons (Fig. 4.12). The phononless emission line coincides in an almost resonant way with the reflection line of the exciton ground state [109], which is another indication of direct electron transition. The lines of phonon replicas form an equidistant series biased toward lower energies from the resonance emission line of excitons. The energy difference between these lines is about 140 meV, which is close to the calculated energy of the LO phonon in the middle of the Brillouin zone [51] and measured in the paper by Plekhanov and O'Konnel-Bronin [245]. Here we also ought to mention an overall increase in the intensity across the entire luminescence spectrum. Evidently, the intensity of a phononless emission line free of excitons (the proof of the existence of quasi-momentum is given below) increases because the rate of emissionless recombination on the pure surface decreases. This seems natural for the surfaces of specimens cleaved in liquid helium because the surface states (as a rule, of extrinsic origin [246, 247]) and their electric fields (recall that the value of the exciton Rydberg energy is relatively low, $E_b = 40\,\mathrm{meV}$ (see above) lead to broadening of the luminescence lines and also to quenching of their intensity, and first of all, quenching of

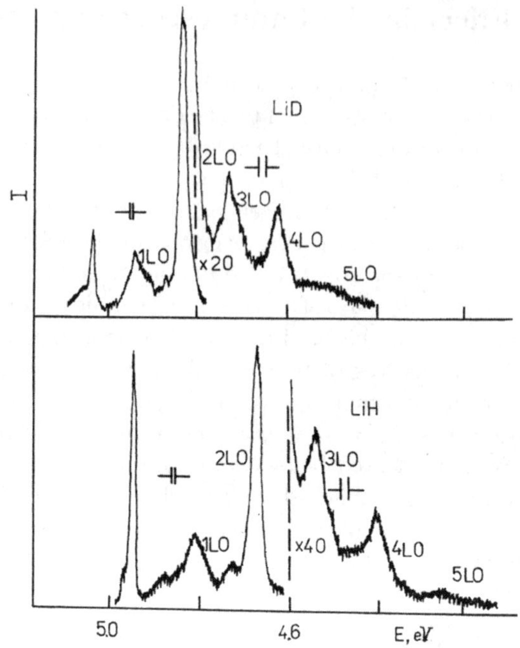

Fig. 4.12. Emission spectra of free excitons at 2 K in LiH and LiD crystals cleaved in liquid helium. The spectral resolution of the instruments is indicated on the diagram

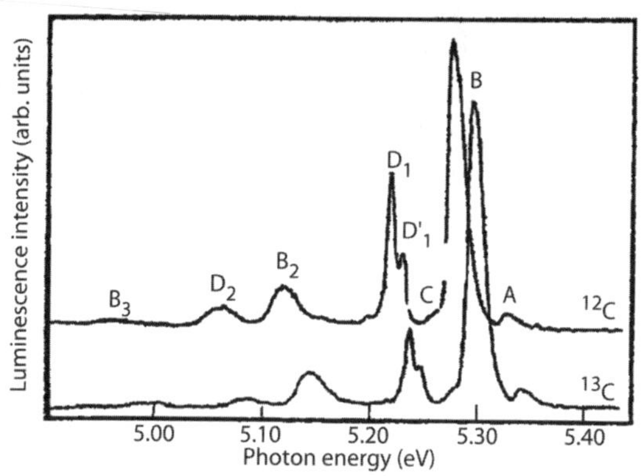

Fig. 4.13. Spectra of the phonon-assisted free exciton cathodoluminescence feature (A, B, and C) and the phonon-assisted bound–exciton features (D) from a natural semiconducting ^{12}C diamond and a ^{13}C synthetic diamond measured at 77 K. (after Collins et al. [248])

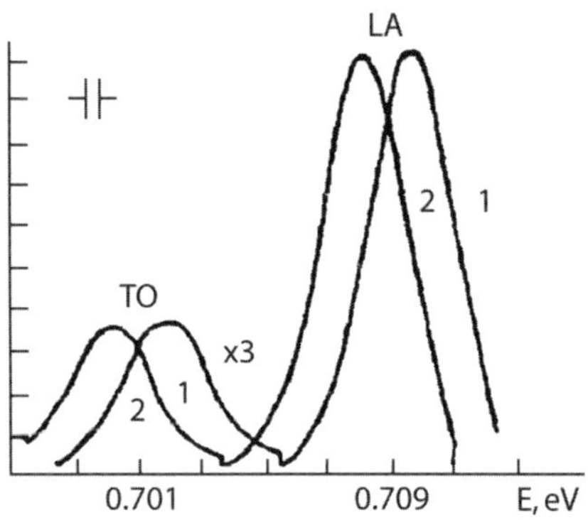

Fig. 4.14. LA- and TO-phonon replicas of electron–hole liquid line emission of ^{76}Ge (1) and ^{72}Ge (2) at 1.7 K. (After Agekyan et al. [88])

the intensity of the zero-phonon line. The isotopic shift of the zero-phonon line of LiH crystals equals that in reflection spectra, 104 meV.

In Fig. 4.12, we see immediately that the structures of the two spectra are the same. The difference is in the distance between the observed lines, as well as in the energy at which the luminescence spectrum begins, and in the half-widths of the line. The first feature is explained by the gradual and smooth decrease of the energy of the LO phonon upon transition from LiH to LiD. This is related to the unimodal nature of LO phonons in mixed LiH_zD_{1-x} crystals at high concentrations of isotopes ($0.4 < x < 1.0$ [25]). This was concluded from the RRLS spectra in the paper cited. The second feature is attributed to phonon renormalization (mainly through interaction with LO phonons [78]) for the energy of band-to-band transitions. The change in E_g is also smooth and continuous though nonlinear (see details in [109] and above).

We should add that isotopic substitution in diamond (Fig. 4.13) [248], ZnO [249] and Ge (Fig. 4.14) [88, 250, 251] also resulted in a shift of the free exciton luminescence spectra. Moreover, isotopic substitution in Ge leads to the shift of the luminescence spectrum and also to the nonlinear concentration dependence of the emission line half-width, which, as in lithium hydride, was attributed to isotopic disordering of the crystal lattice (for details, see [23]).

5 Process of Self-Diffusion in Isotopally Pure Materials and Heterostructures

5.1 General Remarks

Interest in diffusion is as old as metallurgy or ceramics. The first measurement of diffusion in the solid state was made by Roberts-Austen in 1896 [252]. Many measurements, especially of chemical diffusion in metals, were made in the 1930s; the field was reviewed by Mehl [263], Jost [254], and Seith [255]. Diffusion research increased after World War II; the increase was motivated by the connection among diffusion, defects, and radiation damage and was helped by the availability of many artificial radiotracers. These researchers were the first to attempt to identify the basic underlying atomistic mechanisms responsible for mass transport through solids by quantitative investigations and theoretical analysis of the activation energies required for diffusion by exchange, interstitial, and vacancy mechanisms in solids. Prior to this time, there had been little concern with treating diffusional phenomena on a microscopic basis, and most research was concerned with fairly crude observation of overall bulk transfer processes at junctions between regions with strong compositional differences. It was at this time that suggestions for carrying out high-precision, highly reproducible diffusion experiments were first put forward (Slifkin et al. [256] and Tomizuka [257]). The three major factors that determine the quality of a diffusion measurement are

1. the method used,
2. the care taken in the measurement, and
3. the extent to which the material is specified [258].

The most accurate method has, in general, been considered radiotracer sectioning (Tomizuka [257]), and most of this chapter is devoted to this method, especially to points for which special care must be taken: the measurement of temperature, the accuracy of sectioning, and the reproducibility of counting the radioactivity. The importance of specifying the material cannot be overstated. The measured diffusion coefficient depends on the chemistry and structure of the sample on which it is measured. Impurities, nonstoichiometry of compounds, grain boundaries, and dislocations can give apparent values of the diffusion coefficient that are different from, and usually larger than, the true value (see also [259, 260]). The objective of this chapter is to describe some experimental results as well as their theoretical analysis in the last decade. We have organized the chapter around general principles that are applicable to all materials and then listed the particulars (see also [267–271]). The materials we consider are mainly inorganic solids, especially semiconductor and insulator materials. The effects of pressure on diffusion are omitted. Previous reviews covering mainly metals and inorganic materials have been given by Hoffman [261], Tomizuka [257], Cadek and Janda [262], Adda and Philbert [263], Lundy [264], Beniere [265], Nowick [266], and Rothman [267]. Radioactive tracers are essential to many of the experiments described in this chapter [272–275]. Radioactive tracers are hazardous materials, and the experimenter who uses them is under the strongest obligation to avoid exposing his colleagues and contaminating his environment.

5.2 The Relation of Diffusion Experiments to the Mathematics of Diffusion

For measurable diffusion to take place a gradient of some kind is necessary. Diffusion results from the hopping motion of atoms through a solid. The diffusion coefficient D is defined in Fick's first law (Fick [277]),

$$\boldsymbol{J} = -D\boldsymbol{\nabla}C + C\boldsymbol{V} \tag{5.1}$$

where $\boldsymbol{J}$ is the flux of atoms, C their concentration, and $\boldsymbol{V}$ the velocity of the center of mass, which moves due to the application of a force such as an electric field or a thermal gradient (see also [278]). A number of different diffusion coefficients exist, e.g., for the diffusion of a radioactive tracer in a chemically homogeneous solid in the absence of external forces,

$$\boldsymbol{J}* = -D * \boldsymbol{\nabla}C * \tag{5.2a}$$

where the asterisk denotes the radioactive species. For diffusion in a chemical gradient,

$$\boldsymbol{J} = -\tilde{D}\boldsymbol{\nabla}C \tag{5.2b}$$

where $\tilde{D}$ is the interdiffusion or chemical diffusion coefficient. Any of these equations can be combined with the equation of continuity

$$\frac{\partial C}{\partial t} = -\boldsymbol{\nabla} \cdot \boldsymbol{J} \tag{5.3}$$

to yield Fick's second law

$$\frac{\partial C}{\partial t} = \boldsymbol{\nabla} \cdot (D\nabla C), \tag{5.4a}$$

where the mass flow term has been omitted. For a tracer in a homogeneous system,

$$\frac{\partial C^*}{\partial t} = -\boldsymbol{\nabla}^* \cdot \boldsymbol{\nabla}^2 C^*. \tag{5.4b}$$

Equations (5.4a) and (5.4b) describe the types of diffusion experiments discussed in this chapter.

The tracer diffusion coefficient is given also in the atomistic form

$$D^* = \gamma a^2 \Gamma f, \tag{5.5}$$

where γ is a geometric factor, a the jump distance, Γ the atomic jump frequency, and f the correlation factor (see, e.g., [278]). It is thus possible, in principle, to measure D^* by measuring Γ in a resonance experiment [279,280].

We are concerned here with diffusion measurements where the diffusion coefficient is obtained via Fick's second law, i.e., from a solution of the diffusion equation (see also [23]). Fick's second law is used rather than his first because concentrations are easier to measure than fluxes and because D in the solid state is so small that the required steady state is seldom reached. To obtain a solution of the diffusion equation, the initial and boundary conditions (IC and BC) must be known. The IC correspond to the distribution of the diffusing substance in the sample before diffusion annealing, and the BC describe what happens to the diffusing substance at the boundaries of the sample during diffusion annealing. If the experimental IC and BC correspond to the mathematical conditions, the mathematical solution to the diffusion equation $C(x, y, z, t)$ will describe the distribution of the diffusing substance as a function of position in the sample and of annealing time. The diffusion coefficient is finally obtained by fitting the experimentally determined $C(x, y, z, t)$ to the appropriate solution of the diffusion equation with D as a parameter.

Most laboratory experiments are arranged so that diffusion takes place in one dimension. The solution of the diffusion equation is then $C(x, t)$. One most often determines $C(x)$ at constant t, i.e., the concentration distribution in the diffusion direction after diffusion annealing time t. It is also possible to determine $C(t)$ at a constant x (e.g., the concentration at a surface) or $\iint C(x,t)\,\mathrm{d}x\,\mathrm{d}t$ (e.g., the weight gain of a sample as a function of time). The IC, BC, and solutions of the diffusion equation (for D = const) for some common geometries are described below. These and solutions for other cases are given by Crank [281] and Carslaw and Jaeger [282].

(1) *Thin layer or instantaneous source geometry* (Fig. 5.1a). An infinitesimally thin layer ($\ll (Dt)^{\frac{1}{2}}$) of diffusing substance is deposited on one surface of a semi-infinite ($\gg (Dt)^{\frac{1}{2}}$) solid. The initial conditions are

$$C(x, 0) = M\delta(x), \tag{5.6}$$

where δ is the Dirac delta function and M the strength of the source in atoms per unit area. The boundary condition is

$$\partial C(0, t) \partial t = 0, \tag{5.7}$$

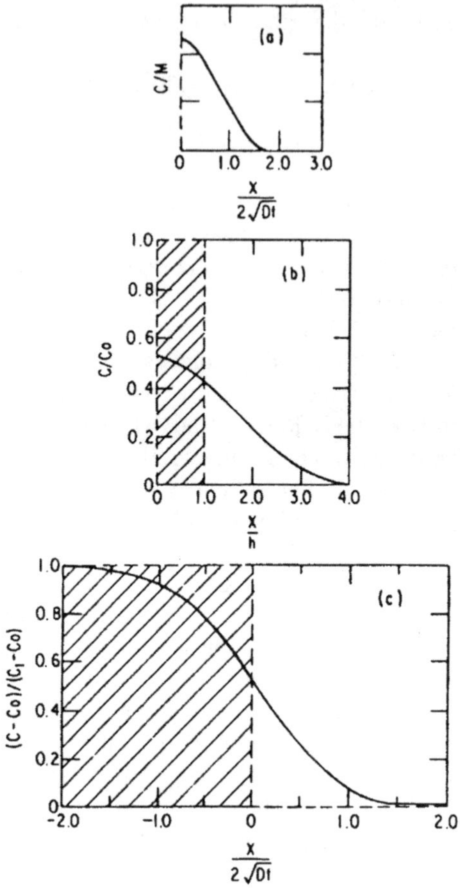

Fig. 5.1. Concentration distributions for different initial conditions. The *otted line* is for $t = 0$ and the *solid line* is for a finite t. **(a)** Thin layer geometry [case (1)]; **(b)** thick layer geometry [case (2)]; *solid curve* for $Dt = h^2$; **(c)** infinite couple [case (3)]: (After Crank [281]]

i.e., there is no flux through the surface. The solution is

$$C(x,t)(M\sqrt{\pi Dt})\exp\left(-\frac{x^2}{4Dt}\right). \tag{5.8}$$

One determines $C(x)$ for constant t.

(2) *Thick layer geometry* (see Fig. 5.1b). Similar to the above, except that the layer thickness h is of the order of the diffusion distance:

$$\text{IC}: C(x,0) = C_0, h \geq x \geq 0,$$

$$C(x,0) = 0, x > h, \tag{5.9a}$$

$$\text{BC}: \partial C(0,t)\partial x = 0. \tag{5.9b}$$

Solution:

$$C(x,t) = \frac{C_0}{2}\left[\operatorname{erf}\left(\frac{x+h}{2\sqrt{Dt}}\right) - \operatorname{erf}\left(\frac{x-h}{2\sqrt{Dt}}\right)\right], \tag{5.10}$$

where

$$\operatorname{erf}(\lambda) = \frac{2}{\sqrt{\pi}}\int_0^{\lambda} \exp(-\eta^2)\mathrm{d}\eta. \tag{5.11a}$$

Measure $C(x)$ for constant t. Note:

$$\operatorname{erf} c(\lambda) \equiv 1 - \operatorname{erf}(\lambda). \tag{5.11b}$$

(3) *Infinite couple* (see Fig. 5.1c). A sample of uniform concentration C_0 is welded to a sample of uniform concentration C_1. The weld plane is situated at $x = 0$. The couple contains a volatile:

$$\begin{aligned} \mathrm{IC} : C(x,0) &= C_1, x < 0, \\ C(x,0) &= C_0, x > 0. \end{aligned} \tag{5.12}$$

Solution:

$$C'(x,t) \equiv \frac{C(x,t) - C_0}{C_1 - C_0} = \left[1 - \operatorname{erf}\left(\frac{x}{2\sqrt{Dt}}\right)\right]. \tag{5.13}$$

Measure $C(x)$ for constant t.

(4) *Vapor–solid couple.* A semi-infinite couple containing a volatile component component is placed in a dynamic vacuum at $t = 0$:

$$\mathrm{IC} : C(x,0) = C_0, x > 0, \tag{5.14a}$$

$$\mathrm{BC} : C(0,t) = 0, t > 0. \tag{5.14b}$$

Solution:

$$C(x,t) = C_0 \operatorname{erf}\left(\frac{x}{2\sqrt{Dt}}\right). \tag{5.15}$$

Exposing a sample initially devoid of a volatile component to a vapor of the volatile component at a pressure in equilibrium with C_0 gives the analogous mathematics:

$$\mathrm{IC} : C(x,0) = 0, x > 0, \tag{5.16a}$$

$$\mathrm{BC} : C(0,t) = C_0, t > 0. \tag{5.16b}$$

Solution:

$$C(x,t) = C_0 \left[1 - \operatorname{erf}\left(\frac{x}{2\sqrt{Dt}}\right)\right] = C_0 \operatorname{erf} c\left(\frac{x}{2\sqrt{Dt}}\right). \tag{5.17}$$

The same equations apply to isotopic exchange between solid and vapor. Measure either $C(x)$ at constant t or integral weight gain (loss)

$$\int_0^\infty \int_0^t C(x,t)\mathrm{d}t\,\mathrm{d}x.$$

(5) *Grain boundary diffusion.* The mathematics in this case are more complicated (see, e.g., [283]), owing to the coupled lattice diffusion, but one still measures $C(x)$ at constant t.
(6) *Exchange experiment* [284]. This technique is used for materials for which a massive sample cannot be prepared. It involves diffusion exchange between an assembly of powder and a gas of limited volume, from which very small aliquots are drawn at different times.

In the first three sample configurations, two bodies of widely different composition are brought into contact. The assumption implicit in the BC is that diffusing material passes across the resulting interface without hindrance, i.e., it is not held up by surface oxides, low solubility, chemical reactions, etc. Nonfulfillment of this condition leads to deviation of the experimental $C(x)$ from the solution of the diffusion equation (for details, see [285]).

In the vapor–solid couple and the exchange experiment, the assumption implicit in the BC is that the surface of the solid equilibrates instantaneously with the gas phase. However, optical measurements of the change in the surface concentration at low temperature have indicated that the attainment of the solid–gas equilibrium can be a slow process (see also [286]).

In this connection, we should add that thin geometry has several advantages. A thin layer can be deposited without straining the sample, which is essential for single crystal samples. A thin layer also allows the use of highly specific radioisotopes and thus measurements of diffusion without a chemical gradient. Diffusion under a large chemical gradient can lead to deformation of the sample and generation of defects (see also [287–290]). For these reasons, thin layer geometry is most often used in experiments in which diffusion is measured to study the fundamentals of diffusion and defect behavior in solids. Such experiments usually concern diffusion as a function of temperature, pressure, or concentration, and small differences in D are not important, in contrast to engineering experiments in which the magnitude of the penetration of one material into another is of interest.

Note that all solutions of the diffusion equation considered above are expressed in terms of the dimensionless variable $\frac{x}{(2\sqrt{Dt})}$. The length $2\sqrt{Dt}$ is a kind of mean penetration distance, and this has to be the same order of magnitude as the characteristic distance associated with an experiment. For a sectioning experiment, the characteristic distance is the section thickness. For ion-beam depth profiling, it is the ion range, etc. [275].

In the ordinary thin-layer sectioning experiment, one wishes to measure diffusion over a drop in specific activity C of $\sim 10^3$; any effects due to diffusion along short-circuiting paths are likely to show up as curvature in the penetration plot over such a range, though they may not be visible if the measurement is only over a factor of 6 in C [291]. Usually, 20 sections suffice to define a penetration plot; from (5.8), the section thickness required for a drop of 10^3 in C over 20 sections is

$$\theta \approx \frac{\sqrt{Dt}}{3.8}. \tag{5.18}$$

A preliminary estimate of D is useful in planning an experiment.

If the isotope decays significantly during the experiment, more radioisotope has to be deposited. Under the conditions of $\theta \approx \frac{\sqrt{Dt}}{3.8}$, the specific activity drops by a factor of ~ 2 per section at the twentieth section. These last points on the penetration plot have the greatest weight in determining D, so the counting statistics must be maintained and the penetration plot extended as far as possible. This implies use of an intense source of radioactivity, on the other hand, too much activity poses an unnecessary health hazard and increases the dead-time correction. The radiotracer may rapidly reach the side surfaces of the sample by surface diffusion or evaporation and then diffuse inward. To keep the diffusion one-dimensional, one removes $\approx 6\sqrt{Dt}$ from the sides of the sample before sectioning.

5.3 The Self-Diffusion Process

As is well known, in all diffusion mechanisms, the atoms under consideration have to jump between different sites (see, e.g., [292]). If the extreme case of coherent tunneling [293] is left aside, the diffusional jumps are assisted by the thermal movement of atoms. In the standard situation, the jump rate is entirely determined by the temperature T (apart from the effects of hydrostatic pressure, which may be incorporated by formulating the theory in terms of enthalpy and Gibbs free energy). For the purposes of this chapter, we may in the first approximation disregard the quantum mechanical contribution to the diffusivity (naturally excluding the self-diffusion in LiH), so that the diffusion coefficient in cubic crystals under standard conditions may be written as an Arrhenius expression

$$D_\alpha = D_{\alpha 0} \exp\left(-\frac{H_\alpha^{\mathrm{M}}}{kT}\right) \tag{5.19}$$

with the preexponential factor

$$D_{\alpha 0} = g_\alpha a_0^2 \nu_{\alpha 0} \exp\left(\frac{S_\alpha^{\mathrm{M}}}{k}\right) . \tag{5.20}$$

Here H_α^{M} denotes the enthalpy and S_α^{M} the entropy of migration, a_0 the lattice constant, and $\nu_{\alpha 0}$ the attempt frequency, k has its usual meaning as Boltzmann's constant, and g_α is a factor that takes into account the geometry of the crystal structure and the atomistic details of the different processes. The subscript α refers to the defect species controlling the diffusion process, i.e., in a direct interstitial mechanism, it indicates the chemical nature, geometrical configuration, etc., of the interstitial involved, whereas in indirect diffusion, it characterizes the intrinsic defects acting as diffusion vehicles. In the latter case, we should write β instead of α if we wish to indicate that these intrinsic defects are monovacancies or monointerstitials.

The tracer self-diffusion coefficient, i.e., the diffusivity of radioactive self-atoms under thermal equilibrium, is given by (see also [292])

$$D^{\mathrm{T}} = \sum_{\beta=I,V} f_\beta\, D_\beta^{\mathrm{SD}} = \sum_{\beta=I,V} f_\beta\, D_\beta\, C_\beta^{\mathrm{eq}} \tag{5.21}$$

where

$$C_\beta^{\mathrm{eq}} = \exp\left(\frac{S_\beta^F}{k}\right) \exp\left(-\frac{H_\beta^F}{kT}\right) \tag{5.22}$$

are the concentrations of self-interstitials ($\beta = I$) and monovacancies ($\beta = V$) in thermal equilibrium. As is clear, in (5.21), contributions by clusters of I or V are neglected. The f_β denote correlation factors, $D_\beta^{\mathrm{SD}} \equiv D_\beta C_\beta^{\mathrm{eq}}$ contributions to the uncorrelated self-diffusion coefficient $\sum_{\beta=I,V} D_\beta^{\mathrm{SD}}$, and S_β^F and H_β^F entropies and enthalpies of formation, respectively. Insertion of (5.19), (5.20), and (5.22) into (5.21) yields

$$\begin{aligned} D^{\mathrm{T}} &= \sum_{\beta=\mathrm{I,V}} D_\beta^{\mathrm{T}} = \sum_{\beta=I,V} f_\beta\, g_\beta\, a_0^2\, \nu_{\beta 0} \exp\left(-\frac{G_\beta^{\mathrm{SD}}}{kT}\right) \\ &= \sum_{\beta=\mathrm{I,V}} D_{\beta 0}^{\mathrm{T}} \exp\left(-\frac{H_\beta^{\mathrm{SD}}}{kT}\right) \end{aligned} \tag{5.23}$$

with the preexponential factors

$$D_{\beta 0}^{\mathrm{T}} = f_\beta\, g_\beta\, a_0^2\, \nu_{\beta 0} \exp\left(-\frac{H_\beta^{\mathrm{SD}}}{kT}\right) \tag{5.24}$$

and the Gibbs free energy of self-diffusion

$$G_\beta^{\mathrm{SD}} = H_\beta^{\mathrm{SD}} - T S_\beta^{\mathrm{SD}} , \tag{5.25}$$

the self-diffusion enthalpy

$$H_\beta^{\mathrm{SD}} = H_\beta^F + H_\beta^{\mathrm{M}}, \tag{5.26}$$

and the self-diffusion entropy

$$S_\beta^{\mathrm{SD}} = S_\beta^F + S_\beta^{\mathrm{M}}. \tag{5.27}$$

The diffusion coefficient D^s of foreign substitutional atoms in thermal equilibrium may be derived from (5.21) or (5.23) by inserting factor h_β under the summation signs. These factors account for the interaction between intrinsic thermal equilibrium defects and substitutional atoms. They depend on temperature and the atomic fraction of the substitutional atoms, unless this is small compared to unity. For a more detailed and complete discussion of this field, we refer the reader to the review by Frank et al. [292] (see also [275]).

Compared with metals, self-diffusion in semiconductors is a very slow process. For the elemental semiconductors, this is illustrated in Fig. 5.2, in

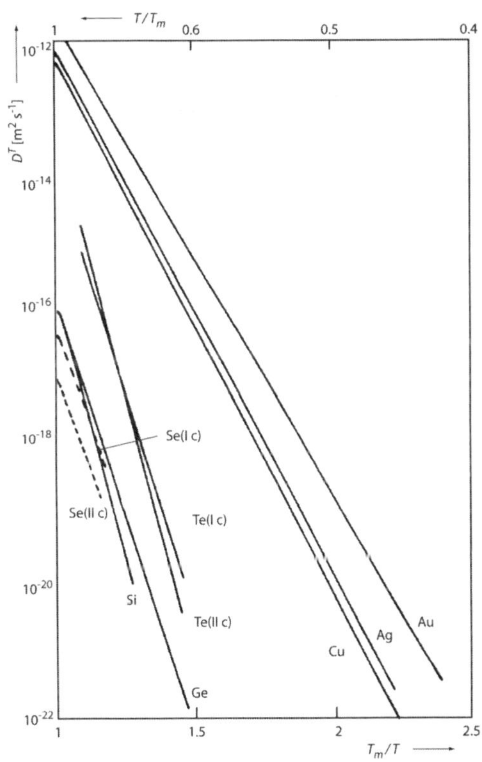

Fig. 5.2. Comparison between the self-diffusivities of the cubic semiconductors Ge and Si (Frank et al. [292]), the trigonal semiconductors Te (Ghoshtagore [294], Werner et al. [295] and Se (Brätter and Gobrecht [296] and the typical metals Cu, Ag, Au (after [292])

which the self-diffusivities of the cubic semiconductors Si and Ge and of the trigonal semiconductors Te and Se are compared with those of typical metals such as Cu, Ag, and Au on a temperature scale normalized to the melting temperature T_{m}. Figure 5.2 reveals the following differences between metals and semiconductors, already emphasized by Seeger and Chik [297].

1. Near melting temperatures, self-diffusion in semiconductors is several orders of magnitude slower than in typical metals.
2. At lower normalized temperatures, the ratio of the self-diffusivities of metals and semiconductors becomes even larger.

Generally speaking, the origin of these differences lies in the homopolar bonding of semiconductors (for details, see [292]).

As is well known, the conventional and well-established techniques of determining the tracer self-diffusion coefficient D^{T} are based on studying the redistribution of radioactive or stable tracers initially deposited on the specimen surface by serial sectioning methods. For radioactive isotopes, the redistribution may be investigated by radiation detection methods; for stable isotopes secondary ion mass spectroscopy (SIMS) may be used (see also below).

5.4 The SIMS-Technique

The most complete description of the experimental technique for studying diffusion processes in solids is the excellent review by Rothman [285]. Here, we briefly discuss the peculiarities of sputtering and SIMS techniques.

We define microsectioning [298] as the cutting of sections a few hundred nanometers or less in thickness, so that the surface on which the tracer is deposited, the "front" surface, is not necessary flat on the scale of $\sqrt{Dt}$, and so the thickness of the individual sections are not determined separately. The isoconcentration contours then follow the contour of the front surface, and one must remove sections parallel to this nonflat surface, rather than parallel to a flat surface. If this condition is met and if the undulations in the front surface are gentle (radius of curvature $\rho \gg \sqrt{Dt}$), one can treat the sections as if they were flat (see, however [299]).

Simple chemical dissolution followed by counting the solution has been used for metals [300], alloys [301], and silicate glasses [302, 303]. One uses a constant volume of solvent in the counting vial, rinses the sample so that the rinse flows into the same vial, and then adjusts the total volume of solution so that it is the same for all sections.

In sputtering, material is removed by ion bombardment owing to the transfer of momentum from the bombarding ions to the atoms of the targets. A depth profile can be constructed by

1. analyzing the sputtered-off material in a mass spectrometer (SIMS);
2. collecting and analyzing the sputtered-off material;

3. determining the concentration of the diffusing material in the remaining surface by, e.g., Auger electron spectroscopy (see, e.g., [304–307]), or
4. counting the residual activity of the entire sample [308].

As a rule, noble gas ions, especially Ar^+, are accelerated to a few hundred electron volt or more, with current densities $\leq 1\,mA\,cm^{-2}$. This is called physical sputtering, in contrast to bombardment with reactive ions, which is called chemical sputtering. Typical removal rates are of the order of $10\,nm\,min^{-1}$ for $1\,mA\,cm^{-2}$ of $500\,eV$ Ar^+ ions. There are two excellent reviews of the subject of sputtering (Behrisch [309] and Chapman [310]), and the reader is referred to them for an understanding of the process. Equipment for sputtering includes a vacuum chamber, pumping equipment, and a controlled gas flow such as a micrometer needle valve. A high-speed pumping system is needed as gas is passed continuously and there are bursts of desorbed gases to cope with. Cold-trapped diffusion pumps, cryopumps, or turbopumps have been used. All sputtering equipment has a gaseous discharge in it. Common glow discharges are not suitable, as too high a gas pressure is required, with resulting low mean free paths and back diffusion of the sputtered atoms. Therefore, either an ion gun or a rf power source is used. Two types of ion sources have been used in sectioning experiments, the custom-modified duoplasmatron of Maier and Schule (see [292]) and a commercially available Kauffman-type gun [311, 312]. Almost any ion source used in ion milling should be usable. The main requirement is that the source puts out ion currents $\geq 1 mA \cdot cm^{-2}$ at $\sim 1\,kV$ over $\sim 4\,cm^2$ in a reasonably uniform beam ($\pm 10\%$ except at the very edge), and that the current stays constant for several hours. The length of a run is limited by life of the filament. In addition to the ion source, chamber, pumps, and valving, one needs a collector and a sample holder. These are usually custom-made. Designs have been given by Gupta and Tsui [313] and Atkinson and Taylor [314] for rf systems and by Mundy and Rothman [312] for ion gun systems. The collector is either a carousel with six Al planchets, which allow six sections to be taken before the chamber is opened [313], or a device like a camera back, on which polyester film is rolled; the latter allows taking 32 sections.

In the SIMS technique, the sample is bombarded by reactive ions, and the sputtered-off molecules are ionized in a plasma and fed into a mass spectrometer. The mass spectrum is scanned, and the ion current for tracer and host atoms can be recorded simultaneously. The beam is swept over the sample and, in effect, digs a crater, whose bottom is more or less flat; an aperture prevents ions originating from the edges of the crater from reaching the mass spectrometer (see Fig. 5.3). The penetration plot is constructed from the plots of instantaneous tracer/host atom ratio versus sputtering time and of distance sputtering time. The distance is obtained by interferometric measurement of the total crater depth under the assumption that material is removed uniformly as a function of time. Large changes in chemical composition in the diffusion direction can invalidate that assumption. The limita-

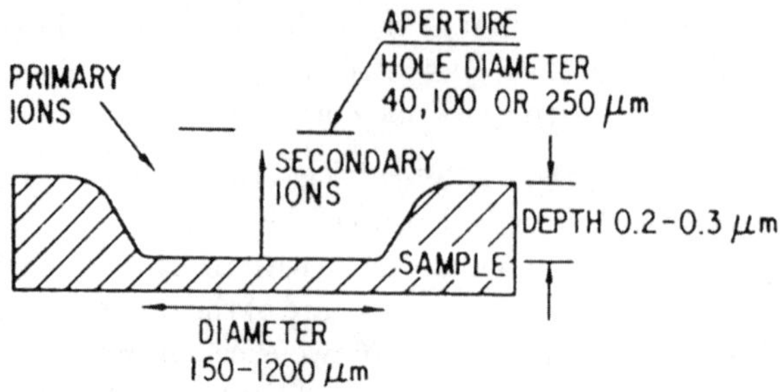

Fig. 5.3. Schematic diagram of crater caused by sputtering in a SIMS apparatus (after Dorner et al. [315])

tions of the SIMS technique have been discussed by Liebl [316] and Reuter and Baglin [317], and a detailed description of its application to diffusion has been given by Seran [318] and Macht and Naundorf [319]. A recent paper [315] shows the quality of results that can be obtained (see also [320]). In general, the resolution is no worse than that obtained by sputtering and counting the sections, and the sensitivity is no worse than that of counting techniques. The major disadvantage of SIMS is its cost. The SIMS apparatus is commercially made (see, e.g., [285] and references therein) but represents a large capital investment. Not withstanding the cost of the apparatus careful controls must be applied to the measurements, and artifacts [321] must be avoided.

If the entire sample is sectioned, the depth of material removed is best determined by weighing on a microbalance. With care, a sample can be weighed to $\pm 3\,\mu g$, which corresponds to $\pm 150\,nm$ for a cross-sectional area of $0.1\,cm^2$ which is about the minimum useful area, and a density of $2\,g\,cm^{-3}$. For larger areas or densities, even better sensitivities are obtained, down to perhaps $\pm 10\,nm$ (for details, see the review [285] and references therein).

5.5 Self-Diffusion of Li and H in LiH Crystals

Self-diffusion is the migration of constituent atoms in materials. This process is mediated by native defects in solids and thus can be used to study the dynamics and kinetics of these defects. The knowledge obtained in these studies is pivotal for understanding many important mass transport processes such as impurity diffusion in materials. Self-diffusion of D(H) and Li in LiH crystals was studied in papers [322–324] and [324, 325], respectively. As shown before, the gas–solid isotope exchange method has been used for measuring self-diffusion coefficients in solids. Two papers [323, 324] reported on a thermogravimetric study of the pressure and temperature dependence of the diffusion coefficient of the deuteride ion in LiH crystals. As is well known, in this method, a crystal of the compound of interest is equilibrated in a furnace with a gas usually containing an isotopic species of the diffusant. The weight change in the crystal due to the permutation process from the gas to the solid is then monitored as a function of time. By assuming that the mass uptake is due to the isotopic exchange process with deuterium gas and subsequent diffusion of the deuteride ion into the crystal and that the rate of the exchange process is diffusion controlled, the mass gain of the crystal is calculated from a solution of Fick's law. The best least squares fit of the data obtained in papers [323, 324] to one-dimensional and three-dimensional models was used to find the diffusion coefficients and the activation energy for the deuteride ion.

According to [323], the solution of Fick's second law [278]

$$\frac{\partial C}{\partial t} = \frac{\partial^2 C}{\partial x^2} \tag{5.28}$$

subject to the boundary conditions

$$\begin{aligned} C(x,0) &= 0, \\ C(0,t) &= C_{\mathrm{s}}, \end{aligned} \tag{5.29}$$

where C_{s} is the surface concentration of the diffusing species and $\lim C(\, ,\mathrm{t}) \to 0$, $x \to 0$ gives, for Q, the total amount of diffusing substance which has entered the solid at time t,

$$Q = 2A_{\mathrm{s}}C_{\mathrm{s}} \left(\frac{D}{\pi}\right)^{\frac{1}{2}} t^{\frac{1}{2}} + B, \tag{5.30}$$

where A_{s} is the total surface area of the crystal, D is the diffusion coefficient, and B a constant that accounts for the initial condition that $Q \neq 0$ at $t = 0$. Q is expressed as the ratio of the number of moles of deuteride ion diffusing to the total moles of LiH in the crystal.

Equation (5.30) may be written as

$$Q = C' t^{\frac{1}{2}} + B, \tag{5.31}$$

where C' is defined as

$$C' = 2A_{\mathrm{s}}C_{\mathrm{s}} \left(\frac{D}{\pi}\right)^{\frac{1}{2}}. \tag{5.32}$$

Assuming that the surface coverage C_{s} can be written in terms of an adsorption isotherm, θ, C' was rewritten as

$$C' = 0.8748 \left(\frac{A_{\mathrm{s}}\theta}{\rho V}\right) D^{\frac{1}{2}}, \tag{5.33}$$

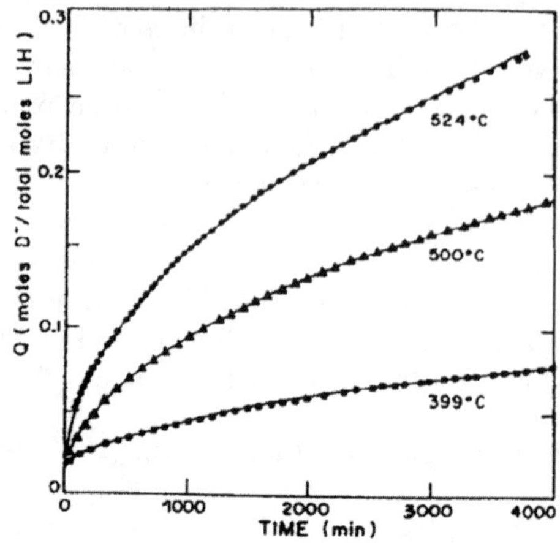

Fig. 5.4. Q as a function of time for data from samples 9, 10 and 13. The *solid curves* are the best least squares fit of (5.31) to the experimental data (after [324])

where ρ *is* the density of LiH at a given temperature, V is the volume of the crystal, and the constant includes the necessary factors for consistency of units. Rearrangement gives

$$C = C' \left(\frac{A_s}{\rho V} \right)^{-1} = 0.8748\,\theta\, D^{\frac{1}{2}} . \tag{5.34}$$

The values for C' were found from the fit of (5.31) to the data at each temperature (see Fig. 5.4). All fits were exceptionally good (see also Table 5.1). Values of C calculated from C' by using the relation given in (5.28) are also given in Table 5.1.

Equation (5.34) may be written in the form

$$C = 0.8748\ \theta\ D_0^{\frac{1}{2}}\ \mathrm{e}^{-\frac{E_a}{2RT}} \tag{5.35}$$

Table 5.1. Summary of one-dimensional semi-infinite solid data (after [324])

Sample no.	T (°C)	$C' \cdot 10^3$ (min^{-1})	$C \cdot 10^4$ (gcm^{-2}min$^{-\frac{1}{2}}$)	E_a (kcal)
8	550	4.74	3.96	22.7±2.8 [a]
9	524	4.37	3.49	22±2 [b]
10	500	2.80	2.52	
11	450	1.50	1.53	
12	410	1.37	0.96	
13	399	1.09	0.86	

[a] E_a as found from the best least squares fit of (5.30) to C with $\theta = 0.80$. The error reported is the 95% confidence level fit to the data

[b] E_a from [326]

by substituting for D,

$$D = D_0\,\mathrm{e}^{-\frac{E_a}{2RT}}, \tag{5.36}$$

where D_0 is a constant and E_a the activation energy.

According to (5.35), the temperature variance of C is determined by the exponential term involving temperature and the temperature dependence of the fractional surface coverage θ. If θ is known as a function of temperature, C may be fit to (5.35) to give D_0 and the activation energy. Lacking knowledge of the exact variation of θ with temperature, (5.29) may be rewritten in the more convenient form

$$\frac{C}{\theta} = 0.8748\,D_0^{\frac{1}{2}}\,\mathrm{e}^{-\frac{E_a}{2RT}} \tag{5.37}$$

and θ is assumed to vary with temperature according to the expected behavior outlined above. The quantity $\frac{C}{\theta}$ was calculated next using $\theta = \frac{T}{(b+T)}$ and the fit of the data to (5.31) found, giving the activation energy as 22.1 ± 3.0 kcal and D_0 as $4.01 \cdot 10^{-3}\,\mathrm{cm^2\,s^{-1}}$. The activation energy thus found is in good agreement with the value of 22 ± 2 kcal obtained by Funkee and Richtering [326] from NMR measurements.

In the three-dimensional bulk diffusion model, the solution of Fick's law [277] for a finite solid

$$\begin{aligned} Q = \theta\Bigg(1 - &\left\{\sum_{u=0}^{\infty} \frac{8}{(2u+1)^2\pi^2} \exp\left[\frac{-(2u+1)^2\pi^2 Dt}{4a^2}\right]\right\} \\ &\times \left\{\sum_{v=0}^{\infty} \frac{8}{(2u+1)^2\pi^2} \exp\left[\frac{-(2v+1)^2\pi^2 Dt}{4b^2}\right]\right\} \\ &\times \left\{\sum_{w=0}^{\infty} \frac{8}{(2mw+1)^2\pi^2} \exp\left[\frac{-(2w+1)^2\pi^2 Dt}{4c^2}\right]\right\}\Bigg) \\ &+ \cdots + B, \end{aligned} \tag{5.38}$$

where the symbols are as previously defined and 2a, 2b, and 2c are the dimensions of the crystal in the x, y, and z directions. The best fit of the experimental data of papers [323,324] gives $D = 2.41 \cdot 10^{-2}\,\mathrm{e}^{\frac{-24.3\cdot 10^3}{RT}}\,\mathrm{cm^2 s^{-1}}$ for the diffusion coefficient as a function of temperature. The diffusion coefficient for the deuteride ion in lithium hydride at 465°C found in paper [323] from the three dimensional model was $(1.9 \pm 0.6) \cdot 10^{-9}\,\mathrm{cm^2 s^{-1}}$ (see also Fig. 5.5). The diffusion coefficient calculated from (5.38) was $1.6 \cdot 10^{-9}\,\mathrm{cm^2 s^{-1}}$. A comparison of tracer diffusion coefficients calculated from one- and three-dimensional models is given in Table 5.2, and a plot of $-\log D$ versus $\frac{10^3}{T}$ is given in Fig. 5.6.

The theoretical activation energies for various migrating species have been calculated by Dellin et al. [327]. They find that activation energies for interstitial H^- diffusion lie in the range 11.5 to 23 kcal, and the calculated activation

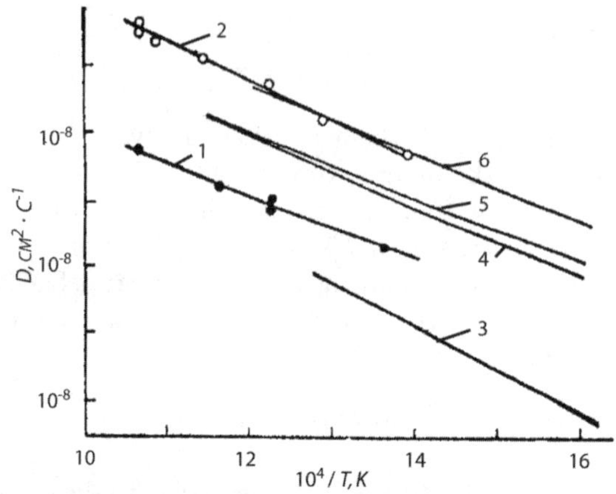

Fig. 5.5. Temperature dependence of self-diffusion coefficients Li (1, 3) and H (2, 4–6) in LiH single crystals; 1, 2: see [325]; 3, 6: see [326]; 4, 5: see [324] (after [325])

energy for H^- vacancy migration is 2.3 kcal. From these calculations, interstitial H^- migration seems possible based on the activation energy of about 24 kcal found by Spencer et al. [324]. However, Dellin et al. [327] in agreement with Pretzel et al. [328] find that interstitial H^- is an unstable species in LiH and thus could not be the difffusing species.

To conclude this section, we should mention once more that the activation energy found by Spencer et al. [324] $\sim$ 24 kcal, is in excellent agreement with the 22 kcal determined for H^- self-diffusion in LiH by NMR [326]. This agreement, plus the consistencies of previous work [323], make D^- vacancy migration still the most likely species and mode of migration (see also [325]).

Table 5.2. Comparison of tracer diffusion coefficients found from one- and three-dimensional models (after [324])

Sample no	T (°C)	$1-D \cdot 10^9$ (cm^2s^{-1})^a	$3-D \cdot 10^9$ (cm^2s^{-1})^b	$3-D^c$
8	550	5.8	8.7	0.2 ± 0.1
9	524	3.7	5.3	
10	500	2.3	3.4	
11	450	0.84	1.1	
12	410	0.34	0.41	
13	399	0.26	0.31	

^{a} Calculated from (5.36).

^{b} Calculated from (5.38).

^{c} Data from [325].

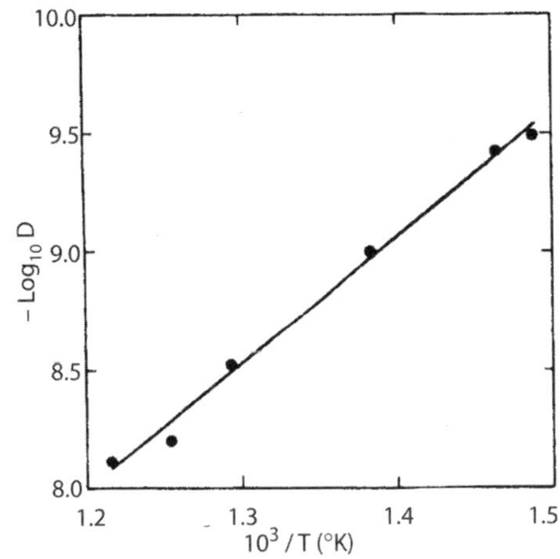

Fig. 5.6. Arrhenius plot of the diffusivity of D^- in LiH in the temperature region $400 - 550°C$ for $\theta = 0.8$. The activation energy determined from this plot is 24.3 ± 2.6 kcal (after [324])

5.6 Self-Diffusion in Intrinsic Ge

In intrinsic germanium, the temperature dependence of the tracer self-diffusion coefficient of the radioactive isotope ^{71}Ge has been measured by several groups [329–334] using different techniques (see Fig. 5.7). Except for the latest experiments, precision grinding techniques were used to remove sections whose thickness was of the order of 1 μm from the diffusion zone of the annealed specimens. As a consequence, the temperature range covered by the earlier experiments is rather limited. Using sputtering technique for serial sectioning [333, 334] extended the range of self-diffusion studies in Ge to diffusivities as low as $10^{-22}\,\mathrm{mm^2 s^{-1}}$.

The overall agreement among Ge self-diffusion data of different authors is good. In the region of overlap, a small difference may be seen between the data of Vogel et al. [333] and those of the earlier workers. We tend to attribute this to problems in determining small diffusion coefficients in the earlier work. Widmor and Gunther-Mohr [331] used Gruzin's or Steigmann's methods, both of which are known to be less reliable than the layer-counting method since these methods require precise knowledge of the absorption coefficient of the radiation involved. In the work of Valenta and Ramasastry [330], the condition $\delta \ll \sqrt{D^{\mathrm{T}}}$ (δ is the thickness of the deposited tracer layer) was not always fulfilled. Since, nevertheless, these authors used the thin-film solution of the diffusion equation to deduce tracer diffusion coefficients, the values obtained are likely to be somewhat larger than the true D^{T} values. As may be seen in Fig. 5.7, the temperature dependence of the D^{T} data on Ge is well described by an Arrhenius law (the preexponential factors D_0^{T} and the self-diffusion enthalpies H^{SD} obtained from measurements of dif-

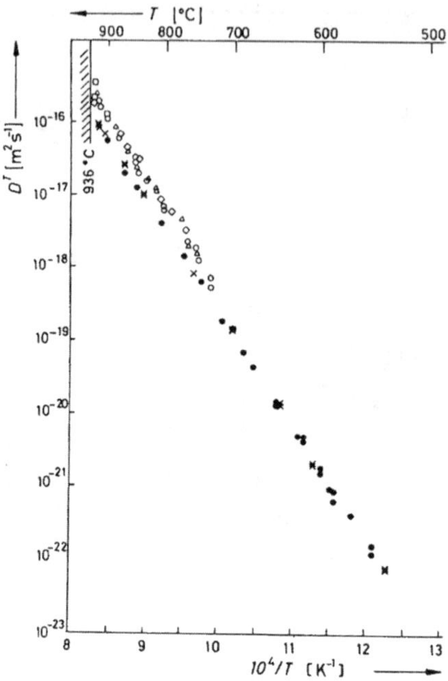

Fig. 5.7. Tracer self-diffusion coefficient of Ge as a function of temperature: ◇ [329]; △ [330]; ○ [331]; □ [332]; • [333]; × [334] (after [292])

ferent authors are compiled in Table 5.3). Seeger and Frank [337] argued that this result may be accounted for in terms of an indirect self-diffusion mechanism involving one type of intrinsic defect. Guided by further observations, they suggest that it is the vacancy mechanism (details see [292]) that controls self-diffusion in Ge. Table 5.3 shows that the preexponential factor D_0^{T} of Ge is considerably larger than the D_0^{T} values typical for metals ($10^{-6}\,\mathrm{m^2s^{-1}} \lesssim D_0^{\mathrm{T}} \lesssim 10^{-4}\,\mathrm{m^2s^{-1}}$) [292]. Arguing that for an ordinary mechanism, the product $f_{\mathrm{v}} g_{\mathrm{v}} a_0^2 \nu_{\mathrm{v0}}$ in (5.18) for $D_{\mathrm{v0}}^{\mathrm{T}} (\equiv D_0^{\mathrm{T}})$ should be of the same order of magnitude for Ge and metals, Seeger and Frank [337] interpreted the large D_0^{T} value for Ge in terms of a large self-diffusion entropy of the vacancy in Ge, $S_{\mathrm{v}}^{\mathrm{SD}} \approx 10\,\mathrm{kcal}$. They suggested that this large $S_{\mathrm{v}}^{\mathrm{SD}}$ value arises from spreading of the vacancy over several atomic volumes.

As we can see from Table 5.3, the published values of fundamental quantities such as the diffusion coefficient vary by several orders of magnitude for various authors (see also [292, 337]). Such a spread in the experimental data makes it difficult to determine conclusively the underlying physical processes. Reliable diffusion data are therefore crucial to clarify the diffusion mechanisms and to accurately determine the corresponding material parameters. The conventional technique (see, e.g., Table 5.3) for determining the self-diffusion coefficient D^{SD} in semiconductors is to deposit a thin layer of radioactive tracer on the surface of the crystal (e.g., $^{71}\mathrm{Ge}$, $^{31}\mathrm{Si}$). In a subse-

Table 5.3. Self-diffusion data for germanium and silicon (after [292])

Element	$D_0^{\mathrm{T}}\,(10^4\,\mathrm{m^2s^{-1}})$	H^{SD} (eV)	Temperature range (K)	Technique	References
Ge	7.7	2.95	1039–1201	SG	[329]
	32	3.1	1023–1143	SG	[330]
	44	3.12	1004–1188	SM + GM	[331]
	10.8	2.99			
	24.8	3.14	822–1164	SS	[333]
	13.6	3.09	808–1177	SS	[334]
	$1.2 \cdot 10^{-3}$	3.05	543–690	SIMS	[346]
Si	1800	4.77	1473–1673	HL	[338]
	1200	4.72	1451–1573	CS	[339]
	9000	5.13	1373–1573	ES	[340, 341]
	1460	5.02	1320–1660	SS	[342]
	8	4.1	1173–1373	R	[343]
	154	4.65	1128–1448	SIMS,^{30}Si	[344]
	20	4.4	1103–1473	R	[345]

SG: sectioning by grinding
SAM: Steigmann's method
GM: Gruzin's method
SS: sectioning by sputtering
HL: hand lapping
CS: chemical sectioning, n activation of ^{30}Si
ES: electrochemical sectioning
R: (p, γ) resonance of ^{30}Si

quent annealing step, the tracers diffuse into the crystal. The depth profile of the tracer atoms is then determined by serial sectioning and measurements of the corresponding radioactivity. There are several experimental difficulties arising from this method (see also [346]).

1. Traditionally, lapping and grinding was used for serial sectioning. This requires that the mean penetration distance $(D^{\mathrm{SD}}t)^{\frac{1}{2}}$ of the tracer atoms during the time t of diffusion anneal has to be in the micron range. Especially in silicon, the large distance and the short half-life (2.6 h for ^{31}Si) limit this method to higher temperatures (larger D^{SD}). Germanium is more convenient in this respect (the half-life of ^{71}Ge is 11.2 days), but it was not until microsectioning techniques (e.g., sputtering) were invented that the measurements could be extended to lower temperatures in recent years (see [285]).
2. Surface effects such as oxidation, contamination, strain, etc. might influence tracer diffusion substantially (e.g., through the formation of intrinsic defects).

Fuchs et al. [346], recently reported results of a very accurate method for measuring the self-diffusion coefficient of Ge which circumvents many of the experimental problems encountered in the conventional methods. These authors used germanium isotopic heterostructures (stable isotopes), grown by molecular-beam epitaxy (MBE) (for details, see [347–349]). As is well known, isotopic heterostructures consist of layers of pure (e.g., ^{70}Ge, ^{74}Ge) or delib-

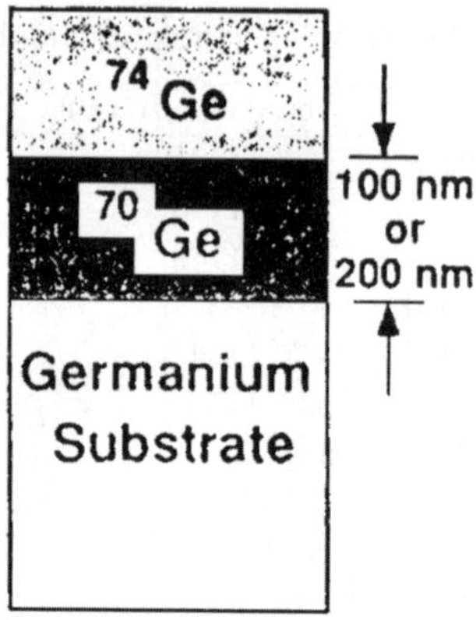

Fig. 5.8. Schematic of the isotopic heterostructure used by Fuchs et al. (after [346])

erately mixed isotopes of a chemical element. Figure 5.8 shows the schematic of the particular samples used by Fuchs et al. [346]. At the interface, only the atomic mass is changing, while (to first order) all other physical properties stay the same. In the as-grown samples, this interface is atomically flat with layer thickness fluctuations of about two atomic ML (monolayer) (for details, see [348]). Upon annealing, the isotopes diffuse into each other (self-diffusion) at a rate that depends strongly on temperature. The concentration profiles in paper [346] were measured with SIMS, after pieces of the same samples were separately annealed at different temperatures. This allows accurate determination of the self-diffusion enthalpy as well as the corresponding entropy. Isotopic heterostructures are unique for self-diffusion studies in several aspects (see also [23]).

1. The interdiffusion of germanium isotopes takes place at the isotopic interface **inside** the crystal, unaffected by possible surface effects (e.g., oxidation, impurities, and strain) encountered in the conventional technique.
2. One sample annealed at one temperature provides five more or less independent measurements (Ge consists of five stable isotopes). Their initial respective concentrations vary for the different layers of the as-grown isotopic heterostructure. After annealing, the concentration profile of each of the five isotopes can be analyzed separately to obtain five data points for each annealing temperature. The samples were cut into several pieces. One piece was kept in paper [346] for reference (as-grown); the other were separately annealed at five different temperatures (543, 586, 605, 636, and 690°C). The temperature controller permitted a variation of 1 − 2°C in the temperature. The concentration depth profiles of all five stable Ge isotopes were recorded with SIMS. The oxygen primary beam had an impact energy of 8 keV per incident ion. The beam was rastered over a

area of about 200 μm^2, and the detected secondary ions were extracted from the central 30 μm diameter region of the crater. The precision of the SIMS data, it was estimated, was within ±5%. The depth resolution of the system was determined from profiles taken from the as-grown samples with an atomically flat interface. What theoretically should be a step function in the concentration profile appeared as a slope of about 4 nm per decade of the measured atomic fraction at the leading edge of a layer and about 16 nm per decade at the falling edge (for details, see [346]).

As is well known, diffusion in crystals occurs through thermally activated jumping between different sites in the lattice [259, 260]. In principle, there are many possibilities for such jumps [substitutional or interstitial sites, vacancies, etc. (for details, see [292])]. In Ge crystals, however, it is known that the only process of significance for the migration of germanium atoms is through the vacancy mechanism (see also [337]) In this case, the self-diffusion coefficient D^{SD} can be written as an Arrhenius expression [346] [see (5.24)]

$$D^{\mathrm{SD}} = gfa^2 \nu_0 \exp\left[\frac{-G^{\mathrm{SD}}}{kT}\right] = D_0 \exp\left[\frac{-H^{\mathrm{SD}}}{kT}\right], \tag{5.39}$$

where G^{SD} is the Gibbs free energy of self-diffusion,

$$G^{\mathrm{SD}} = H^{\mathrm{SD}} = TS^{\mathrm{SD}}, \tag{5.40}$$

H^{SD} is the self-diffusion enthalpy, and the preexponential factor

$$D_0 = gfa^2 \nu_0 \exp\left[\frac{S^{\mathrm{SD}}}{k}\right] \tag{5.41}$$

contains the self-diffusion entropy S^{SD}, the correlation factor f ($f = \frac{1}{2}$ for the vacancy mechanism in the diamond lattice [350]), the attempt frequency ν_0, the geometric factor g ($g = \frac{1}{8}$ for vacancies in Ge), and the lattice constant a; k is Boltzmann's constant (see also Sect. 5.3). The enthalpy H^{SD} and the entropy S^{SD} depend on the formation (subscript F) as well as the migration (subscript M) of the vacancy:

$$H^{\mathrm{SD}} = H^{\mathrm{SD}}_{\mathrm{F}} + H^{\mathrm{SD}}_{\mathrm{M}} \quad \text{and} \quad S^{\mathrm{SD}} = S^{\mathrm{SD}}_{\mathrm{F}} + S^{\mathrm{SD}}_{\mathrm{M}}. \tag{5.42}$$

The quantity which we can extract from the data of paper [346] is primarily the self-diffusion coefficient D^{SD} as a function of annealing temperature T. This was done in the cited paper by fitting experimental depth profiles to theory, with D^{SD} the only fitting parameter. Equation (5.33) then allows us to determine the self-diffusion enthalpy H^{SD}, and the self-diffusion entropy S^{SD} is deduced by using (5.35). Solving Fick's diffusion equation for the specific geometry of samples used in the indicated paper (see Fig. 5.8), these authors obtain the atomic fraction c_i of a given germanium isotope i in terms of error functions (erf) [see (5.38)]:

$$c_i(x) = \left[\frac{c_i^{0,\mathrm{I}} - c_i^{0,\mathrm{II}}}{2} \operatorname{erf}\left(\frac{h/2 + x}{2\sqrt{D^{\mathrm{SD}}t}}\right) + c_i^{0,\mathrm{I}}\right] + \left[\frac{c_i^{0,\mathrm{II}} - c_i^{0,\mathrm{III}}}{2} \operatorname{erf}\left(\frac{h/2 - x}{2\sqrt{D^{\mathrm{SD}}t}}\right) + c_i^{0,\mathrm{III}}\right], \qquad (5.43)$$

where h is the layer thickness (110 or 200 in Ge samples, see Fig. 5.8), and $c_i^{0,\mathrm{I}}$, $c_i^{0,\mathrm{II}}$, and $c_i^{0,\mathrm{III}}$ are the initial concentrations of the isotope i in the enriched ^{74}Ge layer, in the enriched ^{70}Ge layer, and in the substrate, respectively. Figure 5.9 shows the profiles of all five isotopes of an annealed sample (586°C for 55.55 h), together with a fit of the data to (5.43). For clarity, only the fit to the ^{70}Ge profile is shown, but other profiles can be independently fitted as well. The excellent quality of the fit over four orders of magnitude displays the remarkable accuracy of the method used by Fuchs et al. As a reference, the corresponding concentration profiles for the as-grown sample are displayed in Fig. 5.10. The annealing time was purposely chosen so that the plateaus in the annealed samples (around 300 and 100 nm) correspond to the original concentrations in the isotopically enriched layers.

The values for the self-diffusion coefficient D^{SD} obtained at 543, 586, 605, 636, and 690°C are presented in an Arrhenius plot in Fig. 5.11. The lines in Fig. 5.11 represent the results of previous authors [292]. The variation in

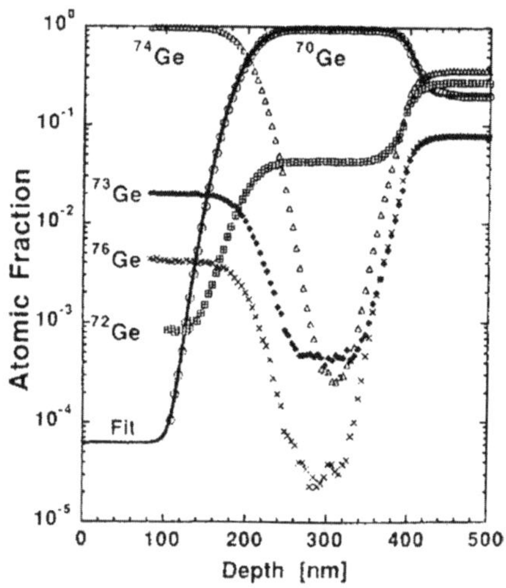

Fig. 5.9. Experimental depth profile of the atomic fraction of ^{70}Ge, ^{72}Ge, ^{73}Ge, ^{74}Ge, and ^{76}Ge (symbols) of a diffusion annealed sample (annealed at 586°C for 55.55 h). The *solid line* is a fit of the ^{70}Ge data of 5.43. For clarity, only the fit to the ^{70}Ge data is shown. (After [346])

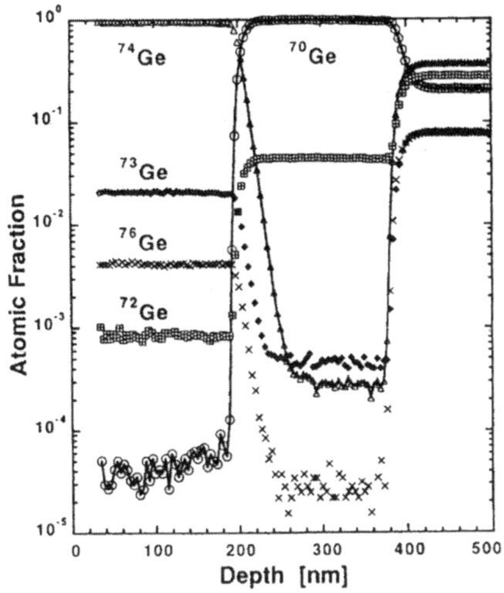

Fig. 5.10. Experimental depth profiles of the same sample as Fig. 5.9, but before annealing (after [346])

D^{SD} obtained from different groups is comparable with the scatter of the data within the work of each of the publications. Fitting the experimental values of D^{SD} to (5.33), Fuchs et al. obtained the self-diffusion enthalpy H^{SD} of 3.0(5) eV. As can be seen from Table 5.3, this in excellent agreement with previously published values of 2.95 – 3.14 eV. The value of the experimental preexponential factor D_0 is $1.2 \cdot 10^{-3}\,\mathrm{m^2s^{-1}}$. This compares to previously published values of $(0.78 - 4.4) \cdot 10^{-3}\,\mathrm{m^2s^{-1}}$. Converting D_0 into the self-diffusion entropy S^{SD} through (5.41), they obtain $S^{SD} \approx 9\,\mathrm{kcal}$ (using $\nu_0 = 8 \cdot 10^{12}\,\mathrm{s^{-1}}$ and $a = 0.565\,\mathrm{nm}$). The self-diffusion entropy for Ge is larger than that for metals (2–4 kcal). As an explanation, Seeger and Frank [337] invoked the idea of extended (spread-out) defects, and Bourgoin and Lanoo [351] proposed that vacancy in Ge is strongly relaxed.

Finally, we want to mention the effect of isotopic mass on the self-diffusion coefficient (see also [292]). The many-body treatment of atomic jump processes leads to an expression for the strength of the isotopic effect in terms of the correlation factor f of (5.35) and the fraction ΔK of the kinetic energy which is associated with the motion in the jump direction [352, 353]:

$$\frac{\left[\dfrac{D_{\mathrm{I}}}{D_{\mathrm{II}}}\right] - 1}{\left[\dfrac{m_{\mathrm{II}}}{m_{\mathrm{I}}}\right]^{\frac{1}{2}} - 1} = f\Delta K. \tag{5.44}$$

In previous Ge self-diffusion experiments, Campbell [332] found $f\Delta K$ values between 0.26 and 0.30, which translates into a ratio of $D_{SD}^{70Ge}/D_{SD}^{74Ge}$ between 1.007 and 1.008 [354]. This small difference, however, is below the precision of

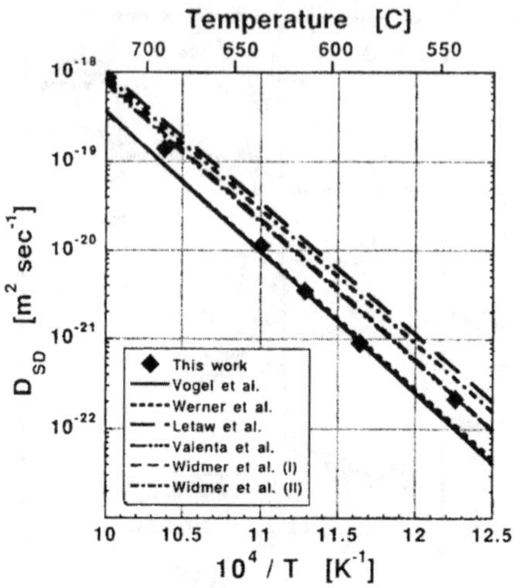

Fig. 5.11. Arrhenius plot of the self-diffusion coefficient as a function of temperature. Data of Fuchs et al. [436] agree favorably well with the most recent data [333, 334]. The older data [329–331] might be less accurate. (After [346])

the Fuchs et al. work. When fitting the experimental depth profiles to (5.43), they could not detect any appreciable difference among the different isotopes. In addition, such small deviations would be insignificant in the Arrhenius plot (logarithmic scale of D^{SD} in Fig. 5.11) for the determination of the self-diffusion enthalpy H^{SD} and entropy S^{SD}.

5.7 Self- and Interdiffusion of Ga and Al in Isotopically Pure and Doped Heterostructures

Self-diffusion is the most fundamental matter transport process in solids. Understanding this process is pivotal to understanding all diffusion phenomena in solids, including those for native defects and impurities. As noted above, compared to metals [355], self-diffusion processes in semiconductors are significantly more complex (see also [254, 255, 278, 356, 357]) due to the much richer spectrum of native defects and to the much larger effects of small concentrations of defects on the Fermi level position and other properties [292]. In III–V compounds, experiments are more difficult because of the high partial vapor pressure of group-V elements and the dependence of native defect species and concentrations on stoichiometry [358].

During the past 30 years, there have been only a few attempts to study Ga self-diffusion directly in GaAs using isotopes [359–361]. Goldstein [359] and Palfrey et al. [360] diffused radioactive ^{72}Ga into bulk GaAs at elevated temperatures and obtained depth profiles of ^{72}Ga by mechanical sectioning

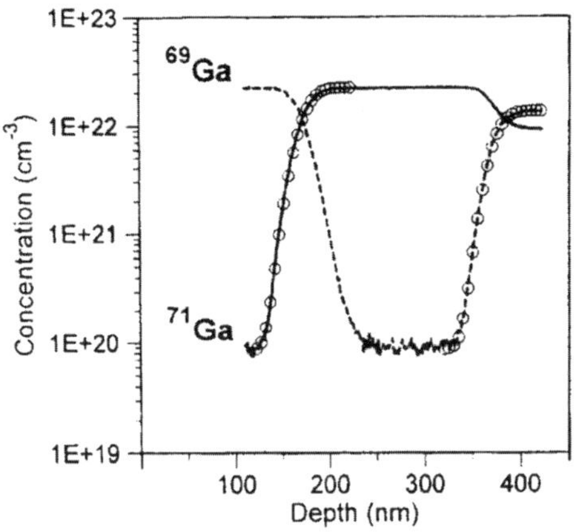

Fig. 5.12. SIMS depth profiles of ^{69}Ga and ^{71}Ga in GaAs isotopic epilayers annealed at 974°C for 3321 s. The *circles* are theoretical fits. (After [363])

and radioactive assaying. With a rather limited temperature range investigated, they reported activation enthalpies of 5.6 and 2.6 eV, respectively. Tan et al. [361] studied the disordering of ^{69}GaAs/^{71}GaAs isotope superlattice structure and found an activation enthalpy of 4 eV. However, arguing that the heavily Si-doped substrates in their samples affected the result, these authors discarded this value in favor of their earlier estimate of a 6 eV activation enthalpy [358]. In view of this controversy and the fact that our knowledge of self-diffusion in GaAs is derived primarily from studies of interdiffusion (see also below) of Ga and Al in $Ga_xAl_{1-x}As$ systems [362,368], it is of great interest to study diffusion of Ga isotopes with some new approaches.

Wang et al. reported results from Ga self-diffusion studies in GaAs [363, 364]. They measured concentration profiles of ^{69}Ga and ^{71}Ga using SIMS, determined the activation enthalpy and entropy by analyzing the diffusion coefficients obtained between 800 and 1225°C, and examined the effects of substrate doping. ^{71}GaAs and ^{69}GaAs layers of 200 nm each were grown using molecular-beam epitaxy (MBE) at 580°C on GaAs substrates of natural isotopic composition (^{69}Ga : ^{71}Ga = 60.2 : 38.8). The nominal isotopic purity in the epilayers was 99.6%.

Figure 5.12 shows the SIMS depth profiles of ^{69}Ga and ^{71}Ga after annealing at $T = 974$°C for 3321 s. Excellent fits over 2.5 orders of magnitude in concentration were obtained for this and all the other depth profiles taken from samples with smooth surfaces.

The characteristic diffusion length R is defined as

$$R = 2\sqrt{Dt}, \tag{5.45}$$

where D (as usual) is the Ga self-diffusion coefficient and t is the annealing time. The Ga self-diffusion coefficient D can be derived from (5.39). An

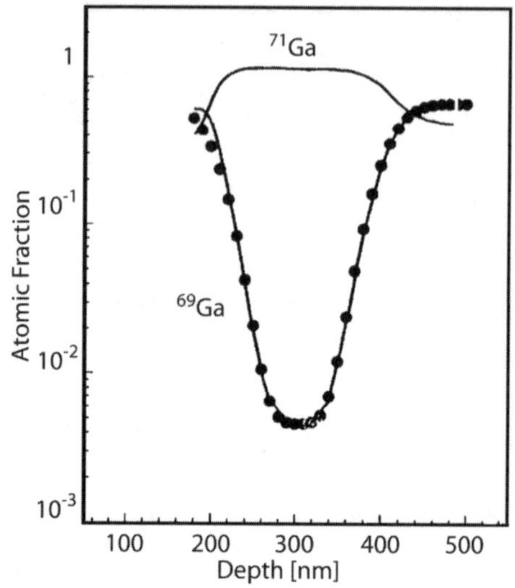

Fig. 5.13. SIMS depth profiles of ^{69}Ga and ^{71}Ga in GaP isotopic epilayers annealed at 1111°C for 231 min. The *filled circles* represent the calculated ^{69}Ga concentration profile. (After [364])

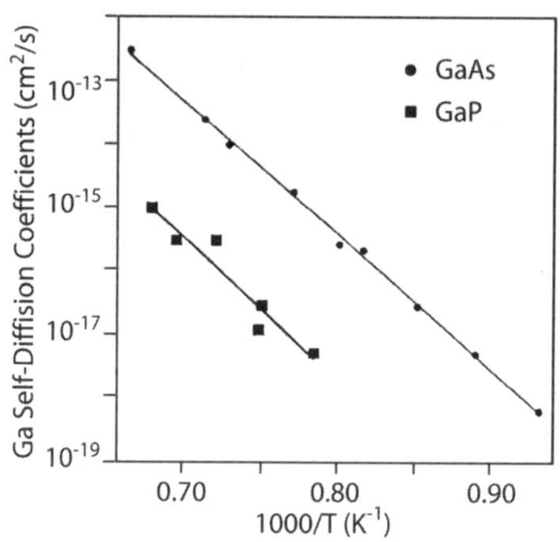

Fig. 5.14. Arrhenius plots of Ga self-diffusion coefficients in GaAs (*filled circles*) and GaP (*filled squares*) (after [364])

Arrhenius plot for D is presented in Fig. 5.14 (see below). The D values span six orders of magnitude in the temperature range from 800 – 1225°C.

Expanding the research on Ga self-diffusion beyond GaAs to other III–V compound semiconductors can be quite instructive in elucidating the microscopic mechanism. This was the main reason for an investigation [364] of Ga in gallium phosphide (GaP) using the same SIMS technique. For the experiment, ^{71}GaP and ^{69}GaP epitaxial layers 200 nm thick were grown by solid source molecular-beam epitaxy (MBE) at 700°C on undoped GaP substrates. The isotopic composition was the same as in GaAs. Figure 5.13 shows the SIMS profiles (*solid lines*) and the calculated $C(x)$ of ^{69}Ga (*circles*) and ^{71}Ga (*continuous line*) in a sample annealed at 1111°C for 3 h and 51 min. Excellent agreement is obtained between the measured and the calculated profiles across two and a half orders of magnitude in concentration. This agreement strongly supports the assumptions made in (5.13). The fitting procedure leads to an accurate determination of D through (5.45).

In Fig. 5.14, D is plotted versus temperature T. The D values span two orders of magnitude in the temperature range from 1000 – 1190°C. They can represented by (5.19). From this, the authors cited determined the activation enthalpy H^{SD} and the preexponential factor D_0 as 4.5 eV and 2.0 cm^{2}s^{-1}, respectively. The self-diffusion entropy can be obtained from (5.41). In GaP, $g \sim 1$, $a = 5.45$ Å, and $\nu_0 = 1.2 \cdot 10^{13}$ Hz. Using these values, Wang et al. [364] obtained $S^{\mathrm{SD}} = 4$ kcal. The Ga self-diffusion coefficients in GaAs from [363] are also shown in Fig. 5.14 for comparison. The activation enthalpy and entropy for GaAs are 4.24 eV and 7.5 kcal, respectively.

As in GaAs, the Ga self-diffusion coefficients in GaP follow an Arrhenius relation described by (5.39), indicating that a single type of native defect is most likely responsible for mediating Ga self-diffusion in GaP over the temperature range studied. In intrinsic GaAs, the defect mediating Ga self-diffusion, it has been ascertained, is the triply negatively charged gallium vacancy acceptor, V_{Ga}^{3-} [358]. Such an assignment may still be premature for GaP. From a recent positron annihilation study in GaP, Krause-Rehberg et al. [365] reported that positron trapping by vacancies behaves similarly in GaP and in GaAs. Vacancies are detected by these authors only in n-type GaP at room temperature, with a detection limit of $2 \cdot 10^{15}$ cm^{-3}. These findings neither support nor exclude the possibility that it is also the acceptor-like V_{Ga} that mediates Ga self-diffusion in intrinsic GaP.

Figure 5.14 shows that the Ga self-diffusion coefficient in GaP is about two orders of magnitude lower than that in GaAs. The decreased cation diffusion in III–V compounds, with phosphorus replacing arsenic as anion, has been previously observed [366, 367]. Interdiffusion of Ga and Al, it was determined, is two orders of magnitude slower in the AlGaInP/GaInP superlattice system than in the AlGaAs/GaAs system (for details, see below). The change has been attributed to the stronger Ga–P bond compared to the Ga–As bond [367]. Although Wang et al. [364] found a higher activation enthalpy in GaP (4.5 eV) than in GaAs (4.24 eV), the difference is not large enough to be outside experimental uncertainty. The difference in the preexponential factor D_0, or the entropy term S^{SD}, between GaP (4 kcal) and GaAs (7.5 kcal) is more reliable. As was shown above, this entropy term is the sum of the formation entropy S^{F} and migration entropy S^{M} for the native defect mediating self-diffusion. S^{F} or S^{M} represents the number of equivalent formation

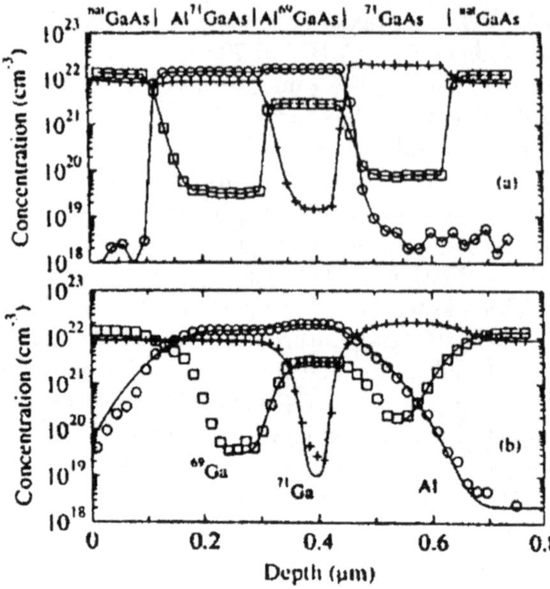

Fig. 5.15. SIMS depth profiles of Al (○), ^{69}Ga (□), and ^{71}Ga (+) in the as-grown $Al^{71}GaAs/Al^{69}GaAs/^{71}GaAs$ heterostructure (b) [see **(a)**] and after annealing at 1050°C for 1800 s [see **(b)**]. *Solid lines* in **(a)** connect the data to guide the eye. *Solid lines* in **(b)** show best fits to the experimental profiles. For clarity, only every fourth data point is plotted in **(a)**, **(b)**. (After [368])

configurations or migration jumps. The significant difference in S^{SD} indicates profound variations in the way that the mediating native defects are formed or migrate in GaP compared to GaAs (see also [23]). The small value of S supports a simple native defect species as the major diffusion vehicle (for details, see [363]).

Bracht et al. used three undoped isotopic heterostructures of $Al_x^{71}Ga_{1-x}As/Al_y^{69}Ga_{1-y}As/^{71}GaAs$ with (a) (X, Y) = (0.41; 0.62), (b) (0.62; 0.85), and (c) (0.68; 0.88), and in addition (d) $AlAs/^{71}GaAs$ layer structure for Ga self- and Al–Ga interdiffusion experiments. The thicknesses of the layers were between 100 and 200 nm. The structures were grown by MBE at about 600°C on a 200 nm thick undoped natural GaAs buffer layer which was deposited on (100) oriented GaAs substrate wafers. A natural GaAs capping layer, about 200 nm thick, was grown on top of the structure to protect the AlGaAs layer from oxidation in air. Concentration profiles of Al, ^{69}Ga, ^{71}Ga, and As in the annealed samples were measured with SIMS using a Cs^+ ion beam with an energy of 5.5 keV. The depth of the craters left from the analysis were determined with a Tencor P-10 surface profilometer. The measured secondary ion counts were converted into concentrations taking into account the count rates obtained on an $Al_{0.56}Ga_{0.44}As$ standard.

Concentration profiles of Al, ^{69}Ga, and ^{71}Ga of the as-grown structure (b) are shown in Fig. 5.15a. Figure 5.15b illustrates the corresponding distribution after annealing of sample (b) at 1050°C for 1800 s. Concentration profiles of ^{69}Ga (see also [368]) which lie within the Al^{71}GaAs and ^{71}GaAs layers, are accurately described by the solution of Fick's law (see above) for self-diffusion across an interface taking into account a concentration-independent diffusion coefficient. The measured ^{69}Ga profiles within the ^{71}GaAs layer of samples (a) to (d) all yield the same Ga self-diffusion coefficient within experimental accuracy, even though the Al concentration in the ^{71}GaAs layer varies from $10^{18}\,\mathrm{cm}^{-3}$ (detection limit) in, e.g., sample d up to $10^{20}\,\mathrm{cm}^{-3}$ in sample b due to in-diffusion of Al from the adjacent AlGaAs layer. The temperature dependence of Ga self-diffusion in $\mathrm{Al}_x\mathrm{Ga}_{1-x}\mathrm{As}$ with X = 0, 0.41, 0.62, and 0.68 is shown in Fig. 5.16. The activation enthalpy H^{SD} of Ga self-diffusion in $\mathrm{Al}_x\mathrm{Ga}_{1-x}\mathrm{As}$ and the corresponding preexponential factor D_0 are summarized in Table 5.4. Recently, Wang et al. [363] determined an activation enthalpy of (4.24 ± 0.06) eV for Ga self-diffusion in GaAs which deviates from the present result of (3.71 ± 0.07) eV. The authors [363] favor the activation enthalpy of 3.71 eV for Ga self-diffusion in GaAs because simultaneous annealing of the former ^{69}GaAs/^{71}GaAs heterostructure with sample a has revealed that the Ga profile near the interface of the ^{71}GaAs layer and the GaAs substrate, which was considered by Wang et al. for the self-diffusion study, deviates from the expected error function solution. This may be caused by surface contamination of the GaAs substrate wafer. The temperature dependence of Ga self-diffusion in AlGaAs reveals that Ga diffusion decreases with increasing Al content whereas H^{SD} appears to be constant within experimental error (see Fig. 5.16 and Table 5.4).

The Al and Ga profiles near the Al^{69}GaAs/^{71}GaAs interface result from Al–Ga interdiffusion. The Al–Ga interdiffusion coefficient $\tilde{D}$ can be expressed as (see [356])

$$\tilde{D} = (X_{\mathrm{Al}}\, D_{\mathrm{Ga}} + X_{\mathrm{Ga}}\, D_{\mathrm{Al}})\, \Phi\, S, \tag{5.46}$$

Table 5.4. Activation enthalpy H^{SD} and natural logarithm of the preexponential factor D_0 for Al and Ga diffusion in $\mathrm{Al}_x\mathrm{Ga}_{1-x}\mathrm{As}$ for intrinsic and As-rich ($p_{\mathrm{As}} \sim$ 1 atm) conditions (after [368])

Element	X of $\mathrm{Al}_x\mathrm{Ga}_{1-x}\mathrm{As}$	H^{SD}	$\ln(D_0/\mathrm{cm}^2\mathrm{s}^{-1})$
Al	0.0	3.50 ± 0.08	-1.77 ± 0.80
Ga	0.0	3.71 ± 0.07	-0.45 ± 0.67
Ga	0.41	3.70 ± 0.21	-1.05 ± 1.98
Ga	0.62	3.60 ± 0.13	-2.51 ± 1.24
Ga	0.68	3.51 ± 0.19	-3.96 ± 1.78
Ga	1.0	3.48 ± 0.23	-4.51 ± 2.16

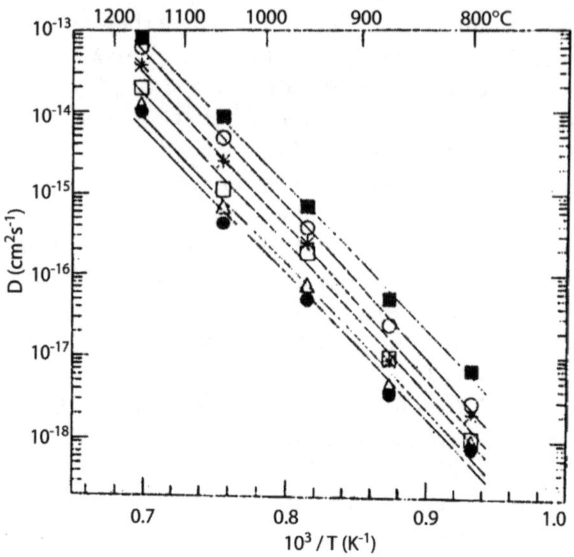

Fig. 5.16. Temperature dependence of the diffusion coefficient D of Ga in $Al_xGa_{1-x}As$ with $x = 0$ (○), 0.41 (*), 0.62 (□), 0.68 (△), and 1.0 (•) and of Al in GaAs (■) (after [368])

where X_{Al} and X_{Ga} are the mole fractions of Al and Ga. D_{Ga} and D_{Al} represent the Ga and Al diffusion coefficients in AlAs and GaAs, respectively. Φ is the thermodynamic factor, and S the vacancy wind factor which takes vacancy interaction and correlation effects into account. For modeling Al–Ga interdiffusion, authors [368] assume the simplest possible values of $\Phi = 1$ (ideal solution) and of $S = 1$. On this basis, Fick's second law was solved numerically. Both measured Al and ^{71}Ga profiles shown in Fig. 5.15b can be described with the composition dependent interdiffusion coefficient $\tilde{D}$ according to (5.46) which takes into account the actual mole fraction of Al and Ga as a function of depth. According to Bracht et al. [368], all interdiffusion profiles of samples a to d, which were annealed at the same temperature, are accurately described by data for D_{Ga} and D_{Al} which are consistent within 40%. The temperature dependence of Al diffusion in GaAs and of Ga diffusion in AlAs is shown in Fig. 5.16. Fitting (5.39) to these results yields data for H^{SD} and D_0 which are listed in Table 5.4.

All values for H^{SD} shown in Table 5.4 lie in the range of (3.6 ± 0.1) eV. Recently, Wee et al. [369] reported an activation enthalpy of (3.6 ± 0.2) eV and a preexponential factor of $0.2\,\mathrm{cm^2s^{-1}}$ for interdiffusion of $Al_{0.2}Ga_{0.8}As/GaAs$ at temperatures between 750 and 1150°C. Their data are consistent with the data of paper [368] on Ga diffusion in AlGaAs and its dependence on Al content. Tan et al. (see [361] and references therein) proposed 6 eV for Ga self-diffusion. This result is based on a compilation of Ga self-diffusion and Al–

Ga interdiffusion data obtained under various experimental conditions which includes AlGaAs/GaAs heterostructures with Al content up to 100%. The activation enthalpy of 6 eV now appears to be questionable since Bracht et al. results unambiguously show that Al–Ga interdiffusion does not represent Ga self-diffusion.

The single activation enthalpy found for Ga self- and Al–Ga interdiffusion suggests that the diffusion is mediated by the same native defect. Vacancies in the sublattice of group-III atoms are assumed to mediate the Ga self- and Al–Ga interdiffusion under intrinsic conditions [370]. In this case, the self-diffusion coefficient is given by [see also (5.39)]

$$D^{\mathrm{SD}} = f_{\mathrm{V}}\ C_{\mathrm{V}}^{\mathrm{eq}}\, D_{\mathrm{V}} \ = \ f_{\mathrm{V}}\ ga^2\nu \exp\left(-\frac{G_{\mathrm{V}}^{\mathrm{SD}}}{kT}\right), \tag{5.47}$$

where f_{V} is a correlation factor and $C_{\mathrm{V}}^{\mathrm{eq}}$ and D_{V} are the thermal equilibrium concentration of vacancies and their diffusivity. $C_{\mathrm{V}}^{\mathrm{eq}}\, D_{\mathrm{V}}$ can be expressed by the geometric factor g, the jump distance a, the jump attempt frequency ν, and the Gibbs free energy $G_{\mathrm{V}}^{\mathrm{SD}}$ of self-diffusion via vacancies. ν is proportional to $\frac{1}{\sqrt{m}}$, where m represents the atomic mass of the jumping atom. Different jump frequencies of ^{27}Al and ^{71}Ga caused by the difference in their masses are proposed as the cause for the experimentally observed higher Al diffusion in GaAs compared to Ga self-diffusion. The experimentally determined ratio between the Al and the Ga diffusion coefficient in GaAs is consistent with $\left(\frac{m_{\mathrm{Ga}}}{m_{\mathrm{Al}}}\right)^{0.5}$. The decreasing Ga diffusivity as increasing Al content in AlGaAs can be interpreted with a different location of the intrinsic Fermi energy level with respect to the vacancy charge transition state [371]. This causes different thermal equilibrium concentrations of vacancies in AlGaAs for different Al compositions.

In the next, we follow the Bracht et al. papers that consider the experimental ratio between Ga diffusion in GaAs and AlAs. This equals the ratio between $C_{\mathrm{V}}^{\mathrm{eq}}\, D_{\mathrm{V}}$ for GaAs and AlAs if the binding energy between Ga and a vacancy in AlAs is negligible. Assuming only a single negatively charged vacancy, the total concentration of vacancies in thermal equilibrium is given by [372]

$$C_{\mathrm{V}}^{\mathrm{eq}} = C_{\mathrm{V^0}}^{\mathrm{eq}}\left[1 + \exp\left(E_{\mathrm{F}}^{\mathrm{in}} - E_{\mathrm{V}-/0}/kT\right)\right], \tag{5.48}$$

where $C_{\mathrm{V^0}}^{\mathrm{eq}}$ is the equilibrium concentration of a neutral vacancy, $E_{\mathrm{F}}^{\mathrm{in}}$ the Fermi level under intrinsic condition, and $E_{\mathrm{V}-/0}$ the acceptor energy level of a singly charged vacancy. The Fermi-level position is given by

$$E_{\mathrm{F}}^{\mathrm{in}} = 0.5\, E_{\mathrm{g}} + 0.75\, kT \ln\left(\frac{m_{\mathrm{V}}^*}{m_{\mathrm{C}}^*}\right), \tag{5.49}$$

where E_{g} is the band-gap energy and m_{V}^* and m_{C}^* are the effective density of state masses for holes and electrons in GaAs and AlAs, respectively [357].

Using (5.48) and (5.49), the ratio between C_V^{eq} in GaAs and AlAs is estimated at $6.5 \cdot \exp\left[\frac{0.077\,\mathrm{eV}}{(kT)}\right]$ assuming that $C_{V^0}^{eq}$ in GaAs and AlAs are similar with respect to the valence-band position of AlAs as an energy reference. Bracht et al. [320] obtain a ratio of 12.8 atm, e.g., 1050°C, which is consistent with their experimental result (details see [368]).

Dopant-enhanced as well as reduced-layer disordering of semiconductor heterostructures are phenomena that have been reported frequently in the literature. Superlattice structures doped with Si during growth [373, 374] or by implantation [375, 376] reveal enhanced Al–Ga interdiffusion after annealing, compared to intrinsic conditions. Conversely, Be doping of Si-doped AlAs/GaAs, it has been found, suppresses superlattice disordering when the Be doping level exceeds that of Si [377,378]. A dopant-induced change in the position of the Fermi level, which strongly alters the concentrations of charged native point defects and hence self- and interdiffusion, is generally considered possible for this diffusion phenomenon [379]. So far, the relative contributions of the various charged native point defects to self- and interdiffusion, which are required to predict the disordering of GaAs-based superlattice structures for different doping levels, are not accurately known. Total energy calculations provide one way to determine the nature of native defects in group III–V compound semiconductors. These calculations predict that the triply negatively charged vacancy in the gallium sublattice (V_{Ga}^{3-}) is the dominant native defect in GaAs, both for intrinsic and n-type doping conditions under an As-rich ambient [380]. On the other hand, the charge state of native defects can be deduced experimentally from the doping dependence of Ga self-diffusion in GaAs. Generally, Al–Ga interdiffusion in AlAs/GaAs superlattices is considered to simulate the self-diffusion process of the group-III atom. Tan and Gösele [381] analyzed data of Al–Ga interdiffusion in Si-doped AlAs/GaAs superlattices [382] and found, in agreement with the theoretical calculations of paper [380], that self-diffusion in the Ga sublattice is mediated by V_{Ga}^{3-}. In contrast to these, seemingly consistent theoretical and experimental studies show that singly negatively charged Ga vacancies V_{Ga}^{-} mediate Ga self-diffusion and Al–Ga interdiffusion under n-type doping conditions [383–385]. Some of the data of Mei et al. [382], which have been used to support the triply charged defect hypothesis [381], can also be described by assuming a singly charged vacancy [384]. Furthermore, Ga self-diffusion in GaAs, it has been proposed, governs by triply charged defects at high Si doping levels, whereas at low doping concentrations, a Fermi-level independent mechanism was assumed [386]. These discrepancies in the doping dependence of group-III atom diffusion in the AlGaAs material system forced Bracht et al. [387] to investigate Ga self-diffusion in undoped, Si-, and Be-doped GaAs with the SIMS technique by using ^{71}GaAs/^{nat}GaAs isotopic heterostructures.

Three Al^{71}GaAs/Al^{69} GaAs/^{71}GaAs/^{nat}GaAs isotopic heterostructures with different Al content and one AlAs/^{71}GaAs/^{nat}GaAs structure were used for the diffusion experiments [387]. One set of samples was kept undoped and another set was doped with either Si or Be, i.e., altogether 12 different isotopic structures were used. Each of these samples contained a ^{71}GaAs/^{nat}GaAs interface which was focused on in this self-diffusion study.

SIMS measurements on the undoped set of isotopic heterostructures have revealed a change in the SIMS sputter rate with increasing Al concentration [388]. Since all samples used for the Ga self-diffusion experiments contained AlGaAs layers, a decrease in the sputter rate with increasing Al content results in thicker AlGaAs layers and thinner GaAs layers. Therefore, a point-by-point correction of the penetration depth of the SIMS profile was performed, taking into account the measured sputter-rate dependence. According to the authors cited, the thicknesses of the AlGaAs and GaAs layers corrected in this way are consistent within 10% or better with the thicknesses determined directly by transmission electron microscopy.

Concentration profiles of ^{69}Ga near the ^{71}GaAs/^{nat}GaAs interface measured with SIMS after annealing of Si- and Be-doped samples are shown in Figs. 5.17a,b and 5.18. The Si and Be profiles revealed a nearly constant dopant concentration across the ^{71}GaAs/^{nat}GaAs interface. A few sufficiently large Si-doped samples were also analyzed with C–V profiling to determine the free carrier concentration after annealing. The carrier concentration profile of one sample is shown in Fig. 5.18 together with the corresponding SIMS profiles of SI and Ga. C–V profiling revealed a Si donor concentration, C_{Si}, of about $2 \cdot 10^{18}\,\text{cm}^{-3}$ along the ^{69}Ga profile. Similar results were obtained from the analysis of other samples. The electron concentration n after annealing is smaller than the C_{Si} of the as-grown structures. This decrease is also apparent in the observed effect of doping on Ga self-diffusion. The hole concentration p of Be-doped samples could not be accurately determined by C–V profiling after annealing. In all cases, the craters left behind by electrochemical etching were nonuniform. Consequently, no reliable data could be obtained for p. SIMS analysis shows that the Be concentration along the Ga self-diffusion profile equals the concentration in the as-grown structure. Therefore, it is assumed that the free carrier concentration p due to Be doping is not significantly affected by thermal annealing. This is supported by the doping dependence of Ga self-diffusion, which is accurately described by taking into account the Be doping level of the as-grown samples (details see [387]).

Ge self-diffusion profiles shown in Figs. 5.17 and 5.18 are accurately reproduced by the solution of Fick's law for self-diffusion across an interface (see above). All experimental ^{69}Ga profiles were described on the basis of (5.43) to take into account the diffusion of ^{69}Ga from the adjacent Al^{69}GaAs layer into the ^{71}GaAs layer. The thickness $d = x_1 - x_2$ determined in this way for the ^{71}GaAs layer is consistent within 10% with the thickness in the as-grown structure. The uncertainty results mainly from the accuracy of the crater depth measurements.

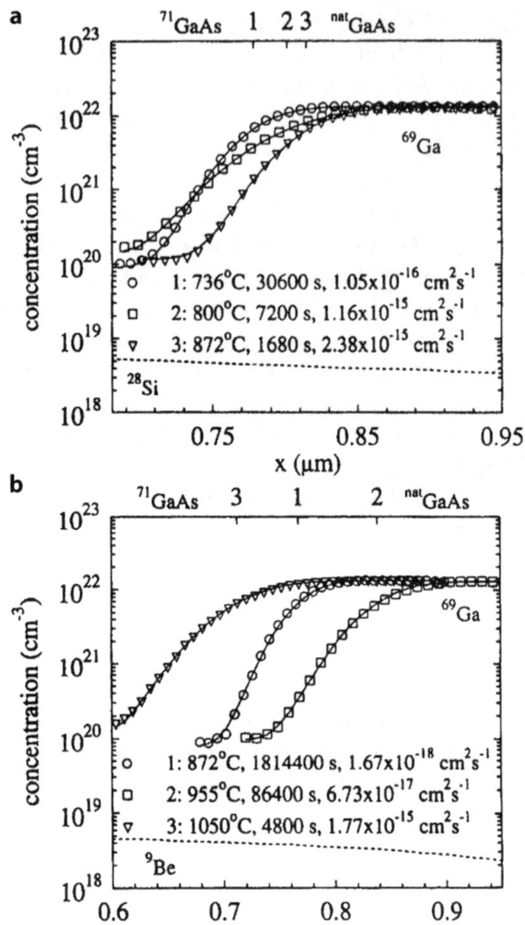

Fig. 5.17. SIMS depth profiles of ^{69}Ga, ^{28}Si and ^{9}Be after annealing of **(a)** Si-doped and **(b)** Be-doped ^{71}GaAs/^{nat}GaAs isotopic structures at various temperatures and times, as indicated. The *solid lines* show the best fits to experimental data. Ga self-diffusion coefficients deduced from each profile are listed in the figure. The numbers 1, 2, and 3 at the top x-axis indicate the position of the ^{71}GaAs/^{nat}GaAs interface of the particular sample. For clarity, only the dopant profile (*dashed line*) from Ga profile 3 is shown. (After [387])

The self-diffusion coefficients D_{Ga} extracted from the analysis of all Ga diffusion profiles in undoped, Si-, and Be-doped isotopic samples are listed in Table 5.5. Each value of D_{Ga} is the average of at least four diffusion coefficients; each of them has been deduced from a different sample. The accuracy given for D_{Ga} represents the standard deviation of all data which belong to the same temperature and doping level. The temperature dependence of D_{Ga} for intrinsic, Si-, and Be-doped GaAs is depicted in Fig. 5.19. Experimental

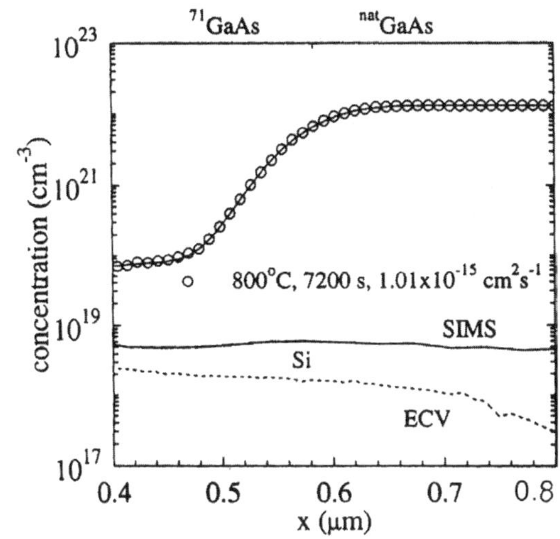

Fig. 5.18. SIMS depth profiles of ^{69}Ga and ^{28}Si of a Si-doped ^{71}GaAs/^{nat}GaAs structure annealed at 800°C for 2 h. The *solid line* shows the best fit to the experimental data which yield the Ga self-diffusion coefficient listed in the figure. The room temperature electron concentration, which was measured by C–V profiling, is shown as a *dashed line*. (After [387])

results [387] for intrinsic conditions are accurately described by (see also [368])

$$D_{\mathrm{Ga}} = 0.64 \exp\left(-\frac{3.71\,\mathrm{eV}}{kT}\right) \mathrm{cm}^2\mathrm{s}^{-1}. \tag{5.50}$$

The *solid lines* in Fig. 5.19 are best fits to the experimental data; they accurately reproduce the temperature dependence of Ga self-diffusion under the different doping conditions.

According to Bracht et al. [387], the Ga self-diffusion coefficient is given by the sum of the transport coefficients of vacancies in various charge states

$$D_{\mathrm{Ga}} = \frac{1}{C_0} \sum_{r=0}^{3} f_r\, C^{\mathrm{eq}}_{\mathrm{V}^{r-}_{\mathrm{Ga}}}\, D_{\mathrm{V}^{r-}_{\mathrm{Ga}}}, \tag{5.51}$$

where C_0, f_r, $C^{\mathrm{eq}}_{\mathrm{V}^{r-}_{\mathrm{Ga}}}$ represent the Ga atom density in GaAs ($C_0 = 2.215 \cdot 10^{22}\,\mathrm{cm}^{-3}$), the diffusion correlation factor, and the thermal equilibrium concentration; and the diffusion coefficient of the vacancy is $D_{\mathrm{V}^{r-}_{\mathrm{Ga}}}$ with the charge $r \in [0, 1, 2, 3]$, respectively. The correlation factor contains information about the microscopic jump mechanism [278]. The different vacancy configurations can introduce energy levels within the energy band-gap of GaAs. Occupation of these energy states depends on the position of the Fermi level. Under extrinsic conditions, i.e., when the hole or the electron concentration introduced by doping exceeds the intrinsic carrier concentration, the Fermi level

Table 5.5. Ga self-diffusion coefficients D_{Ga} in undoped, Si-doped, and Be-doped $^{71}\mathrm{GaAs}/^{\mathrm{nat}}\mathrm{GaAs}$ isotopic heterostructures. The numbers in parenthesis represent the number of samples that were annealed at the particular temperature T and time t. A free electron and hole concentration of both $(3.0 \pm 0.5) \cdot 10^{18}\,\mathrm{cm}^{-3}$ was determined by Hall effect measurements for the as-grown Si- and Be-doped samples. n_i, $E_{\mathrm{F}}^{\mathrm{in}}$ and $\frac{n}{n_i}$ represent the intrinsic carrier concentration [390], the position of the Fermi level under intrinsic conditions [390], and the ratio of the electron concentration in the doped and undoped samples, respectively. (After [387])

T (°C)	t (s)	D_{Ga} ($\mathrm{cm^2s^{-1}}$)	Type of doping	n_i ($\mathrm{cm^{-3}}$)	$E_{\mathrm{F}}^{\mathrm{in}}$ (eV)	$\frac{n}{n_i}$ [a]
1160	270(4)	$(6.12 \pm 0.17) \cdot 10^{-14}$	undoped	$1.56 \cdot 10^{18}$	0.512	1.0
1050	1800(4)	$(4.76 \pm 0.80) \cdot 10^{-15}$	undoped	$7.41 \cdot 10^{17}$	0.545	1.0
1050	3600(1); 4800(4)	$(1.46 \pm 0.20) \cdot 10^{-15}$	Be-doped	$7.41 \cdot 10^{17}$	0.545	0.233
955	12600(1); 15120(3)	$(3.84 \pm 0.61) \cdot 10^{-16}$	undoped	$3.54 \cdot 10^{17}$	0.572	1.0
955	10800(1); 86400(4)	$(8.28 \pm 2.50) \cdot 10^{-17}$	Be-doped	$3.54 \cdot 10^{17}$	0.572	0.116
872	260400(4); 432000(2)	$(2.47 \pm 0.17) \cdot 10^{-17}$	undoped	$1.70 \cdot 10^{17}$	0.595	1.0
872	496800(1); 1814400(4)	$(2.37 \pm 0.59) \cdot 10^{-18}$	Be-doped	$1.70 \cdot 10^{17}$	0.595	0.056
872	1680(5)	$(3.67 \pm 0.93) \cdot 10^{-15}$	Si-doped	$1.70 \cdot 10^{17}$	0.595	14.8
800	1209600(2); 1370700(2)	$(2.84 \pm 0.36) \cdot 10^{-18}$	undoped	$8.32 \cdot 10^{16}$	0.615	1.0
800	7200(4); 10800(1)	$(9.86 \pm 1.65) \cdot 10^{-16}$	Si-doped	$8.32 \cdot 10^{16}$	0.615	26.6
736	–	$1.88 \cdot 10^{-19}$ [b]	undoped	$4.09 \cdot 10^{16}$	0.632	1.0
736	25200(1); 30600(4)	$(1.45 \pm 0.21) \cdot 10^{-16}$	Si-doped	$4.09 \cdot 10^{16}$	0.632	45.9

[a] Data correspond to fit #3 (see Table 5.6)

[b] Calculated with the equation $D_{\mathrm{Ga}} = 0.64 \exp\left(-\frac{2.71\,\mathrm{eV}}{k_{\mathrm{B}}T}\right)\mathrm{cm^2s^{-1}}$

deviates from its intrinsic position. As a consequence, the ratio of the charged to neutral vacancy concentrations is changed [372]. For vacancies which can introduce single (V_{Ga}^{-}), double (V_{Ga}^{2-}), and triple (V_{Ga}^{3-}) acceptor states with energy levels at $E_{\mathrm{V_{Ga}^{-}}}$, $E_{\mathrm{V_{Ga}^{2-}}}$, $E_{\mathrm{V_{Ga}^{3-}}}$ above the valence band edge E_{v}, the ratios are given by

$$\frac{C_{\mathrm{V_{Ga}^{-}}}^{\mathrm{eq}}}{C_{\mathrm{V_{Ga}^{0}}}^{\mathrm{eq}}} = g_{\mathrm{V_{Ga}^{-}}} \exp\left(\frac{E_{\mathrm{F}} - E_{\mathrm{V_{Ga}^{-}}}}{kT}\right), \tag{5.52}$$

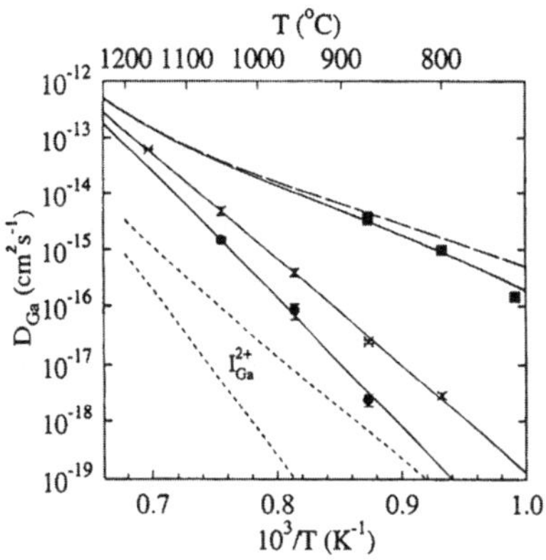

Fig. 5.19. Temperature dependence of the Ga self-diffusion coefficient D_{Ga} in undoped (x), Si-doped (■), and Be-doped (•) GaAs for $P_{\mathrm{As}_4} = 1\,\mathrm{atm}$. The influence of doping on D_{Ga} is best reproduced (see *solid lines*) if the effect of the Fermi-level together with a compensation of Si donors by negatively charged vacancies is taken into account. The *long-dashed line* is expected if the compensation by vacancies is ignored. *Short-dashed lines* represent the contribution of the doubly positive charged Ga self-interstitial I^{2+}_{Ga} to Ga self-diffusion. *Lower dashed line*: I^{2+}_{Ga} contribution for intrinsic conditions and $P_{\mathrm{As}_4} = 1\,\mathrm{atm}$; *upper dashed line*: I^{2+}_{Ga} contribution calculated for a hole concentration of $3 \cdot 10^{18}\,\mathrm{cm}^{-3}$. (After [387])

$$\frac{C^{\mathrm{eq}}_{\mathrm{V}^{2-}_{\mathrm{Ga}}}}{C^{\mathrm{eq}}_{\mathrm{V}^{0}_{\mathrm{Ga}}}} = g_{\mathrm{V}^{2-}_{\mathrm{Ga}}} \exp\left(\frac{E_{\mathrm{F}} - E_{\mathrm{V}^{2-}_{\mathrm{Ga}}} - E_{\mathrm{V}^{-}_{\mathrm{Ga}}}}{kT}\right), \tag{5.53}$$

$$\frac{C^{\mathrm{eq}}_{\mathrm{V}^{3-}_{\mathrm{Ga}}}}{C^{\mathrm{eq}}_{\mathrm{V}^{0}_{\mathrm{Ga}}}} = g_{\mathrm{V}^{3-}_{\mathrm{Ga}}} \exp\left(\frac{E_{\mathrm{F}} - E_{\mathrm{V}^{3-}_{\mathrm{Ga}}} - E_{\mathrm{V}^{2-}_{\mathrm{Ga}}} - E_{\mathrm{V}^{-}_{\mathrm{Ga}}}}{kT}\right). \tag{5.54}$$

The thermal equilibrium concentration of the neutral Ga vacancy $C^{\mathrm{eq}}_{\mathrm{V}^{0}_{\mathrm{Ga}}}$ is independent of the position of the Fermi level E_{F}. However, if E_{F} lies above $E_{\mathrm{V}^{r-}_{\mathrm{Ga}}}$, the formation of negatively charged vacancies is energetically favored, compared to the formation of neutral defects. In (5.51)–(5.53), all degeneracy factors $g_{\mathrm{V}^{r-}_{\mathrm{Ga}}}$, which take into account the spin degeneracy of the defect and degeneracy of the GaAs valence band, have been set to one.

The total concentration of V_{Ga} in thermal equilibrium, $C^{\mathrm{eq}}_{\mathrm{V}_{\mathrm{Ga}}}$, is given by the sum of the corresponding concentrations of the various charge states of gallium vacancies

$$C^{\mathrm{eq}}_{\mathrm{V}_{\mathrm{Ga}}} = \sum_{r=0}^{3} C^{\mathrm{eq}}_{\mathrm{V}^{r-}_{\mathrm{Ga}}} = C^{\mathrm{eq}}_{\mathrm{V}^{0}_{\mathrm{Ga}}} \sum_{r=0}^{3} \frac{C^{\mathrm{eq}}_{\mathrm{V}^{r-}_{\mathrm{Ga}}}}{C^{\mathrm{eq}}_{\mathrm{V}^{0}_{\mathrm{Ga}}}}. \tag{5.55}$$

Based on (5.51)–(5.53), it becomes clear, that $C^{\mathrm{eq}}_{\mathrm{V}_{\mathrm{Ga}}}$ changes when the position of the Fermi level changes. Naturally, the Ga self-diffusion coefficient given by (5.51) may be written as

$$D_{\mathrm{Ga}} = \frac{1}{C_0} C^{\mathrm{eq}}_{\mathrm{V}^0_{\mathrm{Ga}}} \sum_{r=0}^{3} \frac{C^{\mathrm{eq}}_{\mathrm{V}^{r-}_{\mathrm{Ga}}}}{C^{\mathrm{eq}}_{\mathrm{V}^0_{\mathrm{Ga}}}} D_{\mathrm{V}^{r-}_{\mathrm{Ga}}} . \tag{5.56}$$

Recent published molecular dynamic calculations [389] show that the migration enthalpy of Ga is nearly independent of its charge state. For accuracy of migration via second-nearest neighbor hopping, an enthalpy of 1.7 eV was determined for V^0_{Ga} compared to 1.9 eV for V^{3-}_{Ga} [389]. In addition, in paper [388], it was assumed that the entropy change associated with the V^{r-}_{Ga} migration and the correlation factor f_r are similar for all charge states, i.e., $D_{\mathrm{V}^{r-}_{\mathrm{Ga}}} = D_{\mathrm{V}_{\mathrm{Ga}}}$ and $f_r \approx f$. With these assumptions, it follows that [388]

$$D_{\mathrm{Ga}} = \frac{1}{C_0} f \, C^{\mathrm{eq}}_{\mathrm{V}^0_{\mathrm{Ga}}} D_{\mathrm{V}_{\mathrm{Ga}}} \sum_{r=0}^{3} \frac{C^{\mathrm{eq}}_{\mathrm{V}^{r-}_{\mathrm{Ga}}}}{C^{\mathrm{eq}}_{\mathrm{V}^0_{\mathrm{Ga}}}} . \tag{5.57}$$

Taking into account (5.6)–(5.53) and the well-known expression

$$\frac{n}{n_i} = \exp\left(\frac{E_{\mathrm{F}} - E^i_{\mathrm{F}}}{kT}\right) , \tag{5.58}$$

the relationship between the Ga self-diffusion coefficient under intrinsic and extrinsic conditions has the form

$$\frac{D_{\mathrm{Ga}}(n)}{D_{\mathrm{Ga}}(n_i)} = \frac{1 + \sum\limits_{m=1}^{3} \left(\frac{n}{n_i}\right)^m \exp \dfrac{mE^i_{\mathrm{F}} - \sum\limits_{m=1}^{3} E_{\mathrm{V}^{m-}_{\mathrm{Ga}}}}{kT}}{1 + \sum\limits_{m=1}^{3} \exp \dfrac{mE^i_{\mathrm{F}} - \sum\limits_{m=1}^{3} E_{\mathrm{V}^{m-}_{\mathrm{Ga}}}}{kT}} . \tag{5.59}$$

In (5.58) and (5.59), n_i. E^i_{F}, n, and E_{F} define the free electron concentration and the Fermi-level position under intrinsic and extrinsic conditions, respectively. If only one charge state r dominates Ga self-diffusion, (5.59) is reduced to

$$\frac{D_{\mathrm{Ga}}(n)}{D_{\mathrm{Ga}}(n_i)} = \left(\frac{n}{n_i}\right)^r \tag{5.60}$$

with $r \in [0, 1, 2, 3]$. This simplified relationship has been generally used to analyze the doping dependence of Ga self-diffusion and Al–Ga interdiffusion [381]. Bracht et al. [387] prefer the more general (5.59), since this equation takes into account that the charge state of the vacancy mediating Ga self-diffusion may change with doping and temperature.

Fitting of (5.59) to experimental results required data for E^i_{F} and n_i. For these quantities, the authors cited used the data reported by Blakemore [390] which are listed in Table 5.5. The electron concentration n of the Si- and Be-doped GaAs samples were calculated via the charge balance equation yielding [387]

$$n = \frac{1}{2}\left(C_{\mathrm{Si}} - \sum_{m=0}^{3} mC^{\mathrm{eq}}_{\mathrm{V}^{m-}_{\mathrm{Ga}}}\right) + \sqrt{n_i^2 + \frac{1}{4}\left(C_{\mathrm{Si}} - \sum_{m=0}^{3} mC^{\mathrm{eq}}_{\mathrm{V}^{m-}_{\mathrm{Ga}}}\right)} \tag{5.61}$$

and

$$\frac{n_i^2}{n} = p = \frac{1}{2}\left(C_{\mathrm{Be}} - \sum_{m=0}^{3} mC^{\mathrm{eq}}_{\mathrm{V}^{m-}_{\mathrm{Ga}}}\right) + \sqrt{n_i^2 + \frac{1}{4}\left(C_{\mathrm{Be}} - \sum_{m=0}^{3} mC^{\mathrm{eq}}_{\mathrm{V}^{m-}_{\mathrm{Ga}}}\right)}, \tag{5.62}$$

respectively. C_{Si} represents the Si donor concentration and C_{Be} the acceptor concentration due to Be doping. Both concentrations equal the free carrier concentration of $3 \cdot 10^{18}\,\mathrm{cm}^{-3}$, measured at room temperature. Negatively charged vacancies that affect the free carrier concentration are taken into account in both (5.61) and (5.62).

The free carrier concentration given by (5.60) and (5.62) also depends, via (5.51)–(5.53) and (5.58), on vacancy-related energy levels. Therefore, (5.59), (5.61), and (5.62) were solved by Bracht et al. simultaneously. $C^{\mathrm{eq}}_{\mathrm{V}^{m-}_{\mathrm{Ga}}}$ ($r \in [0,1,2,3]$) were calculated with (5.46)–(5.48) using a thermal equilibrium concentration of neutral vacancies which is given by [391, 392].

$$C^{\mathrm{eq}}_{\mathrm{V}^{0}_{\mathrm{Ga}}} = C * P^{\frac{1}{4}}_{\mathrm{As}_4} T^{-\frac{5}{8}} \exp\left(\frac{H^f_{\mathrm{V}^0_{\mathrm{Ga}}}}{kT}\right), \tag{5.63}$$

where C^* is a preexponential factor and $H^f_{\mathrm{V}^0_{\mathrm{Ga}}}$ the formation enthalpy of the neutral vacancy. Equation (5.57) includes the influcnce of the As_4 vapor phase on $C^{\mathrm{eq}}_{\mathrm{V}^0_{\mathrm{Ga}}}$ [391, 392]. So far, equilibrium concentrations of V^0_{Ga} have been estimated with an uncertainty factor of at least 10 [391]. To describe accurately the doping dependence of Ga self-diffusion on the basis of (5.53), $E_{\mathrm{V}^{m-}_{\mathrm{Ga}}}$ ($m \in [0,1,2,3]$) and also C^* were used as fit parameters. The formation enthalpy of neutral vacancies was set to a value of (1.9 ± 0.2) eV. This value equals the difference between the activation enthalpy of (3.71 ± 0.07) eV for Ga self-diffusion and the vacancy migration enthalpy of (1.8 ± 0.2) eV which was deduced in paper [393] from Al–Ga interdiffusion in nonstoichiometric AlAs/GaAs quantum wells [see also (5.50)]. The authors [387] concluded that $H^f_{\mathrm{V}^0_{\mathrm{Ga}}} = (1.9 \pm 0.2)$ eV is fairly reliable.

The temperature and doping dependences of Ga self-diffusion in intrinsic, n-type and p-type GaAs, which were calculated in paper [387] on the basis of the parameters given under fit #3 in Table 5.6, are shown by the *solid*

Table 5.6. Parameter values obtained from fitting (5.59) to the experimental results. Ga self-diffusion is best described by neutral, singly and doubly negatively charged vacancies with relative contributions of the various charge states, that change with temperature and doping. Additionally, a compensation of the Si donor concentration by negatively charged Ga vacancies was taken into account; it becomes especially significant at low temperatures. (After [387])

Fit parameter [a]	Fit #1 [b]	Fit #2 [c]	Fit #3	Fit #4	Fit #5
$E_{V_{Ga}^{-}} - E_V$, (eV)	0.38 ± 0.11	0.42 ± 0.04	0.42	0.42	0.42
$E_{V_{Ga}^{2-}} - E_V$, (eV)	0.72 ± 0.19	0.60 ± 0.04	0.60	0.60	0.60
$E_{V_{Ga}^{3-}} - E_V$, (eV)	1.33 ± 4.33	–	–	–	–
$H^f_{V^0_{Ga}}$, (eV)	1.9	1.9	1.9	1.7	2.1
C^*, (cm^{-3})	0.293	0.293	162 ± 54	18.6 ± 5.8	1385 ± 487

[a] E_V denotes the valence band edge

[b] C^* and $H^f_{V^0_{Ga}}$ according to Tan [391]

[c] V^{3-}_{Ga} contribution ignored; C^* and $H^f_{V^0_{Ga}}$ according to Tan [391]

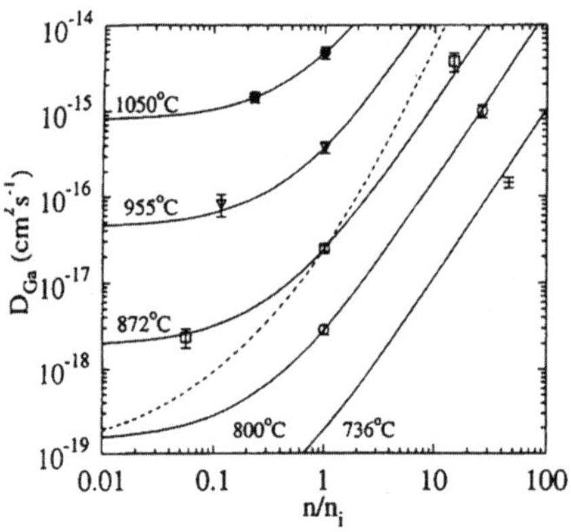

Fig. 5.20. Ga self-diffusion coefficients D_{Ga} versus the ratio $\frac{n}{n_i}$ between the free carrier concentration under extrinsic and intrinsic doping conditions $P_{As_4} = 1\,atm$ and different temperatures as indicated. *Solid lines* were calculated via (5.59) taking into account $E_{V^m_{Ga}}$ given by Fit #3 of Table 5.6. The *dashed line* shows the doping dependence of D_{Ga} at 872°C which was calculated for $E_{V^-_{Ga}} - E_V \approx 0.20\,eV$, $E_{V^{2-}_{Ga}} - E_V \approx 0.52\,eV$, and $E_{V^{3-}_{Ga}} - E_V \approx 0.72\,eV$ according to Baraff and Schlüter [380]. (After [387])

lines in Fig. 5.19. The *long-dashed line* in Fig. 5.19 displays the temperature dependence of D_{Ga} which is expected if compensation via negatively charged vacancies is ignored in the n-type GaAs with $C_{Si} = 3 \cdot 10^{18}\,cm^{-3}$.

A representation of D_{Ga} data [387] versus the ratio $\frac{n}{n_i}$ is given in Fig. 5.20. *Solid lines* have been calculated with (5.59) taking into account the $\frac{n}{n_i}$ values listed in Table 5.5 and the results for $E_{V^{m-}_{Ga}}$ of Fit #3 (see Table 5.6).

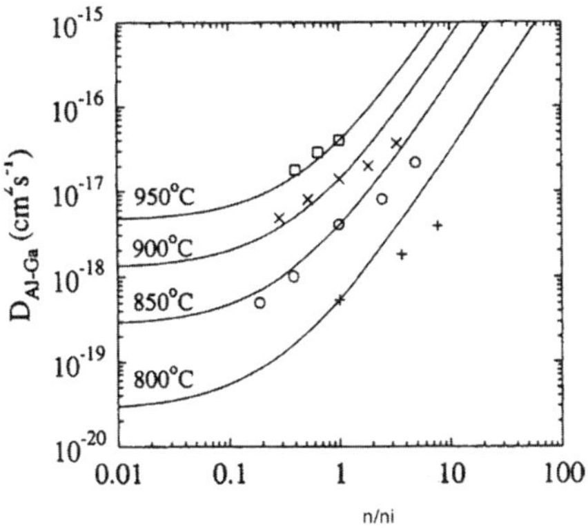

Fig. 5.21. Al–Ga interdiffusion coefficient $D_{\mathrm{Al-Ga}}$ reported by Muraki and Horikoshi [385] versus $\frac{n}{n_i}$. *Solid lines* were calculated with (5.59) taking into account $D_{\mathrm{Ga}}(n_i)$ of Muraki and Horikoshi and the results of Bracht et al. [387] for $E_{\mathrm{V}_{\mathrm{Ga}}^{-}}$. (After [387])

The doping dependence of D_{Ga} at 872°C, which was calculated on the basis of the theoretical results for $E_{\mathrm{V}_{\mathrm{Ga}}^{m-}}$ ($m \in [0,1,2,3]$) reported by Baraff and Schluter [380], is shown in Fig. 5.20 as a *dashed line*. The strong deviation from the data of paper [387] clearly indicates that, in contrast to the theoretical calculations, the triply charged Ga vacancy does not mediate Ga self-diffusion under intrinsic and n-type doping conditions. As noted above, Muraki and Horikoshi [385] studied the Al–Ga interdiffusion of Si- and Be-doped $\mathrm{Al_{0.34}Ga_{0.66}As/GaAs}$ superlattice structures with the help of photoluminescence spectroscopy. They proposed that Al–Ga interdiffusion is mediated by singly negatively charged Ga vacancies in both n- and p-type material. The Al–Ga interdiffusion in these structures with an Al content of 34 atom% is expected to simulate Ga self-diffusion in GaAs. This is supported by recent results obtained in paper [368] on the Al composition dependence of Ga self-diffusion in AlGaAs (see also above) which show that for Al concentrations of 41 atom%, D_{Ga} in AlGaAs is only a factor of 2 smaller than D_{Ga} in GaAs. The interdiffusion coefficients given by Muraki and Horikoshi for different diffusion temperatures and doping levels are illustrated in Fig. 5.21 as a function of the ratio $\frac{n}{n_i}$. Compensation for the Si donors by charged vacancies was also considered by taking into account the equilibrium concentrations of vacancies which, consistent with results of paper [387], were obtained for the particular Si doping levels. The *solid lines* shown in Fig. 5.21 were computed via (5.53) with $D_{\mathrm{Ga}}\left(\frac{n}{n_i}=1\right)$ of Muraki and Horikoshi and the results of Bracht et al. [387] for $E_{\mathrm{V}_{\mathrm{Ga}}^{m-}}$ (see also Fit #3, Table 5.6). No adjustable parameters were used. The dependence on $\mathrm{As_4}$ pressure cancels out because all diffusion experiments were performed under identical conditions. The deviation between the experimental data and the corresponding solid

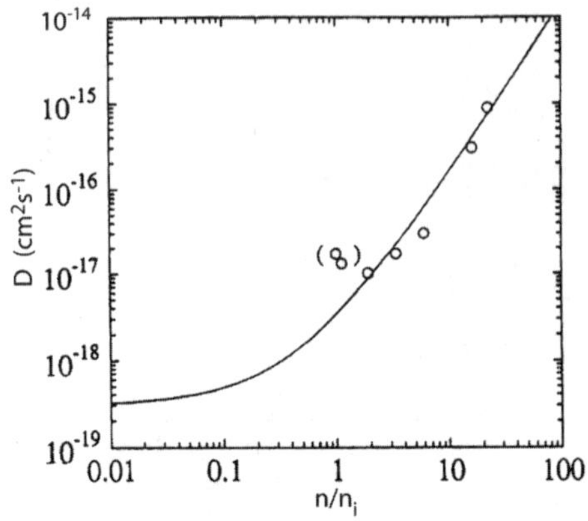

Fig. 5.22. Ga self-diffusion coefficient $D_{\rm Ga}$ reported by Muraki and Horikoshi [385] versus $\frac{n}{n_i}$. The data in parenthesis are not considered reliable. The *solid line* was calculated via (5.53) taking into account the results of Bracht et al. for $E_{V_{\rm Ga}^-}$ and a value of $D_{\rm Ga}(n_i)$ which is a factor of 5 smaller than the corresponding data given by Muraki and Horikoshi. (After [387])

line for 800°C may indicate that the Si donor concentration is additionally reduced by the formation of $\rm Si_{Ga}$ acceptor pairs such as $\rm Si_{Ga}$–$\rm Si_{As}$ complexes or other compensation centers. This is supported by the fact that after annealing, Muraki and Horikoshi observed broad photoluminescent signals in the Si-doped samples which they attributed to deep centers.

Self-diffusion coefficients at 900°C reported by Muraki and Horikoshi are displayed in Fig. 5.22 as a function of $\frac{n}{n_i}$. The *solid line* in Fig. 5.22 was calculated on the basis of the results of Bracht et al. and accurately describes the Ga self-diffusion data for $\frac{n}{n_i} > 1$. Small differences $\left(\frac{n}{n_i} > 1\right)$ between the as-grown and annealed structures can also be caused by SIMS broadening effects associated with a degradation of the surface quality during annealing under As-poor conditions.

Figure 5.23 illustrates the ratio between the reduced interdiffusion coefficients $D_{\rm Al-Ga}(n, P_{\rm As_4} = 1\,\rm atm)$ and $D_{\rm Ga}(n, P_{\rm As_4} = 1\,\rm atm)$ as a function of $\frac{n}{n_i}$. In this double logarithmic representation, the slope of the experimental data equals the exponent r. The best Fit [387] yields $r = 2.3 \pm 0.1$ showing that $V_{\rm Ga}^{2-}$ rather than $V_{\rm Ga}^{3-}$ mediates Al–Ga interdiffusion in agreement with the results of Bracht et al. Based on this reanalysis, the $V_{\rm Ga}^{3-}$ mediated self- and interdiffusion in n-type GaAs and the activation enthalpy of 6 eV for Ga self-diffusion under intrinsic conditions, proposed by Tan and Gösele [381], are incorrect (for details, see [387]).

In summary, the doping dependence of group-III atom diffusion in the AlGaAs material system ([385,386,391,392] and references therein) can be consistently explained with the result presented by Bracht et al. [387]. Neutral, singly and doubly charged Ga vacancies all contribute to self-diffusion in

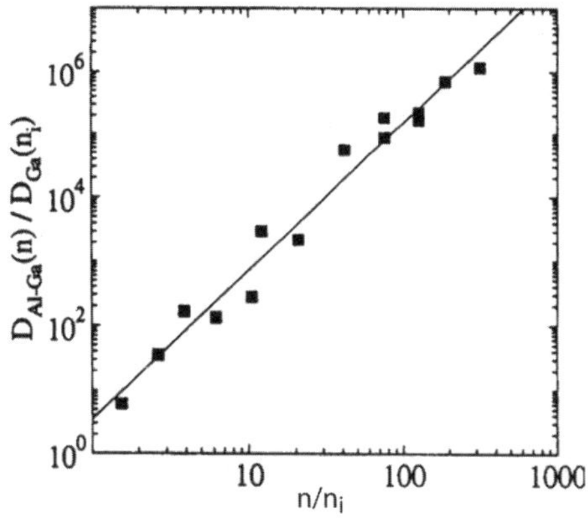

Fig. 5.23. Al–Ga interdiffusion coefficients $D_{\mathrm{Al-Ga}}\,(n)$ of Mei et al. [383] normalized by $D_{\mathrm{Ga}}(n_i)$ [368] as a function of $\frac{n}{n_i}$. The *solid line* represents the best fit which is reproduced by $\frac{D_{\mathrm{Al-Ga}}(n)}{D_{\mathrm{Ga}}(n_i)} = 3.4\left(\frac{n}{n_i}\right)^{2.3}$. (After [387])

undoped, p-type, and n-type material with relative contributions which depend on temperature and doping. The lower power dependence for the doping effects of self-diffusion in samples containing group-VI donors, it is proposed, is caused by the formation of next nearest neighbor complexes between the dopant and the vacancy.

6 Neutron Transmutative Doping

6.1 The NTD Process: A New Reactor Technology

The neutron transmutative doping (NTD) process involves the cooperation of semiconductor materials specialists, device producers, radiation damage and defect specialists, and reactor personnel. Of all possible interactions among these groups, those with the reactor community have traditionally been the weakest. Reactor personnel have, therefore, had the greatest learning curves to overcome. It is to the credit of both the reactor community and the semiconductor industry that these difficulties have been overcome so readily in the few years since 1975 when NTD silicon first appeared on the market. The transmutative doping process simply involves irradiation of an undoped semiconductor with a thermal neutron flux. The major advantages of the NTD process are illustrated schematically in Fig. 6.1. The homogeneity in NTD Si is the result of the homogeneous distribution of silicon isotopes in the target

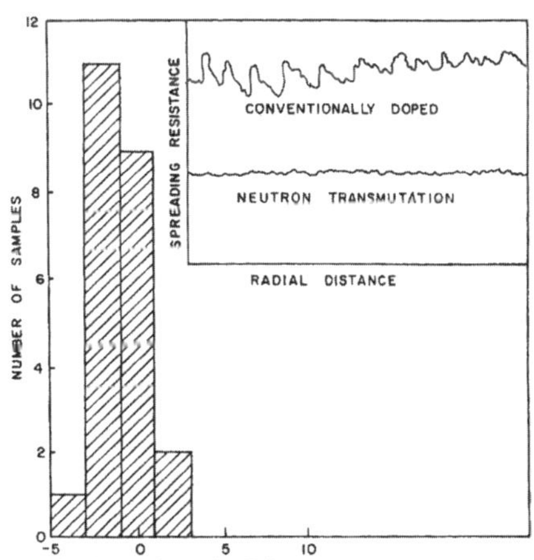

Fig. 6.1. Advantages ofthe NTD process. Histogram of irradiation target accuracy obtained for a commercial sample lot at NURR. The inset is a schematic representation of spreading resistance traces across a wafer diameter for conventionally doped and NTD Si. (After [394])

material and the long range of neutrons in silicon. Doping accuracy is a result of careful neutron flux integration. The material improvements offered by the NTD process form the basis for semiconductor device improvement (for details, see [395]).

As is well known, research reactor facilities provide the best source of thermal neutrons for this purpose at the present time (see, e.g., [6, 48, 49]). These reactors are ideally suited for such projects because they have usually been constructed with sample irradiation as one of the prime design requirements. Although these reactor facilities provide a source of thermal ($E \sim 0.025\,\mathrm{eV}$) neutrons, this thermal flux is always accompanied by a fast neutron component which is not useful in providing doping transmutations, but does produce radiation damage (displacements of atoms from their normal lattice sites) which must be repaired by annealing, the process of heating the irradiated material to temperatures sufficiently high that the irradiation produced defects become mobile and can be removed.

To understand the process further, we must be concerned with the interactions of both thermal and fast neutrons, with the target material to be doped. Because neutrons are neutral particles, their range of penetration in most materials is usually very long. They interact only very weakly with atomic electrons through their magnetic movements. Being neutral, neutrons see no coulombic barrier at the target nuclei, and, therefore even very slow neutrons may reach into the nucleus without difficulty. In fact, the slower the neutron velocity, the greater is the time of interaction between the neutron and the target nucleus. Therefore, we expect that the probability of neutron capture by the target nuclei is enhanced at low neutron energies.

This interaction is described in terms of a capture cross section, σ_{c}, where the number of captures per unit volume, N, is given by

$$N = N_{\mathrm{T}}\sigma_{\mathrm{c}}\Phi, \tag{6.1}$$

where N_{T} is the number of target nuclei per unit volume, σ_{c} the capture cross section and $\Phi = \phi t$ is the influence (flux times time) given in n/cm^2. Figure 6.2 shows the capture cross section as a function of neutron energy for silicon averaged over all three stable silicon isotopes [396]. Similar behavior is found individually for each silicon isotope. It can be seen in Fig. 6.2 that for low energies

$$\sigma_{\mathrm{c}} \sim E^{-\frac{1}{2}} \sim \frac{1}{V}. \tag{6.2}$$

For a given nuclear radius, $\left(\frac{1}{V}\right)$ is proportional to the interaction time. Therefore, the cross section represents the probability of interaction between the nucleus and neutron.

After neutron capture, the target nucleus differs from the initial nucleus by the addition of the nucleon and is a new isotope in an excited state which must relax by emitting energy in some form. This emission is in the form

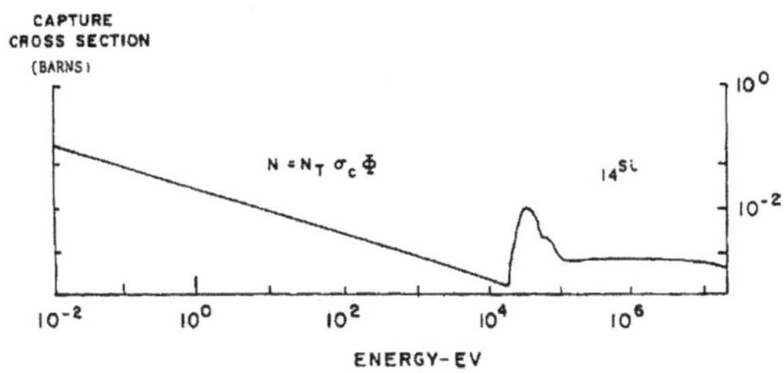

Fig. 6.2. Neutron capture cross section as a function of neutron energy for natural silicon (after [394])

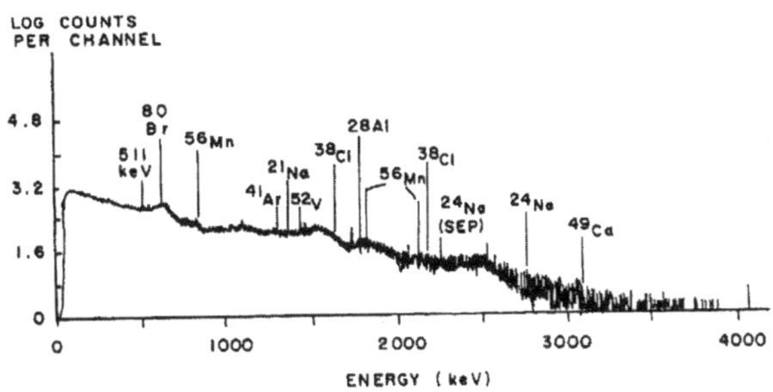

Fig. 6.3. Typical neutron activation analysis (NAA) gamma-ray spectrum to search for trace substances deposited on an air filter after 1 min irradiation at MURR (after [394])

of high-energy electromagnetic radiation (photons) usually called gammas (see, e.g., [397] and references therein). The time for decay of this excess energy by gamma emission can be very short (prompt gammas) or can take an appreciable time in which case a half-life of a factor of 2 can be measured. The gamma emission spectrum is characteristic of the nuclear energy levels of the transmuted target nuclei and can be used as a powerful trace substance technique, called neutron activation analysis (NAA), to detect quantitatively impurity levels as low as 10^9 atoms/cm^3 [398]. A typical trace substance NAA gamma spectrum is shown in Fig. 6.3. Each emission line is characteristic of a particular nuclear transition of a particular isotope. The absorption of a neutron and the emission of gammas is represented by the notation:

$$^{A}\mathrm{X}(n,\gamma)^{A+1}\mathrm{X}, \tag{6.3}$$

where (n, γ) represents (absorption, emission), A is the initial number of nucleons in the target element X before neutron absorption and $A + 1$ is the number after absorption. It is possible for the product isotope $^{A+1}\mathrm{X}$ to be naturally occurring and stable. In many cases, however, the product isotope is unstable. Unstable isotopes further decay by various modes involving the emission of electrons (β decay), protons, α particles, K-shell electron capture or internal conversion until a stable isotopic state is reached (for details, see, e.g., [399]). These decays produce radioactivity and can be characterized by their half-lives $t_{1/2}$. In silicon, three stable target isotopes are transformed by (n, γ) reactions [400] as follows:

$$\begin{aligned}
&(92.3\%)^{28}\mathrm{Si}(n,\gamma)^{29}\mathrm{Si}, \quad \sigma_\mathrm{c} = 0.08\,b; \\
&(4.7\%)^{29}\mathrm{Si}(n,\gamma)^{30}\mathrm{Si}, \quad \sigma_\mathrm{c} = 0.28\,b; \\
&(3.1\%)^{30}\mathrm{Si}(n,\gamma)^{31}\mathrm{Si} \rightarrow {}^{31}\mathrm{P} + \beta^-, \quad \sigma_\mathrm{c} = 0.11\,b; \ (t_{1/2} = 2.62\,\mathrm{h}). \qquad (6.4)
\end{aligned}$$

The relative abundance of each stable silicon isotope is shown in parenthesis (see also [397]). The cross sections are expressed in barns (1 barn = $10^{-24}\,\mathrm{cm}^2$). The first two reactions produce no dopants and only redistribute the relative abundances slightly. The third reaction produces ^{31}P, the desired donor dopant [401], at a rate of about 3.355 ppm per $10^{18}\,n_\mathrm{th}\,\mathrm{cm}^{-2}$ [394]. This production of the ^{30}Si ($N_\mathrm{T} \cong 5 \cdot 10^{22}\,\mathrm{Si} \cdot \mathrm{cm}^{-3} \times 0.031$) is calculated using (5.59).

In addition to the desired phosphorus production reaction and its relatively short half-life for β^- decay, the reaction

$$^{31}\mathrm{P}(n,\gamma)^{32}\mathrm{P} \rightarrow^{32}\mathrm{S} + \beta^-, \quad \sigma_\mathrm{c} = 0.19b \quad (t_{1/2} = 14.3\,\mathrm{d}) \qquad (6.5)$$

occurs as a secondary undesirable effect. The decay of ^{32}P is the primary source of radioactivity in NTD float fine Si. Of course, any undesirable trace impurities in the silicon starting material can lead to abnormally long half-life activities which may require that material be held out of production until exempt limits are reached. These factors have stimulated the subject of radiation protection. Once the dopant phosphorus has been added to a silicon ingot by transmutation of the ^{30}Si isotope, the problems remains to make this radiation damaged and highly disordered material useful from the viewpoint of an electronic device. Several radiation damage mechanisms contribute to the displacement of the silicon atoms from their normal lattice positions (for details, see below):

1. fast neutron knock-on displacements;
2. fission gamma-induced damage;
3. gamma recoil damage;
4. beta recoil damage;
5. charged particle knock-on from (n, p), (n, α), etc. reactions (see details in [397]).

Estimates can be made of the rate at which Si atom displacements are produced by these various mechanisms, once a detailed neutron energy spectrum of the irradiation is known, and these rates are compared to the rate at which phosphorus is produced.

The number of displaced atoms per unit volume per second, N_{D}, is estimated from the equation,

$$\frac{\mathrm{d}N_{\mathrm{D}}}{\mathrm{d}t} = N_{\mathrm{T}}\sigma\phi\nu, \tag{6.6}$$

where N_{T} is the number of target atoms per unit volume, ϕ is the flux of damaging particles, and ν is the number of displacements per incident damaging particle. The cross section for gamma induced displacements in silicon is small, whereas the cross sections for (n,p), (n,α), etc., are of the order of millibarns and have thresholds in the megaelectron volt range. Fast neutron knock-on displacements can be calculated from the elastic neutron scattering cross section, once the reactor neutron energy spectrum is known. Estimates of fission spectra and graphite moderated fission spectra can be found in the literature [402].

Even if the fast neutron damage could be completely eliminated, the recoil damage mechanisms, which are caused by thermal neutron capture, still would produce massive numbers of displacements compared to the number of phosphorus atoms produced. In gamma recoil, a gamma of energy $\hbar\omega$ carries a momentum $\frac{\hbar\omega}{c}$ which must equal the Si isotope recoil momentum MV. The recoil energy,

$$E_{\mathrm{R}} = \frac{1}{2}MV^2 = \frac{1}{2}\frac{(\hbar\omega)^2}{MC^2} \tag{6.7}$$

is, therefore, imparted to the silicon atom of mass M for each gamma emitted. An average over all possible silicon isotope gamma emission and cross sections yields an average recoil energy of 780 eV [403] which is significantly higher than the Si displacement energy. A similar effect is encountered for $^{31}\mathrm{Si}$ β^- decay. The β^- carries a momentum

$$p = \frac{1}{c}\sqrt{E_\beta^2 - (m_0c^2)^2} \equiv MV. \tag{6.8}$$

Therefore,

$$E_{\mathrm{R}} = \frac{1}{2}MV^2 = \frac{1}{2}\left[E_\beta^2 - (m_0c^2)^2\right] / (Mc^2). \tag{6.9}$$

For a β^- emitted with an energy of 1.5 MeV, $E_{\mathrm{R}} = 33.2$ eV, or roughly twice the displacement threshold.

From the above considerations, a very crude estimate of the numbers of displacements per phosphorus atom produced can be made. The results of the estimate are shown in Table 6.1. Though the absolute numbers of displacements should not be taken literally, the relative magnitudes of the amounts of damage produced by these various mechanisms are probably order of magnitude correct. An inspection of Table 6.1 indicates that the gamma recoil mechanism is significant relative to the quantity of phosphorous produced,

Table 6.1. Number of displaced silicon atoms per phosphorus produced for various damage mechanisms shown for an in-core fission spectrum and a graphite moderated spectrum (after [394]))

Damage particle/position	In core	In pool
Fast neutron	$4.06 \cdot 10^6$	$1.38 \cdot 10^4$
Fission gamma	$3.64 \cdot 10^3$	36.4
Gamma recoil	$1.29 \cdot 10^3$	$1.29 \cdot 10^3$
Beta recoil	2.76	2.76
Total Disp/(P)	$4.06 \cdot 10^6$	$1.51 \cdot 10^4$

even in highly moderated reactors. We are led to the inescapable conclusion that transmutative doping will always produce significant amounts of radiation damage that must be repaired in some way. These defects introduce defect levels into the band gap which cause free carrier removal and a reduction in carrier mobility and minority carrier lifetime (see, e.g., [403]).

The defects produced by neutron irradiation are removed by thermal annealing, as discussed previously. It is at this point in the process where disagreement as to the best procedure is likely to be the greatest. The spectrum of possible defect structures and their energetics is impressively large and incompletely understood. Therefore, annealing procedures according to Meese [394] will be based on art rather than exact science. They will also tend to become proprietary for this reason.

This is unfortunate since it is precisely in this area that fundamental knowledge is needed to produce the best possible product. Although carrier concentration and mobility recovery are easily obtainable by various annealing procedures, minority carrier lifetime recovery is very elusive at present.

So, neutron transmutation offers both advantages and disadvantages over conventionally doped silicon (for details, see [397]).

Advantages

1. Precision target doping (= 1% or better)
2. Better axial and radial uniformity
3. No microresistivity structure

Disadvantages

1. Irradiation costs
2. Reduction in minority carrier life-time
3. Radioactive safeguards considerations

The steady growth of the NTD-silicon (and others NTD-semiconductors) market suggests (see also below) that the advantages are outweighing the disadvantages.

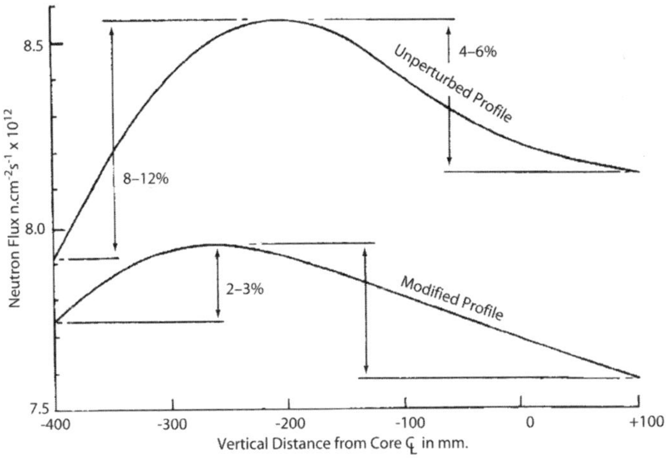

Fig. 6.4. Typical neutron flux profiles (after [404])

6.2 Reactor Facilities for Transmutative Doping

Irradiation of silicon for phosphorus doping has been carried out in the Harwell (England) research reactors since 1975 [404]. At Harwell, silicon is irradiated in the twin, heavy-water, materials-testing reactors DIDO and PLUTO. Both reactors operate continuously throughout the year, and each is available more than 86% of calendar time. As the reactors are D_2O moderated and cooled, the irradiation conditions are particularly good for silicon doping. The ratio of thermal to fast neutrons is in excess of 1000 : 1, which minimizes the damage which has to be removed by annealing. In considering the accuracy that can be achieved in the neutron doping process, it is necessary to consider neutron flux profiles and gradients. Figure 6.4 shows a typical unperturbed flux profile for an irradiation position over a length of 50 cm spaced about the maximum flux value. Note that the maximum and minimum flux values differ by 8 – 12% of the maximum and the gradient at the lower end is particularly steep. To reduce this variation and to smooth the profile, flux flatteners or neutron screens, in the form of stainless steel tubes, are fitted to the facility liners; the result is shown as "modified profile" in Fig. 6.4. As was shown by Smith the severe gradient arising from the 8 – 12% variation has been reduced to 2 – 3%, and the overall variation reduced to 5%.

The "modified profile" can now be examined in more detail, and this is shown in Fig. 6.5. The profile can be considered in terms of "resistivity distribution" which is of interest to the customer and can form the basis for the technical specification of the product. The "Average Target Resistivity" or A.T.R. 50 [404] is, as the name suggests, the average of the resistivity over a length of 50 cm. Achievement of this value is subject to variations arising from irradiation timing, measurement of the mean flux, and the distribution

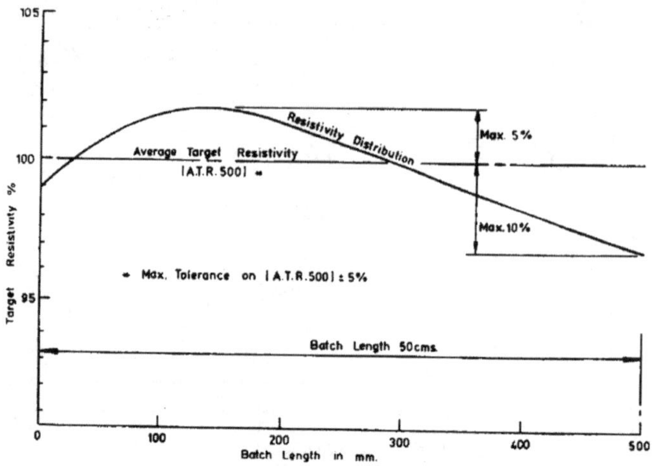

Fig. 6.5. Resistivity variations (after [404])

shape; therefore, the author applied a tolerance of ±5% to the A.T.R. 50 value. The exact shape of the distribution is also subject to variations due to disturbances in the reactor such as control-absorber movements and other irradiations and experiments; according to [404], tolerances are, therefore, also applied to the "Resistivity Distribution" of 5% maximum greater and 10% maximum less than A.T.R. 50.

Researchers at Harwell offer irradiation of volume. Figure 6.6 illustrates this and shows the volume which, for convenience, Smith describes as a "batch." It is a cylindrical volume of 90 mm diameter and 500 mm in length. Although radial gradients are small, crystals are rotated during irradiation, and a maximum variation of ±1% on a diameter of 10 cm (4 inch) is guaranteed. In practice, according to Smith, variations are less than can be measured within the accuracy of a conventional four-point probe. The average annual dose is typically $6.5 \cdot 10^{17}\,\mathrm{n\,cm^{-2}}$, which corresponds to a resistivity of 35 Ω cm. In many of the literature references on neutron doping of silicon, one finds the statement that "three days, or at most a week, after irradiation, silicon is safe to transport and to handle." This is, of course, a relative statement and it is necessary to define what is meant by "safe." The I.A.E.A. publication, "Regulations for the Safe Transport of Radioactive Materials, 1977," states that to qualify as "Exempt" or safe material the following conditions must be met:

1. The radiation level at any point on the external surface of the package shall not exceed 0.1 mRem/h.
2. The non-fixed radioactive contamination of any external surface shall not exceed $10^{-4}\,\mu\mathrm{Ci\,cm^{-2}}$. This level is permissible when averaged over any area of $300\,\mathrm{cm^2}$ of any part of the surface.

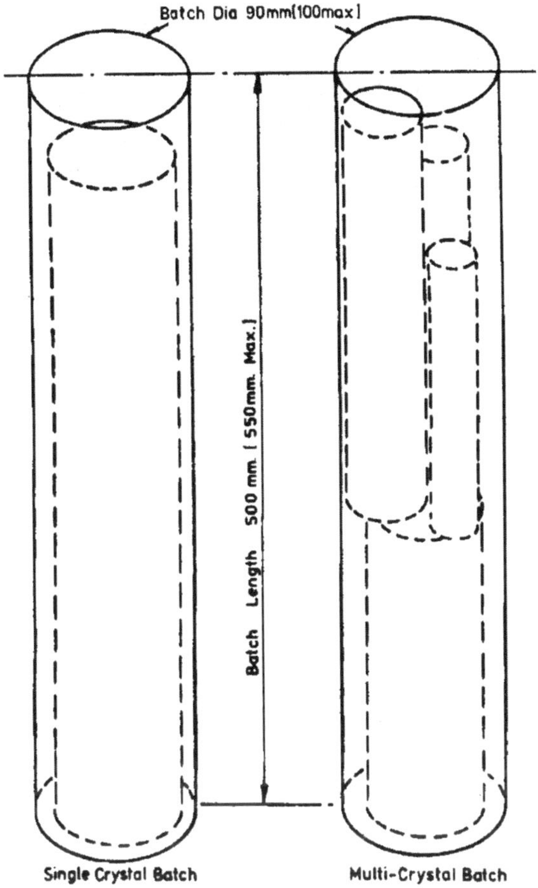

Fig. 6.6. Batch dimensions (after [404])

These are internationally accepted standards for the transport of packages, and Harwell complies with them. To meet this obligation, Smith wrote that each crystal, or piece, of silicon must comply with the following criteria before certification for dispatch:

1. Radiation is less than 0.1 mRem/h, a factor of 5 less than that necessary for the package according to I.A.E.A. Regulations.
2. Contamination is less than $5 \cdot 10^{-5}\,\mu\text{Ci}\,\text{cm}^{-2}$, a factor of 2 less than that necessary for the package.

As a result of applying these low levels, we consider a 3 day delay prior to shipment essential although it is not always possible [404], particularly for material irradiated down to low resistivities (see also [405]).

The tendency of the NTD silicon producers to increase their capacity and the extending range of devices in which NTD silicon is being used calls for special reactor irradiation facilities. The JRC heavy water moderated ESSOR reactor (Ispra, Italy) is described by Bourdon and Restelli [406]. In this article, attention has been especially devoted to obtain an automated operation of the facility and to optimize the characteristics of the irradiation volume. The height of 50 cm, with respect to a core vertical dimension of 150 cm, has been selected to obtain a minimum axial spread of the neutron flux (±4%). A thermal neutron flux of $(2.7-3)\cdot 10^{14}\, n\,\mathrm{cm}^{-2}\mathrm{s}^{-1}$ is available at the irradiation position with a thermal to fast ($>100\,\mathrm{keV}$) neutron flux ratio larger than 400.

The silicon crystals are loaded into a transport unit which can carry up to 100 ingots 77 mm in diameter by 500 mm long (Fig. 6.7). The ingots are protected by a bored plastic capsule which defines the irradiation volume indicated above. The capsules are then loaded into the reactor through a lock between the transport unit and the channel for transfer from the air to the heavy water circuit. Heavy water circulation assures efficient cooling of the silicon ingot during irradiation (the maximum crystal temperature should not exceed 70°C) and induces, by using a suitable capsule shape (see Fig. 6.8), a slow rotation of the ingot to minimize radial dispersion of the neutron flow. At the end of the irradiation, determined by the control system, the capsule is transferred to the transport unit where it stays for at least 4 days before being discharged, thus providing decay time between the irradiation and the extraction. The plastic material for the capsule must be chosen for its mechanical behavior, radiation resistance, heat resistance, and

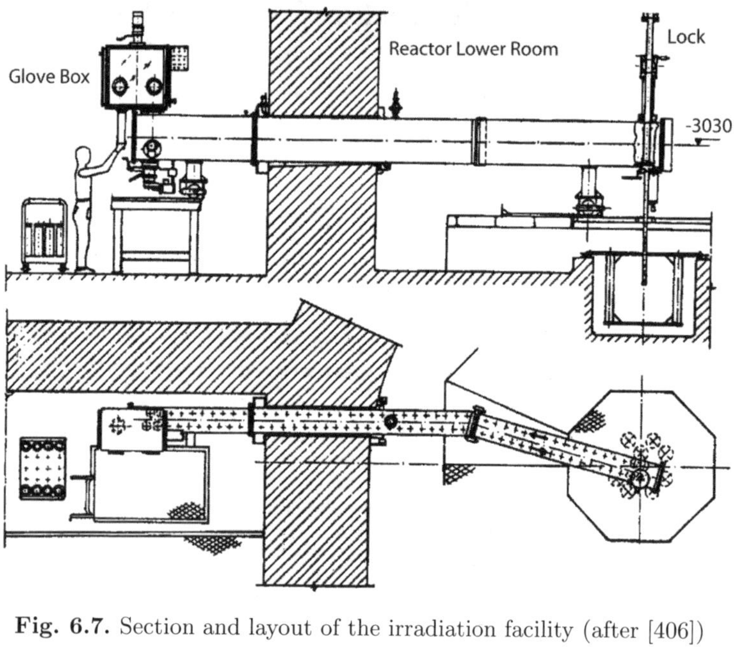

Fig. 6.7. Section and layout of the irradiation facility (after [406])

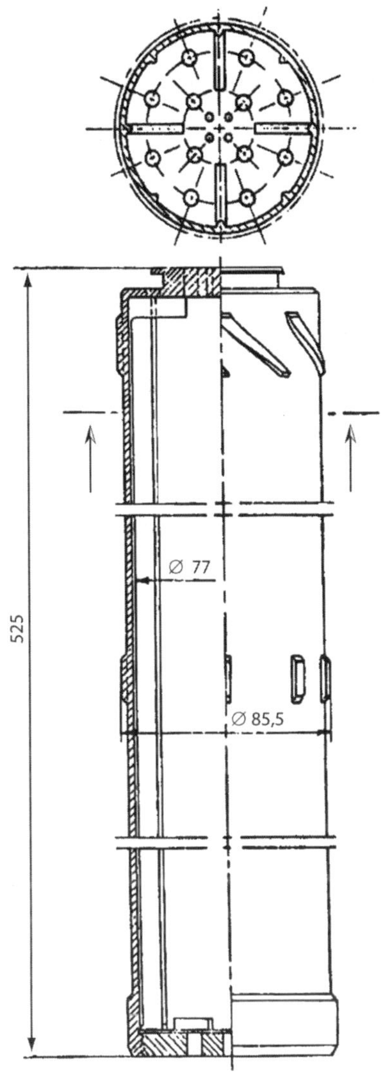

Fig. 6.8. Irradiation capsule (after [406])

low radioactivation. On the basis of the first three requirements, some commercial plastics were selected and samples of each irradiated in the HFR reactor of Petten (Netherland) or in ESSOR, to choose the best material (see also Fig. 6.9). None of the tested materials is completely satisfactory so that Bourdon and Restelli have foreseen the use of Noryl 731 for the preparation of capsules to be used for short irradiation times and polystyrene or PPO for long irradiation times.

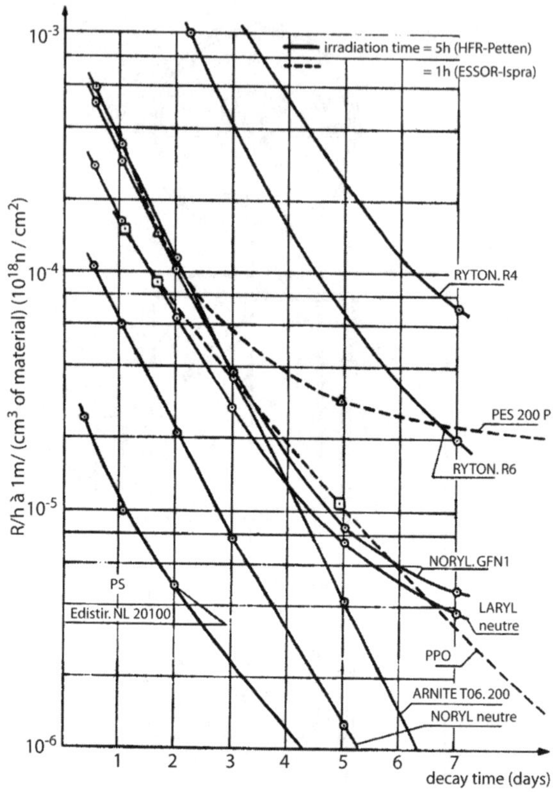

Fig. 6.9. Radioactivation of some plastic materials (after [406])

The high thermal to fast neutron flux ratio assures fewer lattice defects, as demonstrated by the fact that thermal annealing for 5 min at $750-800°C$ has been found sufficient to achieve complete recovery of the final resistivity (see Fig. 6.10).

Concluding this paragraph, we should remember once more the advantages offered by the ESSOR reactor. These advantages are high thermal neutron flux density, very uniform and easily controllable over a large irradiation volume, and a good thermal to fast neutron flux ratio in conjunction with the possibility of irradiating a capsule immersed in D_2O which enssures efficient cooling of the Si crystal.

The General Electric Test Reactor (GETR) (Schenectady, N.Y.) was designed and constructed to provide large irradiation volumes outside the reactor pressure vessel in a surrounding water pool. The thermal neutron flux available for silicon irradiation spans four decades, 10^{11} to $> 10^{14}$ nv (see Fig. 6.11). The large irradiation volume permits the inclusion of flux flattening and spectral softening devices, if desired [407]. As is well known, NTD silicon offers significant technical advantages over chemically doped silicon

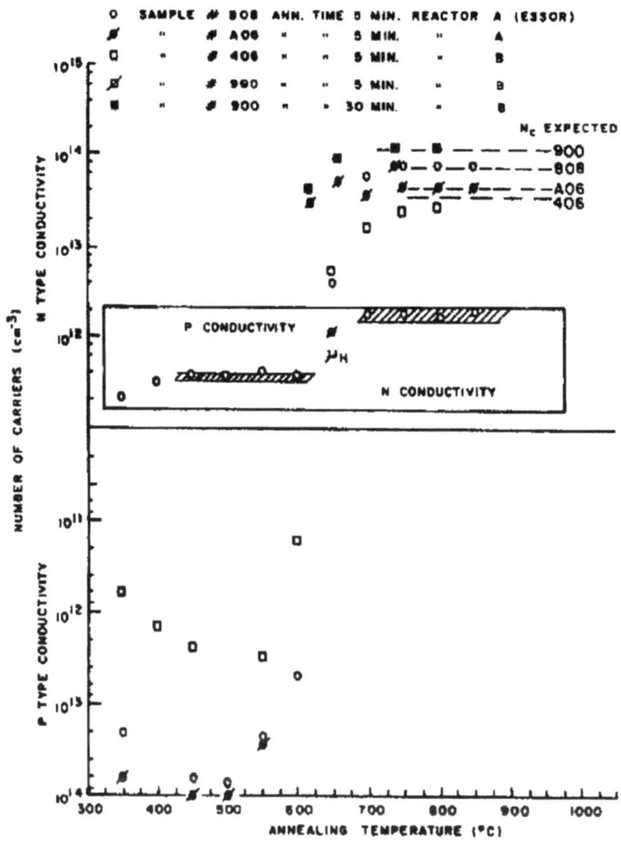

Fig. 6.10. Number of carriers (and Hall mobility) versus annealing temperature (isochronal anneals) (after [406])

(see also [408]). In particular, NTD silicon has a more uniform phosphorus concentration across the radius of an ingot or wafer than chemically doped material. The uniformity approaches 1% for a 3-inch diameter wafer activated in the 50 MW GETR. The availability of such uniform NTD silicon would make it possible to manufacture higher power density thyristors for high-voltage applications (for details, see below).

A program of silicon irradiation is being carried out at the National Bureu of Standards (NBS) (Washington, DC) 10 MW, heavy water moderated reactor. A plan view of the NDBS reactor core showing several of the irradiation locations is shown in Fig. 6.12. A set of five pneumatic rabbit tubes, useful for irradiating silicon chips to analyze for impurities or to study irradiation damage, provide a range of thermal fluxes from $2 \cdot 10^{11}$ n cm^{-2}s^{-1} (copper–cadmium ratio of 3400) to $6 \cdot 10^{13}$ (copper–cadmium ratio of 46) [410]. Those researchers interested in long-term silicon doping irradiation can currently

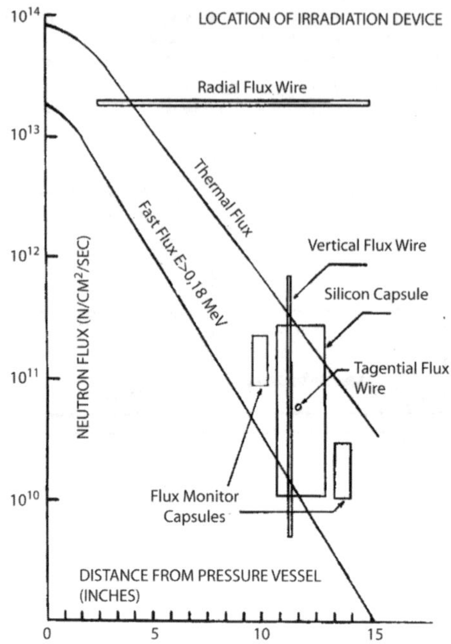

Fig. 6.11. GETR 50 MW neutron flux pool position Z-6 (after [407])

use two vertical facilities designated G2 and G4 [409]. Both facilities are D_2O filled and are completely isolated from the reactor coolant. Since they are isolated, encapsulation of the silicon is unnecessary, and only aluminum hardness is needed to hold the sample. The G 2 tube will accept samples up to 1.6 inches in diameter and has a neutron flux at the core midplane of $1.1 \cdot 10^{14}$ (copper–cadmium ratio of 55) [409]. A vertical flux profile of this facility is depicted in Fig. 6.13. Irradiation of samples in G 2 for periods of 1 day to 6 weeks has been done for Oak Ridge National Laboratory (ORNL). The predicted phosphorus doping rate of $7.5 \cdot 10^{13}$ atom $cm^{-3}h^{-1}$ yielded a concentration in excellent agreement with that measured by ORNL. The G 4 tube is located at the center of the reactor core and will accept samples up to 3 inches in diameter. Its neutron flux has roughly the same shape as that in G 2 but is about 28% greater. A 1-kilogram silicon sample has been irradiated in G 4 to a measured phosphorus concentration of $1.4 \cdot 10^{17}$ atoms cm^{-3} (for details, see [409]).

The absolute differential neutron-energy spectrum for the low-temperature, fast-neutron irradiation facility in the CP-5 reactor was determined by means of a 20-foil activation technique by Kirk and Greenwood [411]. Figure 6.14 shows a simplified schematic of VT 53, the cryogenic fast-neutron irradiation facility at CP-5 (for a more detailed description of this equipment, see paper [411] and references cited therein). The elements silicon, nickel,

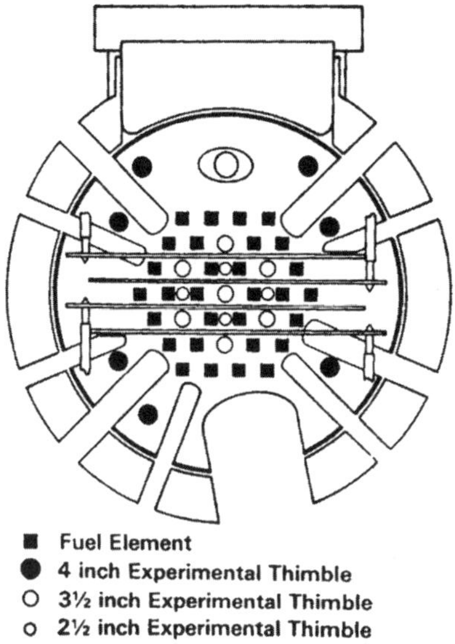

Fig. 6.12. Plan view of NBSW reactor core and irradiation facilities (after [409])

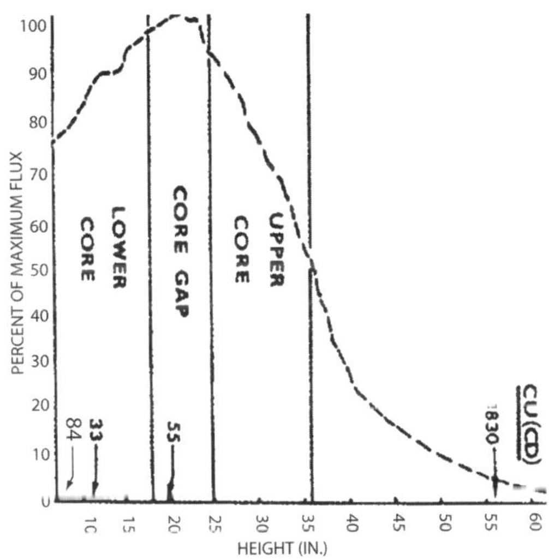

Fig. 6.13. Vertical flux profile of position G 2 (after [409])

niobium, and gold were selected to illustrate the results of these calculations. Table 6.2 gives the group distributions (partially collapsed from the original, more detailed group structure) of primary recoils and the total primary recoil cross-sections for these four elements. Also displayed in this table (for gold only) is an error of one standard deviation for recoils in each group, based on the covariant error matrix generated during neutron spectrum error analysis

Table 6.2. Group distributions of recoils and spectrum averaged cross sections in gold, niobium, nickel, and silicon irradiated in the VT 53 fast neutron energy spectrum (after [411])

	Si	Ni	Nb	Au	
Primary recoil/ Energy group	PRD	PRD	PRD	PRD	±
0.5 (eV ↓)	2.4	4.0	2.6	11.8	1.4
5–10	0.4	1.7	1.2	5.1	0.9
10–20	0.6	2.9	1.9	5.6	1.0
20–50	1.5	6.2	4.3	6.7	0.6
50–100	1.9	6.8	3.9	5.4	0.3
100–200	2.4	8.7	4.9	7.2	0.6
200–400	2.7	10.1	6.1	9.4	0.9
400–600	1.8	8.0	4.7	6.4	0.7
600–800	1.5	6.8	4.1	4.7	0.6
800–1000	1.3	4.4	3.5	3.7	0.6
1–1.5 (keV ↓)	2.8	5.4	7.2	6.6	1.4
1.5–2	2.4	3.9	5.5	4.2	1.1
2–3	3.9	5.3	8.4	5.4	1.5
3–5	6.7	5.8	11.5	5.8	1.3
5–7	5.5	3.6	7.2	3.0	0.5
7–10	6.9	3.9	6.7	2.6	0.4
10–20	18.8	5.7	9.0	3.9	0.4
20–40	18.1	3.3	4.7	2.1	0.1
40–60	6.7	1.3	1.5	0.3	0.02
60–80	3.5	0.8	0.6	0.08	0.004

PRD: Primary recoil distributions (%)

Spectrum averaged cross sections (barns)					
σ_{elastic}	2.86	8.82	6.77	9.35	±12%
$\sigma_{\text{inelastic}}$	0.05	0.12	0.40	0.80	±9%
$\sigma_{\text{n,2n}}$	0.0001	0.001	0.001	0.002	21%
$\sigma_{\text{n,p}}$	0.002	0.02	0.0001	–	–
$\sigma_{\text{n},\alpha}$	0.001	0.001			
σ_{total}	2.91	8.96	7.17	10.15	±11%

by the SANDANL code. To facilitate comparison among the four elements, Fig. 6.15 shows the integral distribution of primary recoils (see also [412]). Using the Robinson analytical approximation [413] to the Linfhard et al. [414] theory of electronic energy losses, it is also possible to calculate the damage energy distribution and spectrum averaged total damage energy cross sections [415, 416]. The number of Frenkel defects (interstitial and vacancy pairs) (see also [403]) produced by a primary recoil of energy T is generally

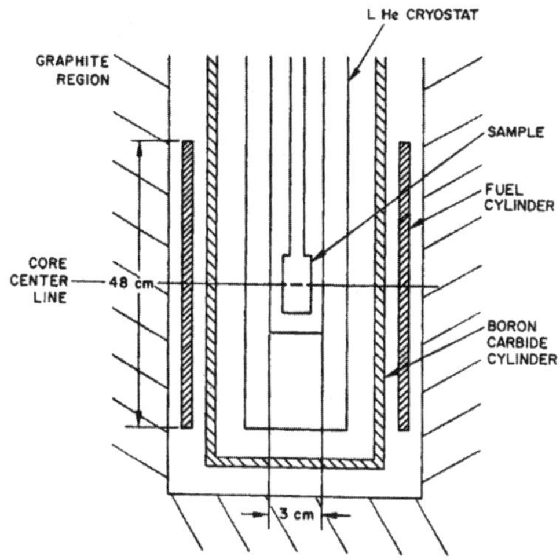

Fig. 6.14. Simplified schematic of the cryogenic fast-neutron facility in CP-5 (VT 53) (after [411])

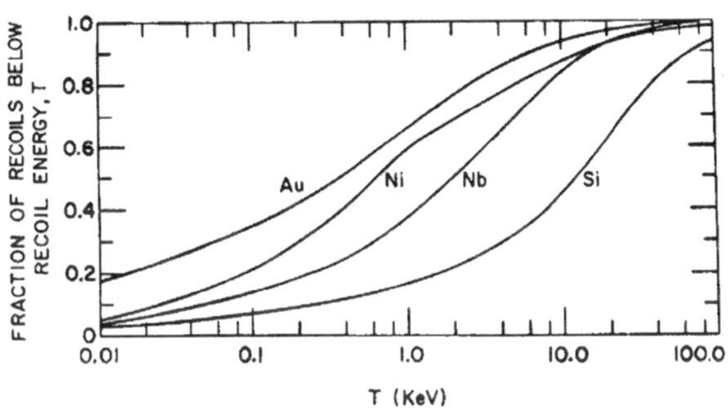

Fig. 6.15. Integral distributions of primary recoils in Au, Nb, Ni, and Si irradiated in VT 53 (after [411])

proportional to the damage energy available from this recoil, which is just the total recoil energy, T, minus the electronic energy losses at this recoil energy. The distribution of damage energy over the recoil energy groups thus gives a good indication of how the Frenkel defects are distributed with primary recoil events. As an example, Table 6.3 gives the distributions of damage energy in recoil energy groups and the spectrum averaged damage energy cross sections for the same four elements shown in Table 6.2. Figure 6.16 graphically illustrates the corresponding integral damage energy distributions.

The integrated neutron flux determined by Kirk and Greenwood ($2.2 \cdot 10^{16}\,\mathrm{n/m^2s} \pm 13\%$, for $E_\mathrm{n} > 0.1\,\mathrm{MeV}$) for the low-temperature, fast-neutron facility in CP-5 is 70% greater than that determined less accurately earlyier [417]. The amount of ^{235}U burn-up in the fuel cylinder during this

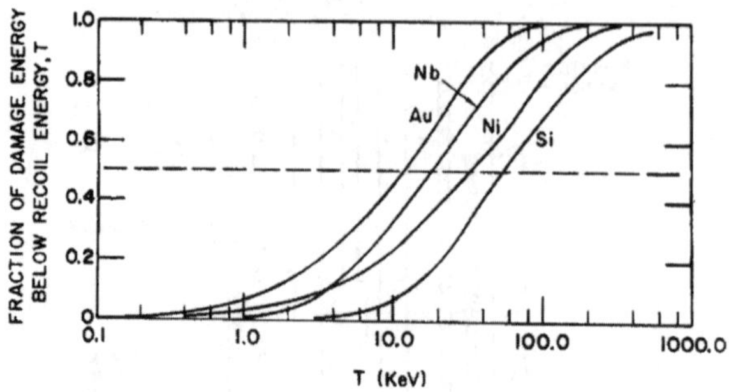

Fig. 6.16. Integral damage energy distributions for Au, Nb, Ni, and Si irradiated in VT 53 (after [411])

Table 6.3. Group distributions of damage energy and spectrum averaged damage energy cross sections ($\langle \sigma T_d \rangle$) in gold, niobium, nickel, and silicon irradiated in the VT 53 fast-neutron energy spectrum (after [411])

Primary recoil/ Energy group	Si DED	Ni DED	Nb DED	Au DED
0–5 (eV ↓)	0.00	0.001	0.001	0.01
5–10	0.0002	0.003	0.002	0.02
10–20	0.0005	0.01	0.006	0.04
20–50	0.003	0.05	0.03	0.11
50–100	0.008	0.1	0.05	0.2
100–200	0.02	0.3	0.14	0.5
200–400	0.04	0.6	0.3	1.2
400–600	0.05	0.8	0.4	1.4
600–800	0.06	1.0	0.5	1.4
800–1000	0.06	0.8	0.6	1.5
1–1.5 (keV ↓)	0.2	1.4	1.6	3.6
1.5–2	0.2	1.4	1.7	3.2
2–3	0.5	2.6	3.6	5.7
3–5	1.3	4.3	7.6	9.5
5–7	1.6	4.1	7.1	7.4
7–10	2.8	6.2	9.2	8.9
10–20	12.7	14.7	20.7	22.6
20–40	12.6	10.7	10.9	6.2
60–80	8.9	9.5	6.3	2.0

DED: Damage energy distribution (%)

Spectrum averaged damage energy cross sections (keV-barns)				
$\langle \sigma T_d \rangle$	42.8	36.0	33.9	19.5 (±8.7%)

time is not known with certainty. However, based on a comparison of resistivity damage rate measurements in copper during a comparable period of time [418, 419], the burn-up is about 5% for an 8-year period. Thus, it is reasonable to assume that the flux (corrected for burn-up) in this facility has remained constant in time within the uncertainty of the present measurement. It is found that the major cause of the difference between the flux measurements is improvement in the accuracy of the cross-sectional data on which these flux determinations are based. Therefore, any use of data from previous experiments in this facility will employ the presently determined neutron spectrum and integrated flux values. One caution that is perhaps obvious should be noted. In comparing experimental data from different neutron irradiation facilities, one must be careful when using integrated flux values that have been determined at different times, since cross-sections have changed with time (for details, see [411, 420]).

6.3 Nuclear Reaction Under the Influence of Charged Particles

According to the modern concept (see, e.g., [399]), the nuclear reactions leading to the formation of doped impurities, may proceed under the influence of charged particles (protons, deuterons, α particles, etc.), neutrons and γ quantites. In such a case, the energy of bombarding the nuclear particle must be sufficient to overcome the mutual Coulomb particle and nuclear repulsion. And in the case of a following particle flying out from an excited nuclear compound the other charged particle should also obtain, in a nuclear compound the energy sufficient to overcome this Coulomb barrier. If r_{n} is a nuclear radius and e is the electron charge, the barrier height [397] has the following relation:

$$B_{\mathrm{b}} = \frac{Z_x Z_{\mathrm{a}} e^2}{r_{\mathrm{n}}} \simeq \frac{Z_x Z_{\mathrm{a}}}{A^{\frac{1}{3}}}\,\mathrm{MeV} \tag{6.10}$$

and reaches $5 - 10\,\mathrm{MeV}$ for a light nucleus, $10 - 20\,\mathrm{MeV}$ for medium nuclei and $20 - 30\,\mathrm{MeV}$ for heavy nuclei [397, 399].

Historically the first sources of charged particles (α particles) were the radioactive elements of Ra, Rn, Po, Pu and others, which in radioactive decay emit $10^{10} - 10^{11}$ α particles in 1 s from 1 g of emitter [421] with energies of $4 - 8\,\mathrm{MeV}$ [399, 421]. The main question is how the beam of α particles passes on its own energy to the atoms of irradiated material. It appears that, owing to the Coulomb interaction of particles with matter, the kinetic energy of the charged particles is spent on ionizing and exciting atoms of the irradiated substance. Estimate show [399] that the cross section of the ionization process ($\sim 10^{-21}\,\mathrm{cm}^2$) is approximately 1000 times larger than the cross section of the nuclear interaction ($\sim 10^{-24}\,\mathrm{cm}^2$). Strictly speaking, for a monoenergetic beam of particles, R_0 (the distance of a particle run) has the meaning of the median distance relative to which it has a Gauss distribution. During one act of ionization, the particle energy is diminished approximately 3.5 eV [399], and the probability of nuclear interaction depends on the particle energy at the moment the particle and nucleus collide. The number of nuclear interactions in a thin layer dx at depth x from the surface of the target has the form

$$\mathrm{d}\nu = \eta(x)N\sigma(x)\mathrm{d}x \simeq \eta_0 N\sigma(x)\mathrm{d}x, \tag{6.11}$$

where N is concentration of the target nucleus; η_0 and η are the particle beams on the surface and in depth x, respectively; $\sigma(x)$ is the cross section of interaction. The full number of interactions on the depth of a substance, whose thickness is more than R_0, is

$$\nu = \eta_0 N \int_0^{R_0} \sigma(x)\mathrm{d}x = \eta_0 N \int_0^{E_0} \frac{\sigma(E)\mathrm{d}E}{\left|\frac{\mathrm{d}E}{\mathrm{d}x}\right|}, \tag{6.12}$$

where E_0 is the initial energy of the particle. The yield of the nuclear reaction is determined by the portion of particles that undergoes the nuclear interaction:

$$V(E) = \frac{\nu}{\eta_0} = N \int_0^{E_0} \frac{\sigma(E)\mathrm{d}E}{\left|\frac{\mathrm{d}E}{\mathrm{d}x}\right|}. \tag{6.13}$$

From the last formula, it follows that the yield of the nuclear reaction at the energy E of a charged particle is determined by the cross section $\sigma(E)$ and the specific ionization of the particle $\frac{\mathrm{d}E}{\mathrm{d}x}$. And, vice versa, the functions $\frac{\mathrm{d}E}{\mathrm{d}x}$ and $\frac{V}{E}$ from (6.13) allow us to find the cross section of interaction:

$$\sigma(E) = \frac{1}{N}\frac{\mathrm{d}V}{\mathrm{d}E}\left|\frac{\mathrm{d}E}{\mathrm{d}x}\right|. \tag{6.14}$$

It is known (see, e.g., [399]) that for a charged particle, the magnitude $\frac{\mathrm{d}E}{\mathrm{d}x}$ is proportional to the square root of the particle charge, the concentration of electrons in the substance (n_e), some function of the velocity $f(v) \sim \frac{1}{v}$, and does not depend on particle mass:

$$\frac{\mathrm{d}E}{\mathrm{d}x} \sim Z^2 n_e f(v). \tag{6.15}$$

The dependence $\frac{\mathrm{d}E}{\mathrm{d}x}$ is permitted to recount the data on the motion of one concrete particle in a concrete substance and on the motion of another particle in another substance (see also [398]). So far $\frac{\mathrm{d}E}{\mathrm{d}x} = f(E)$; then, taking the integral, we can obtain the full run of the particle:

$$R = \int_0^{E_0} \frac{\mathrm{d}E}{f(E)}. \tag{6.16}$$

For example, α particles ($R_{\alpha,x}$) in the substance X_Z^A can be defined on the run in air ($R_{\alpha,\mathrm{air}}$) with the assistance of the empirical formula [399]

$$R_{\alpha,x}(E) = 0.56 R_{\alpha,\mathrm{air}}(E) A^{\frac{1}{3}}, \tag{6.17}$$

where $R_{\alpha,\mathrm{air}}$ in centimeter and relates in air at a temperature of 15°C and ambient pressure, $R_{\alpha,x}$ will be obtained in mg/cm^2.

The flow of protons is connected with the flow of α particles by another formula, which works at $E \geq 0.5\,\mathrm{MeV}$:

$$R_\mathrm{p} = 1.007 R_\alpha(3.972)E - 0.2, \tag{6.18}$$

where $R_\alpha(3.972)$ is the flow of α particles with the energy of 3.972. The flow of other charged particles with mass M_x (exclude electrons) is connected with the flow of protons in

$$R_x(E) = \frac{M_x}{M_\mathrm{p}} R_\mathrm{p}\left(\frac{M_\mathrm{p}}{M_x}E\right). \tag{6.19}$$

Owing to the fast retardation of charged particles, they can dope a thin layer with an ununiform distribution of doped impurities in the depth. The possible reactions obtained with charged particles are described by Smirnov [397], where the half-time decay was indicated. To conclude of this section, we note that regular experience in this field is absent at the present time.

6.4 Nuclear Reaction Under the Action of the γ-Rays

The reactions of (γ, n), (γ, p) and (γ, α) belong to the reactions of splitting nuclei irradiated by γ rays. These reactions are endoenergetical and have some energetic threshold. In the common case, the probability of a (γ, n) reaction is more than the probability of (γ, p) and (γ, α) reactions. The energy of γ rays $\sim$ 10 MeV is called an (γ, n) reaction. At the $E_\gamma \sim 100\,\mathrm{MeV}$, reactions that create several particles ($\gamma, 2n, \gamma, pn$, etc.) are possible [397]. Inasmuch as impurities can be produced in practically any substances with the assistance of photonuclear reactions, therefore they are of interest for semiconductor doping. The main reason for this circumstance is their large penetrating capability compared to charged particles [397]. This circumstance can guarantee uniform doping of a large volume substance with some impurity atoms. The perspective of this direction at the present time is also supported by the possibility of receiving of γ quanta with energy that retards electron emission. For that, monoenergetical electrons with energy of 25 – 60 MeV

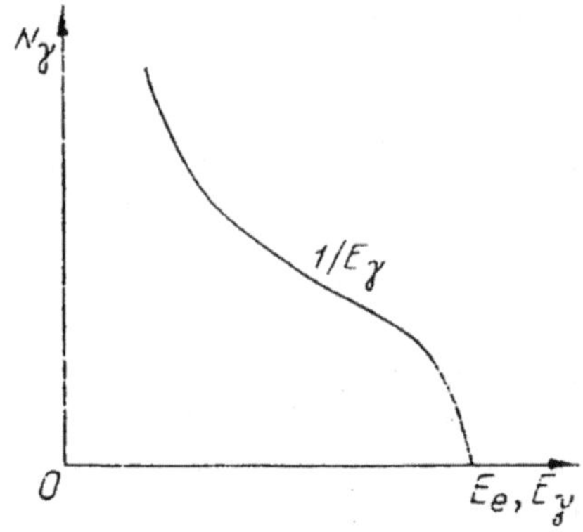

Fig. 6.17. The spectrum of γ emission is created by electrons retarding with energy E_e on the target of the accelerator (after [397])

received by the accelerator, are directed at the target from heavy metals (Pb, Bi, W, U etc.). As a result, the retarding electrons create a continuous spectrum of γ emission. The maximum energy of this continuum is equal to the kinetic energy of electrons E_e and the intensity of the γ emission, approximately inversely proportional to the γ quanta of energy (see Fig. 6.17). The increase in using γ emission for irradiating Si and other semiconductors has been noted [422].

6.5 Nuclear Reactions Under the Influence of Neutrons

The main hope in the field of nuclear doping of semiconductors at the present time is placed on neutrons (see also [423]), inasmuch as these are neutral particles and they possess large penetrating capability. Furthermore, neutrons can interact practically with all nuclei. At the present time, there are different sources of neutrons with different intensity accepted and approached (see above). The history of techniques by which the first neutrons beams were produced can be found in [397]. The main source of neutron beams is still nuclear reactors of different types (see also above). A typical neutron spectrum of the research water–water reactor WWR is given in review [424]. In this spectrum, there are neutrons with energy of $0.01 \leq E \leq 10^7$ eV. Upon the irradiation of a neutron beam, the cross section of nuclear compound σ_x depends on the degree of resonance to the intrinsic magnitude of the energy levels of the nuclear compound.

For neutrons with energy $E_n \sim 10^{-2}$– 10^4 eV [424], we can write $\sigma(n, \gamma)$ for the cross section of the nuclear reaction

$$\sigma(n,\gamma) = \sigma_0 \left(\frac{E_0}{E_n}\right)^{\frac{1}{2}} = \sigma_0 \left(\frac{v_0}{mv}\right), \tag{6.20}$$

where σ_0, v_0 and E_0 are some constants of the cross section of the nuclear reaction, velocity, and neutron energy taken as a starting point. As is well known, NTD thermal neutrons are of greater interest. In this interval, there are almost all numbers of neutrons in the spectrum of a nuclear reactor which is described by the Maxwell distribution of the velocities:

$$n(v) = \frac{4}{\sqrt{\pi}} \left(\frac{M_{\mathrm{n}}}{2kT} \right)^{\frac{3}{2}} v^2 \exp \left(-\frac{M_{\mathrm{n}} v^2}{2kT} \right), \tag{6.21}$$

where, as usual, k is the Boltzmann constant and T is temperature.

The average cross section of the spectrum $n(v)$ provides

$$\bar{\sigma} = \frac{\displaystyle\int \frac{\sigma_0 v_0}{v} n(v) v \mathrm{d}v}{\displaystyle\int n(v) v \mathrm{d}v} = \frac{\sigma_0 v_0}{\bar{v}}, \tag{6.22}$$

where $\bar{v} = \sqrt{\frac{8kT}{\pi M_{\mathrm{n}}}}$ is the average velocity. It is convenient to use the next formula for the more probable velocity of neutrons:

$$v_{\mathrm{mp}} = \sqrt{\frac{2kT}{M_{\mathrm{n}}}} = \bar{v} \frac{\sqrt{\pi}}{2} = \frac{\bar{v}}{1.228}. \tag{6.23}$$

This corresponds to the maximum distribution of $n(v)$, taking into account that $v_0 = v_{\mathrm{mp}}$ and $\sigma_0 = \sigma_{\mathrm{mp}}$ and at ambient temperature ($T = 293\,\mathrm{K}$), $v_{\mathrm{mp}} = 2200\,\mathrm{m/s}$. In such a case, we obtain the relation connected with the middle cross section (n, γ) reaction of the thermal neutrons where the magnitude of the cross section is of the neutrons usually measurable by the more probable velocity

$$\sigma - \frac{\sigma_{\mathrm{mp}} v_{\mathrm{mp}}}{\bar{v}} = \frac{\sigma_{\mathrm{mp}}}{1.128} = \frac{\sigma_{2200}}{1.128}. \tag{6.24}$$

The energetic dependence on the cross section for Si can be described by the following approximate equation [425]:

$$\sigma = \sigma_{\mathrm{a}} + \sigma_{\mathrm{s}} = \sigma_{\mathrm{a}} + \sigma_{\mathrm{fa}}(1 - \mathrm{e}^{-2\mathrm{w}}) = \frac{0.8}{\sqrt{E_{\mathrm{n}}} + 2.25\,(1 - \mathrm{e}^{C E_{\mathrm{n}} T})}, \tag{6.25}$$

where E_{n} is the energy of neutrons (MeV); σ_{a} is the cross section of absorption which is equal $160 \cdot 10^{-3}\mathrm{b}$ at $E_{\mathrm{n}} = 25\,\mathrm{MeV}$; σ_{s} is the cross section of the scattering; σ_{fa} is the cross section of free atom which is equal to 2.25 b; $\mathrm{e}^{-2\mathrm{w}}$ is the Debye–Waller factor; and $C = 1.439 \cdot 10^{-5}$ is the normalized constant of the dependence $\sigma(E)$ on the magnitude $\sigma = 0.55\,\mathrm{b}$ at $E_{\mathrm{n}} = 50\,\mathrm{MeV}$ at 300 K. Calculations based on (6.19), of the energetic dependence of the cross section of Si at different temperatures as well as some experimental data are shown in Fig. 6.18. As is seen for polycrystalline Si (curve 1), the cross

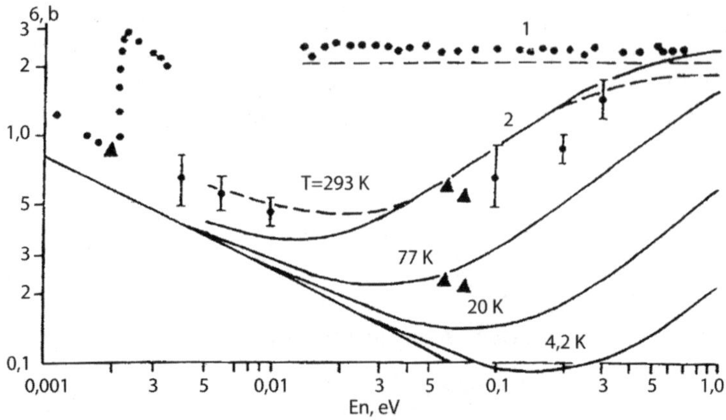

Fig. 6.18. The dependence of the calculated cross section of thermal neutrons in polycrystalline (1) and monocrystalline (2) Si at different temperatures (after [397])

section in the thermal field does not depend on energy, and for monocrystals (curve 2), this dependence is changed according to the law $\frac{1}{v}$ and depends on temperature. The last fact indicates that the NTD of different semiconductors and one semiconductor with different impurities can irradiate at different temperatures (for details, see [397]).

6.6 The Influence of Dopants

According to the common scheme of nuclear reactions, the primary products of (n, γ) reactions present the nuclear $X_{\mathrm{Z}}^{\mathrm{A}+1}$ per one atomic unit of heavier outcomes. These can be heavier isotopes of the same or neighboring elements in the periodic table. If these isotopes have isotopes of irradiated material in a natural mixture and appear stable, then nuclear transformation leads to a change in the initial concentration of stable isotopes in the irradiated substance. In such a case, this process does not create impurities of neighboring elements in the periodic table. The possibility of exchanging electrophysical properties with the assistance of impurities which are created by nuclear doping with irradiation of neutrons is proven for numerous semiconductors (see, e.g., Table 2.6 in [397]). In this way, different properties of semiconductors can be connected with the nature and concentrations of the predominant impurities after respective annealing, for example, Sn and Te in InSb [426–428], Ge and Se in GaAs [429–431], in CdS [432], and phosphorus in Si [400, 427], [433–436], [428], etc.

To estimate the character of the distribution of impurities from the thickness of doped materials, we will use the fact that the attenuation of the intensity of a collimated neutron flux by the layer of substance of thickness

x is confirmed by the well-know law

$$I = I_0 \exp(-N\sigma x) = I_0 \exp(-\mu x) = I_0 \exp\left(-\frac{x}{l_1}\right), \qquad (6.26)$$

where N is the amount of atoms of irradiated material per cm^3; l_1 is the median length of absorption [437] connected with the macroscopic absorption coefficient of the material $\mu = N\sigma$ next relation

$$l_1 = \frac{1}{N\sigma} = \frac{1}{\mu}, \qquad (6.27)$$

where l_1 also characterizes the layer thickness of the material on which the neutron flux and, respectively, the impurity concentration is diminished by $e = 2.72$ times.

The relations mentioned above are correct, if we assume that the neutron cross section of scattering is small compared to the absorption cross section. When there are both absorption and scattering of neutrons, it is necessary to use the conclusions of the common theory of neutron diffusion [437]:

$$I = I_0 \exp\left(-\frac{x}{L}\right), \qquad (6.28)$$

where L is the diffusion length.

For compound semiconductors having various atoms, the attenuation of the emission beam can be considered an additive property of the medium. Taking into account this fact, we represent (6.20) in the following form:

$$I = I_0 \exp\left(-\frac{\mu}{\rho} m\right). \qquad (6.29)$$

Here ρ is the density of the substance, $\frac{\mu}{\rho}$ is the mass coefficient of the attenuation, and m is the mass of substance with a cross section of $1\,\mathrm{cm}^2$ and thickness x. Then, for a compound substance, we can write

$$\frac{\mu}{\rho} = \sum_i C_i \left(\frac{\mu}{\rho}\right)_i, \qquad (6.30)$$

where C_i is the weight concentration of the ith element of the mixture. Equations (6.20)–(6.24) also apply to the attenuation of a narrow beam of γ quanta. In this case, the linear coefficient of attenuation is determined (see also [421] and references therein) from the sum of the contributions of photo and Compton effects and also from the creation of electron–positron pairs, e.g.,

$$\frac{1}{\mu} = N(\sigma_{\mathrm{ph}} + \sigma_{\mathrm{c}} + \sigma_{\mathrm{p}}), \qquad (6.31)$$

Table 6.4. The efficiency of the attenuation of neutrons and γ rays by different semiconductor materials (after [397])

Material	density, $\rho\left(\frac{g}{cm^3}\right)$	SN $N\sigma$ (cm^{-1})	SN l_n (cm)	SN L (cm)	GR $\frac{\mu}{\rho}\left(\frac{cm^2}{g}\right)$	GR μ (cm^{-1})	GR l_γ (cm)
Si	2.42	0.008	125.0	22.2	0.024	0.058	17.2
Ge	5.46	0.25	4.0	4.7	0.39	0.213	4.7
GaAs	5.4	0.36	2.8	-	0.039	0.210	4.8
InSb	5.78	7.0	0.14	-	0.052	0.391	3.43
CdS	4.82	115.0	0.01	-	0.045	0.217	4.6

SN: Slow neutrons

GR: for γ rays with energy $E = 20 - 30\,\mathrm{MeV}$

where σ_{ph}, σ_{c}, and σ_{p} are the cross sections of the irradiated processes of the interaction of γ quanta per atom of the substance. The efficiency of neutron absorption and γ ray absorption of different semiconductor materials in one side irradiation is reflected in Table 6.4. The magnitude of $\frac{\mu}{\rho}$ and, respectively, values of l_{n} and l_γ are calculated using (6.24) and the knowledge of the cross section of neutrons [438] and extrapolation data of the mass coefficients of absorption of γ radiation for some elements. A comparison of the data in Table 6.4 shows that for neutrons, the increasing absorptive ability from Si to CdS decreases rather sharply in the thickness of the semiconductor material layer at the boundary where the distribution of impurities can be regarded homogenous. In Si, the natural doping is inhomogeneous caused by the absorption and scattering of neutrons; it is small and for ingots with a diameter of 50 mm, does not exceed 1% [401], for crystals with a diameter of 80 mm, the relation of concentrations in the center (C_{min}) and at the ends (C_{max}) of the ingot $\frac{C_{\mathrm{min}}}{C_{\mathrm{max}}} \simeq 0.956$ [439] (for details, see [397]).

6.7 Atomic Displacement Effects in NTD

Recoil atoms from γ rays or from particle emissions after thermal neutron capture and recoil atoms from elastically and inelastically scattered fast neutrons produce atomic displacements in solids [440]. The relative importance of the displacement damage produced by thermal and fast neutrons can be estimated by using isotope concentrations, capture or scattering cross sections, and recoil energies. Such an estimate has been made for silicon, and the results are presented in Table 6.5. The first three columns list the silicon isotopes, isotope concentrations, and cross sections for thermal neutron capture [440] and for fast neutron scattering [441]. The probability of interaction (the product of isotope concentration and capture or scattering cross section) is listed in column 4. The energies available from silicon recoils for producing atomic displacements are given in column 5. For thermal neutrons, these

Table 6.5. Energy available for producing displacement damage in silicon by neutron transmutative doping (after [441])

(a) Thermal neutron capture recoil (Energy into displacement, $\nu = 780\,\text{eV}$)

Silicon isotope	Concentration (10^{22}cm^{-3})	Cross section (10^{-24}cm^2)	$N\sigma$ (cm^{-1})	$\nu N\sigma$ (eV/cm)
$^{28}\text{Si}\ (n,\gamma)^{29}\text{Si}$	4.61	0.08	0.0037	2.88
$^{29}\text{Si}\ (n,\gamma)^{30}\text{Si}$	0.23	0.27	0.00062	0.49
$^{30}\text{Si}\ (n,\gamma)^{31}\text{Si}$	0.15	0.12	0.00018	3.51
β (1.5 MeV)$\rightarrow$ 2.61 h $\rightarrow^{31}$P			Total	3.51

(b) Fast neutron knock-on recoil (Average recoil energy of 50 keV assumed, $\nu = 25\,\text{keV}$)

All	5	3 (avg)	0.15	$3.8 \cdot 10^3$

energies were obtained from the product of the probability of thermal neutron capture and the average recoil energy [440] of 780 eV. Kirk and Greenwood [411] used 474 eV for the average silicon recoil energy, but the lower energy only emphasizes further the dominance of the neutrons in producing displacement damage. The details of the damage produced by recoiling ^{31}Si and ^{31}P atoms may, however, be important in determining the lattice location of ^{31}P introduced by NTD. Column 5 of Table 6.5 shows that the energy available for displacements from an incident fast neutron is 10^3 times that from an incident thermal neutron. Fast neutrons will, therefore, dominate the displacement damage until thermal-to-fast ratios exceed 1000 : 1 (see also above). Thermal neutron capture cross sections for germanium and gallium arsenide [441] are much larger than those for silicon (see also Table 6.4). Consequently, displacement damage by thermal neutrons relative to fast neutrons is expected to be more important in these materials than in silicon. An atom recoiling in a host material creates a high-defect density (cluster of defects) along the recoil track [403]. The upper part of Fig. 6.19 illustrates clusters formed by recoil tracks calculated for a 50-keV silicon atom recoiling in silicon [442] and for a 10-keV germanium atom recoiling in germanium [443]. The track in silicon is a side view and the authors [442] emphasize the high damage density in subclusters expected near the end of the range for each recoiling silicon atom. These subcluster regions are similar to the capture–recoil damage regions produced by thermal neutron capture. Therefore, the kind of damage regions produced by thermal neutrons is included in fast neutron damage. The track in germanium is an end view where the open circles, which represent interstitials, are concentrated on the periphery of the track. The lower part of Fig. 6.19 shows the results of calculating the depth distribution of displacement damage by averaging over a large number of tracks for silicon atoms recoiling with energy E_r in silicon [444]. The peak damage density first increases with E_r but then decreases because the

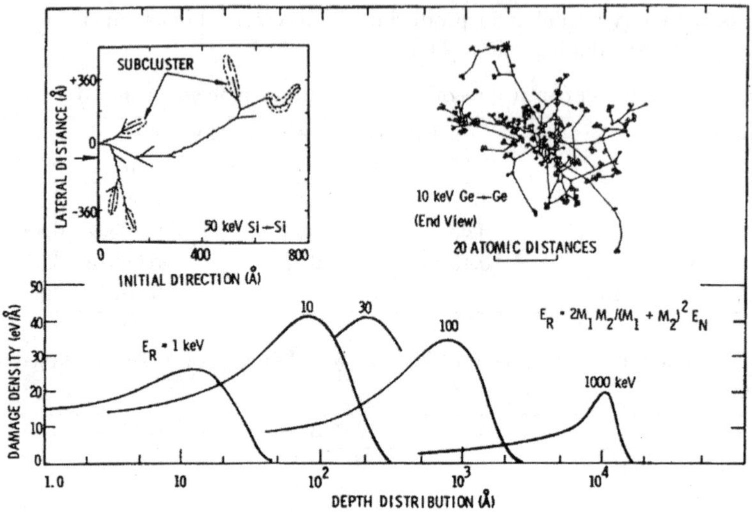

Fig. 6.19. Shown at the upper left is a track for a 50-keV Si atom recoiling in Si according to a calculation by Van Lint et al. [442]]. Upper right is a displacement cascade for a 10 keV Ge atom recoiling in Ge according to calculations by Yoshida [443], where the *open* and *solid circles* represent interstitials and vacancies, respectively. The lower part of the figure shows the energy deposited per Å for different Si recoil energies E_r, according to the formulations of Brice [444]. (After [441])

damage clusters become more diffuse. Knock-on recoils in reactor-neutron irradiations have an average $E_r \approx 50$ keV. The areas under the curves give the energy spent in collision processes and is ≈ 25 keV for $E_r = 50$ keV. The other half of E_r is spent in ionization [444]. Assuming 25 eV/displacement, there is sufficient energy to produce 1000 displacements.

One of the most graphic early examples of defect clusters produced by neutron bombardment of semiconductors was obtained by Bertolotti et al. [445] using etched surface replication transmission electron microscopy (TEM). Figure 6.20 is a sketch taken from such results obtained on 14 MeV neutron irradiated silicon [446]. Most people would agree that the central region is probably due to the core of the displacement-damage clusters. There is less agreement on the interpretation of the outer zone. Direct TEM measurement [447, 448] indicate a strained region around the damage core so that strain-induced differential etching may have caused the outer zone observed in papers [445,446]. A model by Nelson [449] postulates trapping of mobile defects by damage clusters so that the differential etching may have been caused by an excess of trapped vacancies or interstitials. However, the interpretation that has been most extensively used to explain the experimental data is due to Gossick [450] and Crawford and Cleland [451]. In this model, the outer zone represents a space charge region surrounding a p-type germanium. The dam-

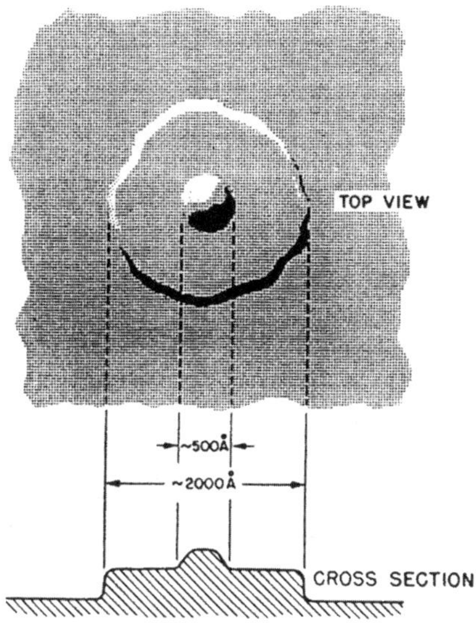

Fig. 6.20. Sketch of a region observed on an etched surface of Si following 14-MeV neutron irradiation. The central core (500 Å) is attributed to recoil damage. (After [441])

age cluster in silicon is nearly intrinsic so that a space charge region would be found in both n- and p-type silicon. The space charge model has been used to interpret minority carrier lifetime data [452], changes in carrier concentration and carrier mobility [453], the light sensitivity of neutron-produced electrical changes [453], photoconductivity [454], and EPR (electron paramagnetic resonance) observations on specific defects in neutron irradiated silicon (see also [428] and references therein).

Thus, calculations of energy deposition into displacement processes show that fast neutrons dominate defect production in NTD processing of silicon unless thermal-to-fast neutron ratios exceed 1000 : 1. Defect clusters are produced by silicon atom recoils from fast neutron collisions. Using an experimental value for the energy needed per unit volume to form amorphous material, it is argued in paper [441] that amorphous zone formation in silicon NTD is highly unlikely.

6.8 Experimental Results

6.8.1 Ge

As noted above, neutron transmutation (NT) is especially intriguing for semiconductors for several reasons. First, the NT process can create new elements removed by just one atomic number. Considering for the moment the elemental group-IV semiconductors Ge and Si, this means that the donors As and

P will be created, respectively, following neutron capture and β-decay of isotopes of these semiconductor elements. The new elements are, of course, the prototypical donors. Neutron capture leads to NTD. Second, the number of new atoms ${}^{A+1}_{Z+1}N$ created is simply (see also Sect. 6.1)

$$ {}^{A+1}_{Z}N = n\sigma_n {}^{A}_{Z}N, \tag{6.32}$$

where n is the total neutron flow (cm^{-2}), σ_n the cross section for thermal neutron capture (cm^2), and ${}^{A}_{Z}N$ the atom concentration of the specific isotope in the given isotope mixture (cm^{-3}) (either natural or man-made). Considering that the values of σ_n lie in the $10^{-23} - 10^{-24}$ cm^{-2} range (see above), its recognized that very large neutron flows are required to transmute a significant number of atoms of one element into another. Whereas this may pose problems to fulfill the medieval alchemist's dream, it is perfectly suited for the person who wants to dope semiconductors. With the thermal neutron flows available in modern nuclear reactors (see also Sect. 6.2), one can dope Ge up to the metal–insulator (MI) transition (2 to $3 \cdot 10^{17}$ cm^{-3}), and Si can be doped with phosphorus to several times 10^{15} cm^{-3} [428]. As will be shown below, this is due to the small atom concentration of ${}^{30}_{14}$Si and the modest value of the thermal capture cross section. Third, there are elements that have light isotopes which upon neutron capture, transmute to a lower Z element either by electron capture or by positron decay. In this case, acceptors are created. A classical case is the transmutation of ${}^{70}_{32}$Ge into ${}^{71}_{31}$Ga.

The main advantage of the NTD method, as we know at present, is precision doping which is connected with the linear dependence of concentration of doping impurities on the dose of neutron irradiation. Such dependence is frequently observed in different experiments (see, e.g., [455–460]). As an example, Fig. 6.21 shows the dependence of the concentration of doped phosphorus on the dose of irradiation of a Si crystal in a nuclear reactor. This dependence was measured with the help of the Hall effect [435]. However, at a large dose of neutron irradiation, nonlinear dependence is observed. Figure 6.22 shows the results of paper [462] where the deviation from linear law at a large dose of neutron irradiation of the sample of ^{74}Ge which was annealed after irradiation at $T = 460°$C was observed different times (see also the caption of Fig. 6.22). A more amazing effect was observed at the second irradiation of the samples of ^{74}Ge previously strongly doped with As by the NTD method. Instead of an expected increase in the concentration of free charges (electrons) n, a decrease in n is observed. This decrease was directly proportional to the neutron irradiation dose of the ^{74}Ge crystals. Both effects analyzed in detail in papers [459, 462].

The transmutation of stable germanium isotopes via the capture of thermal neutrons is well understood. Table 6.6 contains all the information relevant to NTD of germanium. The paper of Haller et al. [455] quoted the values of the thermal neutron capture cross section σ_n of three sources [427,436,463]. The information from Table 6.6 permits computing the acceptor and donor

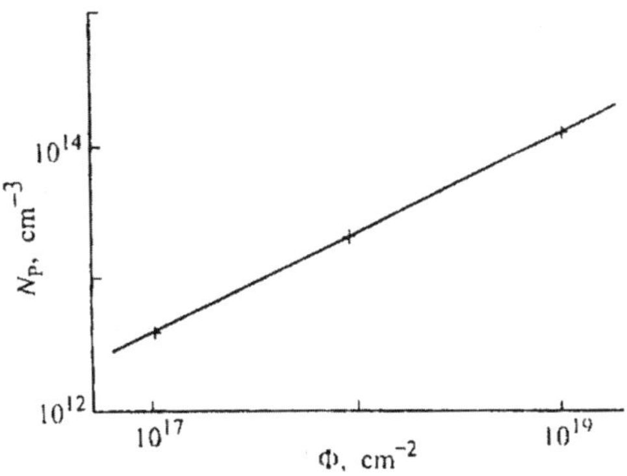

Fig. 6.21. The dependence of phosphorus atom concentration on neutron irradiation dose of Si crystals followed by annealing at 800°C for 1 h. The dependence was measured by the Hall effect. (After [461])

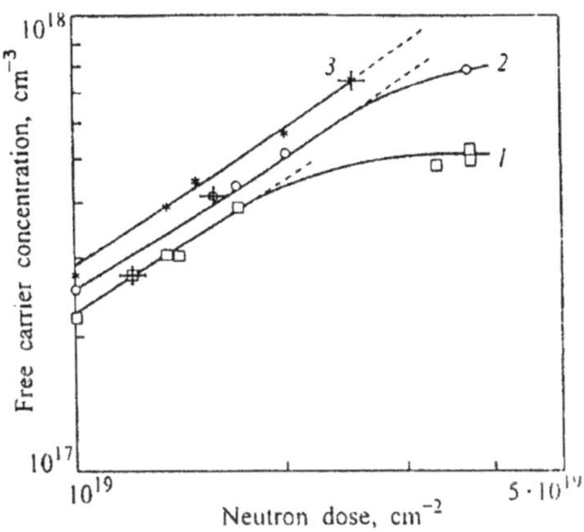

Fig. 6.22. The dependence of the concentration of free electrons in ^{74}Ge NTD on the irradiation of the thermal neutron dose followed for annealing at 460°C by (1): 24, (2): 50, and (3): 100 h (after [462])

Table 6.6. Characteristics of the transmutative process of germanium (after [455])

Isotopic	abundance (%)	NCCS (barn)	NCDR	Dopant type
$^{70}_{32}$Ge	20.5	3.4; 3.2; 3.25	$^{70}_{32}$Ge$(n,\gamma)\rightarrow^{71}_{32}$Ge $\rightarrow^{71}_{31}$Ga	p
$^{72}_{32}$Ge	27.4	0.98; 1.0; 1.0	$^{72}_{32}$Ge$(n,\gamma)\rightarrow^{73}_{32}$Ge	
$^{73}_{32}$Ge	7.8	14.0; 14.0; 15.0	$^{73}_{32}$Ge$(n,\gamma)\rightarrow^{74}_{32}$Ge	
$^{74}_{32}$Ge	36.5	0.62; 0.5; 0.52	$^{74}_{32}$Ge$(n,\gamma)\rightarrow^{75}_{32}$Ge $\rightarrow$ $^{75}_{33}$As	n
$^{76}_{32}$Ge	7.8	0.36; 0.2; 0.16	$^{76}_{32}$Ge$(n,\gamma)\rightarrow^{77}_{32}$Ge $\rightarrow$ $^{77}_{33}$As $\rightarrow^{75}_{34}$Se	n

NCCS: Neutron capture cross sections

NCDR: Neutron capture and decay reactions

concentrations for a known neutron exposure. Not only are these concentrations important, but the ratio of the sum of all minority dopants (donors) and the sum of all majority dopants (acceptors), i.e., the compensation K, are crucial for low-temperature conduction. For germanium, one obtains K from the following equation (see also [455]):

$$K = \frac{\left(\sum \text{donors} \cdot \text{cm}^{-3}\right)}{\left(\sum \text{acceptors} \cdot \text{cm}^{3}\right)} = \frac{(N_{\text{As}} + N_{\text{Se}})}{N_{\text{Ga}}}. \tag{6.33}$$

The substituted selenium impurities are double donors providing two electrons for compensation. Therefore, they are counted twice in the sum of donors. Using the different values for σ_{n}, one finds K ranging from 0.322 to 0.405 for crystals with negligible initial donor and acceptor concentrations. It would be of great help for both the basic understanding of hopping conduction [464] as well as for the application of neutron-transmutation-doped germanium as, for example, bolometer material [455], if these cross sections could be accurately evaluated in one or more well-characterized nuclear reactors (see above).

To obtain the above K values and thus take full advantage of NTD, Haller et al. chose the purest available Ge crystals as starting materials. Germanium is, in this respect, ideally suited for NTD because it can be purified at present to concentrations of $\lesssim 10^{11}\,\text{cm}^{-3}$ (see e.g., [465]). Such low concentrations are negligible when compared with the dopant concentrations in the low $10^{16}\,\text{cm}^{-3}$ range after NTD. According to Haller et al. [455], the concentrations of electrically inactive impurities such as hydrogen, carbon, oxygen, and silicon can be as high as $10^{14}\,\text{cm}^{-3}$. Of all the isotopes of these impurities, only $^{30}_{14}\text{Si}$ transmutes to an electrically active impurity, phosphorus, a shallow donor. Because there is only one silicon atom in every $4.4 \cdot 10^{8}$ germanium atoms and only 3% of all silicon atoms is $^{30}_{14}\text{Si}$ which has a neutron capture cross section much smaller than the germanium isotope cross sections, Haller et al. estimated that less than one phosphorus donor is produced for every 10^{11} gallium majority acceptors during the NTD process. These authors concluded that ultrapure germanium crystals are a virtually perfect starting material. For the NTD study, they chose an ultrapure germanium single crystal which they grew at the crystal growth facility described earlier (see also [455, 465] and references therein).

The measured resistivities (ρ) in paper [455] are presented in Fig. 6.23. The results of these measurements yield the mobility μ:

$$\rho = (p\mu e)^{-1},$$
$$R_{\text{H}} = (pe)^{-1},$$

and

$$\mu = \frac{R_{\text{H}}}{\rho}, \tag{6.34}$$

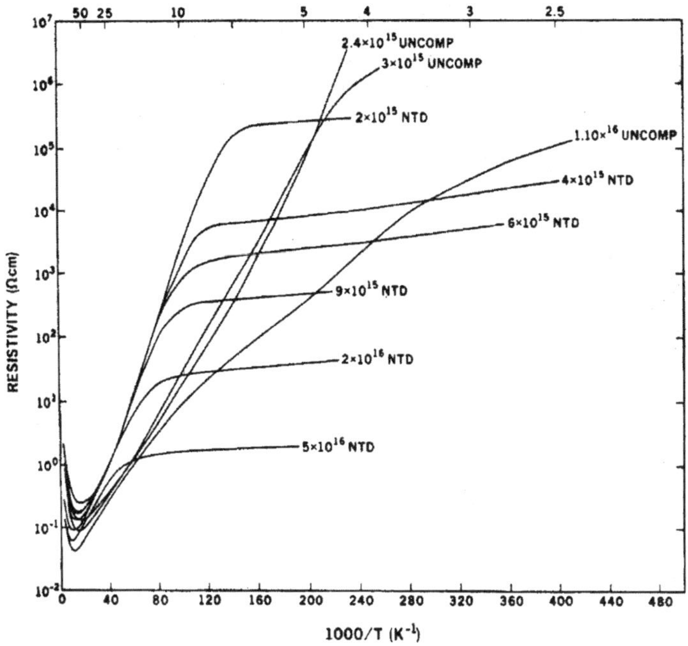

Fig. 6.23. Resistivity as a function of $\frac{1000}{T}$ for NTD and uncompensated germanium samples. Each curve is labeled by the gallium concentration obtained by either NTD or melt doping. (After [455])

where p is the free hole concentration, e is the charge of the electron, and R_{H} is the magnitude from the Hall measurement.

The mobility values are only useful down to the temperature where hopping conduction sets in. The mobility values agree well with published values for melt-doped material in the temperature range above the hopping regime. This indicates that the concentration of residual radiation damage or other free-carrier scattering centers must be very small. Figure 6.23 shows the log (resistivity) versus $\frac{1000}{T}$ dependence for six NTD germanium samples. The number next to each curve corresponds to the acceptor (gallium) concentration in each sample. For comparison, Haller et al. also measured gallium-doped germanium samples that have extremely small values of K. These so-called uncompensated samples were cut from crystals that were doped in the melt, grown in the ultrapure germanium crystal-growing equipment, and not NTD doped. The compensating donor concentration in these crystals is estimated at less than 10^{11} to $10^{12}\,\mathrm{cm}^{-3}$. The resulting K is of the order of 10^{-4} to 10^{-5}. The resistivity-temperature dependence of these NTD samples is characterized by three regimes. At high temperatures (room temperature down to about 50 K), the resistivity decreases because the carrier mobility increases. Below about 50 K, carrier freeze-out begins and reduces the free hole concentration rapidly. The slope of the freeze-out in highly compensated material is proportional to the acceptor binding energy $E_A - E_V \simeq 11\,\text{meV}$. At still lower temperatures, the appearance of hopping conduction causes the resistivity to increase only very slowly. All six NTD germanium samples show these three resistivity regimes very clearly. The low-compensated samples show different $\log(\rho)$ versus $\frac{1}{T}$ dependences. A third conduction mechanism has been proposed for such material [466]. It is based on the idea that carriers can "hop" from a neutral to a neighboring neutral acceptor thereby forming a positively charged acceptor. The NTD process in high-purity germanium leads to a fixed compensation which in turn results in a certain slope of the $\log(\rho)$ versus $\frac{1}{T}$ dependence for a given neutron exposure.

Metal–Insulator Transition

In the next part of this paragraph, we briefly discuss the metal–insulator transition (MIT) [456, 464, 467, 468] in transmuted Ge. In the literature, there is an intensive debate whether MIT is a phase transition of first or second order and what the experimental conditions are to obtain it at finite temperatures and in a real (disordered) system (see, e.g., [457, 458, 460, 469]). If the MIT is a second-order phase transition, a further challenge is the solution of the so-called puzzle of the critical index, μ, for the scaling behavior of the metallic conductivity near the MIT, i.e., just above the critical impurity concentration N_c and a small compensation, K. According to the scaling theory of the MIT for doped semiconductors [468], the conductivity at zero temperature $\sigma(0) = \sigma(T \to 0)$, when plotted as a function of impurity concentration N, is equal to zero on the insulating side of the MIT and remains finite on the metallic side, obeying a power law in the vicinity of the transition,

$$\sigma(0) \propto \left[\left(\frac{N}{N_c}\right) - 1\right]^{\mu}, \tag{6.35}$$

where N_c is the critical impurity concentration of the given system and μ is the critical conductivity exponent. The value of μ, determined experimentally, is compared with theoretical predictions. Until now, $\mu \approx 0.5$ has been obtained with nominally uncompensated semiconductors (Si : P [470], Si : As [471, 472], Ge : As [473], Si : B [474]), whereas $\mu \approx 1$ has been found with compensated semiconductors (Ge : Sb [475], Si : P, B [476], Ge : Ga, As [477]), and amorphous alloys [478–480]. Exceptions are uncompensated Ge : Sb with $\mu \approx 1$ [456] and Ga_xAr_{1-x} amorphous alloys with $\mu \approx 0.5$ [481]. As was shown in [457], the value $\mu \approx 0.5$ obtained with simple systems like uncompensated semiconductors turns out to be inconsistent with the theoretical prediction [80, 467, 468]. In his original theory, Mott considered only the electron–electron (e^-–e^-) interaction (Mott transition) and predicted a discontinuous transition of $\sigma(0)$ at N_c [483]. Although there is much evidence for the importance of e^-–e^- interactions, no experimental observation of such an abrupt transition has been reported. Anderson's ideas of the MIT is based solely on the disordered potential arising from randomly distributed dopants (Anderson transition) [484]. This led to the development of the well-known "scaling theory" which predicted $\mu \approx 1$ for three-dimensional systems (see also [468] and references therein). More recently, higher order calculations of the scaling theory (exclusively with disorder and no interactions) predict $\mu \approx 1.3$ [485], and more importantly, this value, it is shown, is independent of time reversal invariance [486] and of the strength of spin–orbit interactions [487] (see also [458]). Therefore, it is clear that the effect of disorder alone cannot explain the experimental results of $\mu \approx 0.5$ or 1. Chayes et al. combined the theories of Mott and Anderson and successfully set the lowest limit of $\mu > \frac{2}{3}$ [488]. This result permits $\mu \approx 1$ obtained with compensated semiconductors and amorphous alloys. However, there still is no theory that convincingly explains $\mu \approx 0.5$ found for uncompensated semiconductors.

Even with today's advanced semiconductor technology, melt-doping of bulk semiconductors always leads to inhomogeneous dopant distributions due to impurity segregation and striation during crystal growth [469]. In papers [457, 469] these difficulties have been overcome by applying the NTD technique to chemically pure, isotopically enriched ^{70}Ge [457] and ^{74}Ge [469] crystals. The ^{70}Ge crystal of isotopic composition $\left[^{70}\mathrm{Ge}\right] = 96.2\,\mathrm{atom\%}$ and $\left[^{72}\mathrm{Ge}\right] = 3.8\,\mathrm{atom\%}$ was grown [457] using the Czochralski method developed for ultrapure Ge [465]. The as-grown crystal free of dislocations, was p-type with an electrically active net-impurity concentration less than $5 \cdot 10^{11}\,\mathrm{cm}^{-3}$. In paper [469], isotopically engineered germanium was used which was grown from pure ^{74}Ge, enriched to 94%, or by adding a controlled portion of Ge with a natural isotopic content to the ^{74}Ge material. In this way, both the doping as well as the compensation homogeneous due to the NTD and the compensation by controlled mixtures of ^{74}Ge and ^{70}Ge which transmute to ^{75}As donors and ^{71}Ga acceptors. Four series of n-type NTD Ge with different K were grown [469]. The values of K are proportional to the product of the isotopic abundance and the thermal neutron cross section of all isotopes producing impurities (see above: $K = \frac{N_{\mathrm{Ga}}}{N_{\mathrm{As}}+N_{\mathrm{Se}}}$), whereas the impurity concentration is additionally proportional to the irradiation dose. A very small fraction of ^{72}Ge becomes ^{73}Ge which is stable, i.e., no other acceptors or donors are introduced. NTD is known [395, 489] to produce the most homogeneous, perfectly random dopant distribution down to the atomic level. Figure 6.24 shows the temperature dependence of the resistivities (ρ) of 14 insulating samples in the range $N = 0.16 - 0.99\,N_{\mathrm{c}}$ for NTD ^{70}Ge : Ga crystals. The analogous picture for NTD ^{74}Ge : Ga is shown in Fig. 6.25. All curves become linear only when $\ln \rho$ is plotted against $T^{-\frac{1}{2}}$ in good agreement with the theory of variable range hopping conduction for strongly interacting electrons [464]:

$$\rho = \rho_0 \exp\left(\frac{T_0}{T}\right)^{\frac{1}{2}}, \tag{6.36}$$

where ρ_0 is a prefactor and T_0 is given by

$$T_0 \approx \frac{2.8\, e^2}{k(N)\xi(N)}, \tag{6.37}$$

where $k(N)$ and $\xi(N)$ are the dielectric constant and localization length depending on N, respectively. Moreover, $k(N) \propto \left[\frac{N_c}{(N_c - N)}\right]^s$ and $\xi(N) \propto \left[\frac{N_c}{(N_c - N)}\right]^\zeta$ as N approaches N_c from the insulating side, so that T_0 becomes [464]

$$T_0 = A\left[(N_c - N)\, N_c\right]^\alpha . \tag{6.38}$$

Here, $\alpha = +\zeta$ is determined experimentally [457].

Figure 6.26 shows the dependence of T_0 as a function of $N_d = \frac{n}{(1-K)}$ for different K of ^{74}Ge : Ga [469]. A earlier these authors used the intersection point of these dependencies with the X-axis as a tool for the determination of $N_c(K)$. The left half of Fig. 6.27 shows the experimentally determined T_0 versus [Ga] (*filled diamonds*) together with the result of a three-parameter fitting using A, N_c, and α as variables in (6.38) (*solid curve*) [457]. These authors deduced [Ga] for samples using the following equation: $[^{71}\text{Ga}]\ (\text{cm}^{-3}) = 0.1155 \cdot n\,(\text{cm}^{-2})$, since the precise neutron flow used in each irradiation was known. The best fit of T_0 with (6.38) was obtained with the values $\alpha = 1.03 \pm 0.038$ and $N_c = (1.855 \pm 0.012) \cdot 10^{17}\,\text{cm}^{-3}$.

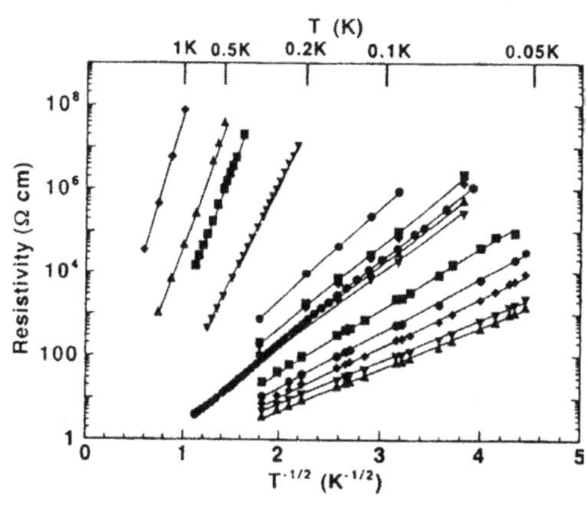

Fig. 6.24. The logarithm of resistivity plotted as a function of $T^{-\frac{1}{2}}$ for 14 insulating NTD ^{70}Ge : Ga samples. The Gallium concentrations from top to bottom in units of $10^{16}\,\text{cm}^{-3}$ are 3.02, 8.00, 9.36, 14.50, 17.17, 17.52, 17.61, 17.68, 17.70, 17.79, 17.96, 18.05, 18.23, and 18.40. (After [457])

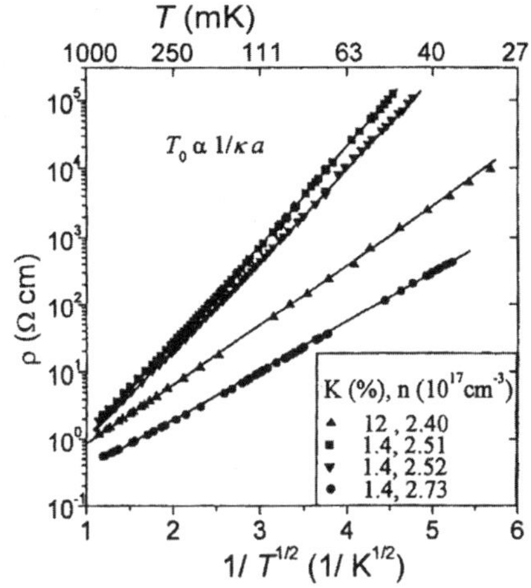

Fig. 6.25. Typical temperature dependences of resistivity as a function of $T^{-\frac{1}{2}}$ for four samples NTD ^{70}Ge : Ga crystals (after [469])

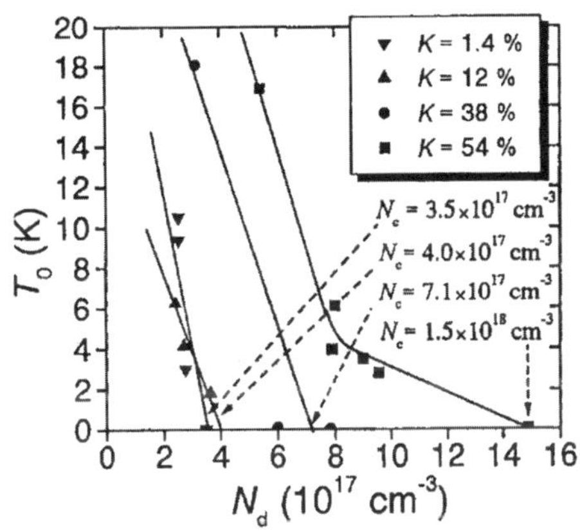

Fig. 6.26. Determination of N_c from the extrapolation of $T_0 \to 0$ in the range $T_0 > T$ (after [469])

A much larger value of $\alpha \approx 2$ has been reported for Ge : As using only three samples: the highest N is far from the transition $0.56\,N_c$ [490]. In paper [457], $\alpha = 1$ obtained with 14 homogeneously doped samples of [Ga] $= 0.16 - 0.99\,N_c$, all demonstrating the lh $\rho \propto T^{-\frac{1}{2}}$ dependence, i.e., this data set should be considered the first reliable determination of the hopping conductivity exponent α for a particular semiconductor system.

Figure 6.28 shows the conductivity σ according to the results of paper [457] for 10 metallic samples plotted against $T^{\frac{1}{2}}$. Extrapolation of each curve to $T = 0\,$K, i.e., the determination of the zero-temperature conductivity $\sigma(0)$, yields a very small error since the dependence of σ on T for all samples

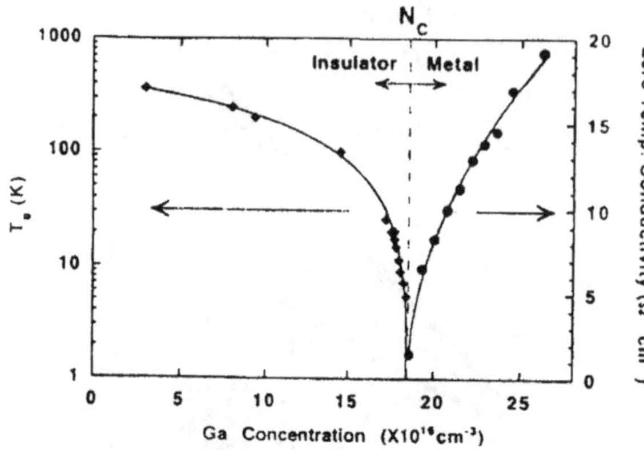

Fig. 6.27. The **left side** shows (^{70}Ge) as a function of Ga concentration (◇). The *solid curve* is the best fit obtained with (6.38) (with $\alpha \approx 1$). The **right side** shows the zero-temperature conductivity σ(○) obtained from the extrapolation in Fig. 6.28 for the metallic samples as a function of Ga concentration (•). The *solid curve* is the best fit obtained with (6.35). (After [457])

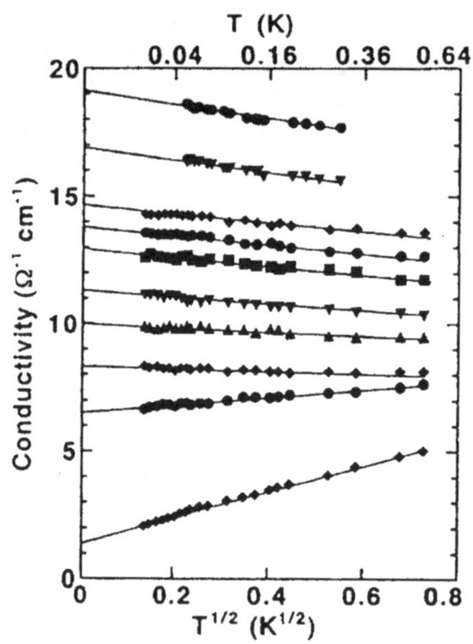

Fig. 6.28. Conductivity plotted as a function of $T^{\frac{1}{2}}$ for 10 metallic NTD ^{70}Ge : Ga samples. *Solid lines* indicate extrapolation to $T = 0\,\mathrm{K}$. Gallium concentrations from top to bottom in units of $10^{16}\,\mathrm{cm}^{-3}$ are 18.61, 19.33, 20.04, 20.76, 21.47, 22.19, 22.90, 23.62, 24.50, and 26.25. (After [457])

is very weak. The right half of Fig. 6.27 shows $\sigma(0)$ as a function of [Ga] (*filled circles*) together with a fit obtained by the scaling expression (6.35) (*solid curve*). The values of the parameters determined in paper [457] from this fit are $\mu = 0.502 \pm 0.025$ and $N_c = (1.856 \pm 0.003) \cdot 10^{17}\,\mathrm{cm}^{-3}$. This value of $\mu \approx 0.5$ is for uncompensated Ge : Ga semiconductors with high confidence, since the two values of N_c obtained from the scaling of T_0 (6.38) and $\sigma(0)$ (6.35) agree perfectly (for details, see also [460, 469]).

Neutral-Impurity Scattering

The low-temperature mobility of free carriers in semiconductors is mainly determined by ionized- and neutral-impurity scattering. The ionized-impurity scattering mechanism has been extensively studied (see e.g., [491] and references therein), and various aspects of this process are now quite well understood. Scattering by neutral impurities (see also [423]) is much less than by ionized centers, i.e., its contribution is significant only in crystals with low compensation and at very low temperatures where most of the free carriers are frozen at the impurity sites. The availability of highly enriched isotopes of Ge which can be purified to residual dopant levels $< 10^{12}\,\mathrm{cm}^{-3}$ has provided the first opportunity to measure neutral-impurity scattering over a wide temperature range. Three Ge isotopes transmute into shallow acceptors (Ga), shallow donors (As), and double donors (Se) (see also above):

$$\begin{aligned}
&{}^{70}_{32}\mathrm{Ge} + n \rightarrow {}^{71}_{32}\mathrm{Ge}_{\mathrm{EC}(t_{1/2}=11.2\,\mathrm{days})} \rightarrow {}^{71}_{32}\,\mathrm{Ga} + \nu_e,\\
&{}^{74}_{32}\mathrm{Ge} + n \rightarrow {}^{75}_{32}\mathrm{Ge}_{\beta^-(t_{1/2}=82.2\,\mathrm{min})} \rightarrow {}^{75}_{32}\,\mathrm{As} + \beta^- + \bar{\nu}_e,\\
&{}^{76}_{32}\mathrm{Ge} + n \rightarrow {}^{77}_{32}\mathrm{Ge}_{\beta^-(t_{1/2}=11.3\,\mathrm{h})} \rightarrow \beta^- + \bar{\nu}_e + {}^{77}_{32}\,\mathrm{As}_{\beta^-(t_{1/2}=38.8\,\mathrm{h})}\\
&\qquad \rightarrow {}^{77}_{32}\mathrm{Se} + \beta^- + \bar{\nu}_e.
\end{aligned} \tag{6.39}$$

The isotopes ^{72}Ge and ^{73}Ge are transmuted into the stable ^{73}Ge and ^{74}Ge, respectively. Controlling the ratio of ^{70}Ge and ^{74}Ge in bulk Ge crystals allows fine-tuning the majority – as well as the minority carrier concentration. Currently, this is the best method for varying the free-carrier concentration independently of the compensation ratio. As opposed to other doping methods, NTD yields a homogeneous, perfectly random distribution of dopants down to atomic levels [492]. Thus, isotopically controlled crystals offer a unique possibility of studying systematically the scattering mechanism of charge carriers in semiconductors. Extensive Hall-effect and resistivity measurements from room temperature down to 4.2 K yielded very accurate free-carrier concentrations and mobilities as a function of temperature, and doping levels were done in papers [354, 493, 494]. Itoh et al. performed temperature-dependent Hall measurements on four different p-type and two different n-type Ge crystals. The n-type crystals were obtained through NTD of isotopically enriched ^{74}Ge and the p-type crystals correspondingly from NTD of isotopically enriched ^{70}Ge. The neutron cross sections for the neutron capture of the isotope

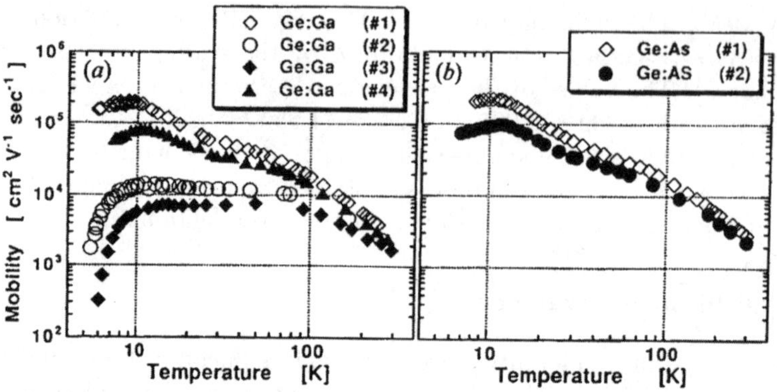

Fig. 6.29. Temperature dependence of the carrier mobility of **(a)** p-type and **(b)** n-type NTD Ge crystals (after [354])

Table 6.7. Carrier concentration of the Ge crystals used in the work of Itoh et al. [354]

p-type		$N_A - N_D$	N_D	$K = \frac{N_D}{N_A}$
	Ge : Ga #1	$3.1 \cdot 10^{14}$	$3 \cdot 10^{12}$	$9 \cdot 10^{-3}$
	Ge : Ga #2	$7.7 \cdot 10^{15}$	$9 \cdot 10^{13}$	$1.2 \cdot 10^{-2}$
	Ge : Ga #3	$1.7 \cdot 10^{16}$	$2 \cdot 10^{14}$	$1.2 \cdot 10^{-2}$
	Ge : Ga #4	$1.0 \cdot 10^{15}$	$1.2 \cdot 10^{13}$	$1.2 \cdot 10^{-2}$
n-type		$N_D - N_A$	N_A	$K = \frac{N_A}{N_D}$
	Ge : As #1	$3.5 \cdot 10^{14}$	$8.5 \cdot 10^{12}$	$2.4 \cdot 10^{-2}$
	Ge : As #2	$1.2 \cdot 10^{15}$	$1.2 \cdot 10^{13}$	$1.0 \cdot 10^{-2}$

for these irradiations were determined as $\sigma_c(^{70}\mathrm{Ge}) = 2.5(5) \cdot 10^{-24}\,\mathrm{cm}^2$ and $\sigma_c(^{74}\mathrm{Ge}) = 0.6(1) \cdot 10^{-24}\,\mathrm{cm}^2$ by Itoh et al. [495]. To remove structural defects due to unintentional irradiation with fast neutrons, all samples had to be thermally annealed at 650°C for 10 s in a rapid thermal annealer. The Hall mobility obtained from the conductivity and free-carrier concentration data (listed in Table 6.7) are displayed in Fig. 6.29. A magnetic field of 3 kG was used, that is, for the temperature range of interest for neutral impurity scattering, the high-field limit $\mu B \gg 1$ is satisfied and the Hall mobility can be equated to the drift mobility.

Itoh et al. [354] analyzed the mobility data of Fig. 6.29 in terms of scattering of the carriers from phonons (μ_1), ionized impurities (μ_i), and neutral impurities (μ_n) assuming

$$\frac{1}{\mu} = \frac{1}{\mu_1} + \frac{1}{\mu_i} + \frac{1}{\mu_n}. \tag{6.40}$$

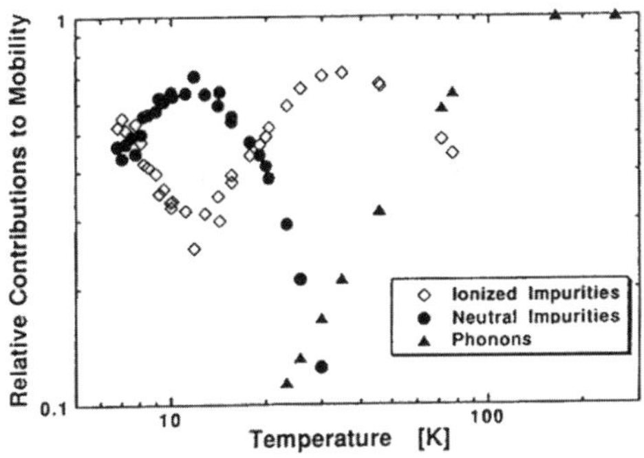

Fig. 6.30. Temperature dependence of the relative contributions to mobility. Note that mobility is dominated by neutral impurity scattering below 20 K (^{70}Ge : Ga #2 crystal). (After [354])

To extract the neutral impurity scattering contribution, they subtracted $\frac{1}{\mu_1}+\frac{1}{\mu_i}$ from the measured $\frac{1}{\mu}$. The relative contributions of phonon scattering $\left(\frac{1}{\mu_1}\right)$, ionized impurity scattering $\left(\frac{1}{\mu_i}\right)$, and the resulting neutral impurity scattering $\left(\frac{1}{\mu}-\frac{1}{\mu_1}-\frac{1}{\mu_i}\right)$ are plotted in Fig. 6.30 (data Ge : Ga #2). For $T > 80$ K, phonon scattering is the dominant scattering mechanism. On comparison of Figs. 6.29 and 6.30, it becomes clear that the "dip" in the carrier mobility around 50 K is caused by scattering from ionized impurities that dominate the scattering of carriers between 20 and 80 K. The flattering and saturation of the mobilities below 20 K originate from neutral impurity scattering, which can only be observed in crystals with very high crystalline quality and low compensation like isotopically enriched NTD Ge crystals used in paper [493].

The experimental data obtained in [493] allow these authors quantitative comparison with theory. According to Erginsoy [496], the inverse relaxation time τ^{-1}, the scattering rate, for neutral-impurity scattering is

$$\tau^{-1}=\frac{20kN_{\mathrm{N}}\hbar^{3}}{m^{*2}e^{2}}, \tag{6.41}$$

where k is the dielectric constant, e is the electron charge, N_{N} is the neutral-impurity concentration, and m^* is the electron effective mass. Equation (6.35) can be considered only a first-order approximation because the prefactor 20 is an empirically determined constant and only the lowest partial wave is taken into account in the phase-shift calculation (see also [497,498]). McGill and Baron [499] used the following equation for $\tau_{\mathrm{neutral}}^{-1}$:

$$\tau_{\mathrm{neutral}}^{-1}=\frac{4\pi N_{\mathrm{N}}\hbar e}{2km^{*}E_{\mathrm{B}}}\sum_{l=0}^{\infty}\frac{(l+1)}{4w^{\frac{1}{2}}}\left[3\sin^{2}\left(\delta_{\mathrm{l}}^{-}-\delta_{\mathrm{l+1}}^{-}\right)+\sin^{2}\left(\delta_{\mathrm{l}}^{+}-\delta_{\mathrm{l+1}}^{+}\right)\right], \tag{6.42}$$

where E_B is the binding energy of the scattering centers, $w \equiv \frac{E}{E_B}$ and E is the incident electron energy, and δ_l^+ and δ_l^- are the lth partial shift for the singlet and triplet states, respectively. The authors [499] graphically showed the accurate $\tau_{\text{neutral}}^{-1}$ as a function of w for neutral-impurity scattering in semiconductors. This result has been considered an appropriate model for neutral-impurity scattering in semiconductors and has been discussed in detail in many standard textbooks (see, e.g., [500]).

Meyer and Bartoli [501] reevaluated this task and obtained an analytical expression that is essentially the same as the graphical solution of the authors of paper [499] but covers a wider incident-electron energy range:

$$\tau_{\text{neutral}}^{-1} = \frac{A(w)kN_{\text{N}}\hbar^3}{m_{\text{H}}^{*\,2}e^2}, \tag{6.43}$$

with

$$\begin{aligned} A(w) = \frac{35.2}{w^{\frac{1}{2}}} & \frac{\left(1+e^{-50\,\text{W}}\right)\left(1+80.6\,\text{W}+23.7w^2\right)}{\left(1+41.3w+133w^2\right)} \\ & \times \left[\frac{1}{w}\ln\left(1+w\right) - \frac{\left(1+0.5w-1.7w^2\right)}{\left(1+w\right)^3}\right]. \end{aligned} \tag{6.44a}$$

Here m_{H}^* is the hydrogenic effective mass given by

$$m_{\text{H}}^* = \frac{E_{\text{B}}k^2m_0}{E_{\text{H}}}. \tag{6.44b}$$

In 6.44b m_0 is the electron rest mass, and $E_{\text{H}} = 13.6\,\text{eV}$ is the binding energy of hydrogen. In the total-mobility calculation; Itoh et al. [493] employed a standard relaxation-time approximation. This approach is valid because it is limited to low temperatures ($T < 25\,\text{K}$), where the inelastic optical–phonon deformation-potential scattering is negligible. Three scattering mechanisms are considered: neutral-impurity, ionized-impurity, and acoustic-phonon deformation-potential scattering. The neutral-impurity scattering contribution was calculated using both (6.41) and (6.43), so the models of Erginsoy and Meyer and Bartoli can be compared with the experimental results of paper [493]. The concentration of neutral-impurity centers as a function of temperature $N_{\text{N}}(T)$ in each sample is given by

$$N_{\text{N}}(T) = N_{\text{MJ}} - N_{\text{MN}} - n(T). \tag{6.45}$$

Here N_{MJ}, N_{MN} and $n(T)$ are the majority-impurity, minority-impurity, and free-carrier concentrations, respectively. For ionized-impurity scattering, Itoh et al. employed the Brooks–Herring expression [501, 502]:

$$\tau_{\text{ion}}^{-1} = \frac{\pi N_{\text{I}}e^4\left(k_{\text{B}}T\right)^{-\frac{3}{2}}x^{-\frac{3}{2}}}{\left(2m^*{}_{\text{con}}\right)^{\frac{1}{2}}k^{\frac{1}{2}}}\left[\ln\left[1+\frac{4x}{a}\right] - \frac{\frac{4x}{a}}{1+\frac{4x}{a}}\right], \tag{6.46}$$

where

$$a = \frac{2\pi\hbar^2 e^2 n}{m^* k k_B^2 T^2}, \tag{6.47}$$

$x = \frac{E}{k_B T}$ (E is the incident electron energy), m^*_{con} is the average conductivity effective mass, and N_I is the ionized-impurity concentration. The temperature-dependent N_I in each sample is given by

$$N_I(T) = n(T) + 2N_{MN}. \tag{6.48}$$

For acoustic-phonon deformation-potential scattering [503]

$$\tau_{ac}^{-1} = B_{ac}\,(m^*_{con}T)^{\frac{3}{2}}\,x^{\frac{1}{2}}, \tag{6.49}$$

where the constant B_{ac} has well-established values for n- and p-type Ge, as shown in Table 6.8. Having found τ^{-1} of all three scattering mechanisms, Itoh et al. calculated an average $\langle\tau\rangle$ using the Maxwell–Boltzman integration:

$$\langle\tau\rangle = \frac{4}{3\sqrt{\pi}} \int_0^\infty \frac{x^{\frac{3}{2}}\exp(-x)}{\tau_{ac}^{-1} + \tau_{ion}^{-1} + \tau_{neutral}^{-1}}\mathrm{d}x. \tag{6.50}$$

Finally, the total mobility μ_{tot} was given by

$$\mu_{tot} = \frac{e\,\langle\tau\rangle}{m^*_{con}}. \tag{6.51}$$

All parameters required for the mobility calculations are well known for Ge (see Table 6.8). The only unknown material parameters at this point are sample-dependent N_{MJ}, N_{MN} and $n(T)$ in (6.45) and (6.48). All three parameters, as will shown below, can be determined precisely for each sample

Table 6.8. Parameters used in the total-mobility calculations (after [493])

	Ga:As (n-type)	Ge:HGa (p-type)
k	16	16
m^*_{con}	$0.12\,m_0$	$0.28\,m_0$
B_{ac} [a]	$1.08 \cdot 10^{10}\,g^{\frac{3}{2}}\,K^{-\frac{3}{2}}$	$9.50 \cdot 10^{8}\,g^{\frac{3}{2}}\,K^{-\frac{3}{2}}$
B_B (theoretical)	12.5 meB [b]	11.2 meV[c]

[a] The values of B are determined experimentally using ultrapure n- and p-type Ge of $N_{MJ} \sim N_{MN} \sim 3 \cdot 10^{11}\,\mathrm{cm}^{-3}$

[b] Altarelli, M., Hsu, W.Y., Sabatini, R.A., J. Phys. C **10**, L605 (1977)

[c] Baldareshi, A., Lipari, N.O., in Proc. 13th Int. Conf. Phys. Semicond. (Fumi, F.G. (ed.), North-Holland, Amsterdam, 1976) p. 595

by performing variable-temperature Hall-effect measurements. Consequently, all mobility calculations are performed without any adjustable or scaling parameters. The experimental curves are fitted with the following standard semiconductor statistics [504], which describes the temperature dependence of the free-carrier concentration in semiconductors doped by shallow majority impurities N_{MJ} and compensated by minority impurities N_{MN}:

$$n(T) = \frac{2\left(N_{\mathrm{MJ}} - N_{\mathrm{MN}}\right)}{\left\{\left[1 + \left(\frac{N_{\mathrm{MN}}}{gN_{\mathrm{B}}}\right)^{\left(\frac{E_{\mathrm{MJ}}}{k_{\mathrm{B}}T}\right)}\right]\right.} + \frac{2\left(N_{\mathrm{MJ}} - N_{\mathrm{MN}}\right)}{\left.\sqrt{\left[1 + \left(N_{\mathrm{MN}} g N_{\mathrm{B}}\right)^{\left(\frac{E_{\mathrm{MJ}}}{k_{\mathrm{B}}T}\right)}\right]^2 + \left(\frac{4}{gN_{\mathrm{B}}}\right)\left(N_{\mathrm{MJ}} - N_{\mathrm{MN}}\right)^{\left(\frac{E_{\mathrm{MJ}}}{k_{\mathrm{B}}T}\right)}}\right\}}, \tag{6.52}$$

where $g = \frac{1}{2}$ ($g = 4$) is the spin degeneracy for a donor (acceptor), N_{B} is the effective conduction- (valence-valence density of states) band density of states, and E_{MJ} are the experimentally determined ionization energies: 14 and 11.07 meV for As and Ga, respectively (Table 6.8).

Figure 6.30 shows the relative strength of scattering from ionized and neutral impurities. There is only a relatively small temperature region in which the scattering from neutral impurities dominates. This range extends to higher temperatures as the free-carrier concentration is increased. The calculated "transition temperatures" above which the ionized impurities are the main scattering centers (see also [505]) compare very well with the experimental results of Itoh et al. [493] (see also Fig. 6.31).

We now turn attention to the low-temperature regime where mobilities are dominated by neutral-impurity scattering. Figure 6.31 directly compares the experimental results with theoretical total-mobility curves calculated. For each sample, two theoretical total-mobility curves are calculated: one uses Erginsoy's model (6.35), and the other uses the Meyer and Bartoli model (6.37). Strikingly good agreement was obtained between the experimental and theoretical mobilities calculated with the Meyer and Bartoli model for all samples (see Fig. 6.31).

To demonstrate the importance of homogeneous dopant distribution, Itoh et al. performed the same study on samples cut from Ge, Ga crystals grown by the conventional Czochralski method, where Ga impurities were introduced into the Ge melt during crystal growth. These authors observed deviations of measured mobility from the theoretical calculations, which are most likely due to inhomogeneous Ga impurity distributions in melt-doped Ge. Only the use of NTD semiconductors with randomly distributed dopants allows an accurate test of neutral-impurity scattering models.

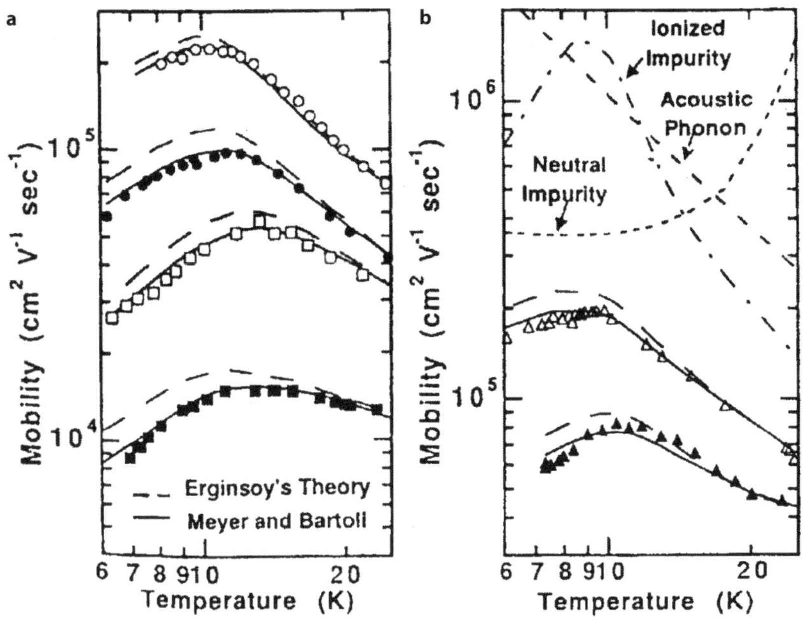

Fig. 6.31. Data points represent experimentally measured carrier mobility in **(a)** four ^{74}Ge : As and **(b)** two ^{70}Ge : Ga samples. A direct comparison of the theoretically calculated mobility using Erginsoy's model (*broken line*) and Meyer and Bartoli (*solid line*) is shown for each sample. The contributions of the different scattering mechanisms to the total mobility of the ^{70}Ge : Ga – one sample – are shown in the upper half of **(b)**. (After [493])

6.8.2 Silicon

It is well known that doping silicon single crystals by incorporating impurities from the melt during solidification in most cases leads to an inhomogeneous distribution of impurities in the solids [401, 482]. This is due to the fact that nearly all impurities in silicon have thermal equilibrium distribution coefficients much less than unity and that the solidification or crystal grown at each position of the interface is characterized by a different state of thermal inequilibrium leading to distribution coefficients that continuously change in space and time and result in nonuniform impurity distribution [408, 506]. In actual crystal production, the nonuniformity is further increased by lack of control of exactly constant melt volume and feed of the doping impurity. The most widely used elements in doping silicon are boron and phosphorus. Boron has a distribution coefficient between 0.9 and 1 which makes a doping uniformity of ±10% easily obtainable (see, e.g., [482]). The thermal equilibrium distribution coefficient of approximately 0.3 for phosphorus leads in general to the previously mentioned large doping variations on both a macroscale (center to periphery) and a microscale (striations). No other n-type doping element has a larger distribution coefficient. Because fast diffusing p-type dopants (Ga, Al) are available, because electron mobility is greater than hole mobility, and because contact alloying technology is reasonable, n-type silicon is generally used for solid-state power devices [408,507]. With avalanche breakdown voltages determined from areas with lower resistivities, use of a conventionally doped material results in hot-spot formation prior to breakdown and too high a forward voltage drop that leads to excessive heat dissipation because of a safe punch-through design [401,482,508].

Phosphorus doping of NTD was suggested by Lark-Horovitz [426] and Tanenbaum and Mills [400] for homogeneity and has been applied to high-power thyristor manufacturing in [401,482,507]. Hill et al. [507] demonstrated how such homogeneous phosphorus doping may result in a "theoretical design" possibility for high-power components (see also below).

The process used for fractional transmutation of silicon into phosphorus and thereby performing n-type doping

$$ {}^{30}_{14}\mathrm{Si}(n,\gamma){}^{31}_{14}\mathrm{Si}^{\beta^-}_{2.62\,\mathrm{h}} \rightarrow {}^{31}_{15}\,\mathrm{P} \tag{6.53} $$

was first pointed out by Lark-Horovitz in 1951 [426]. Apart from special applications [509] and research, the above process was, however, not used to any extent until the early 1970s, at which time manufacturers of high-power thyristors and rectifiers for high-voltage direct current transmission lines, in particular, initiated usage of the transmutative doping process [439,507,510]. The reasons for not using the neutron doping method throughout the 1960s may be found in the lack of a processing technology which could benefit from more uniform doping, insufficient availability of high resistivity starting material, and the lack of nuclear reactors with irradiation capacities in excess of that needed for testing fuel and materials for nuclear power stations.

Let us, for the following discussion, assume that completely uniform neutron doping may be accomplished. The homogeneity of the doped silicon is in this case determined by the background doping, i.e., the distribution of impurities in the starting material, where the net impurity concentration may be of either donor or acceptor type. Let us further, for simplicity, consider starting material of one conductivity type and assume complete n-type conduction after irradiation and annealing. With C_{S} the net impurity concentration of the starting material and C_{D} the resulting donor concentration after irradiation, for both n- and p-type material,

$$ C_{\mathrm{D}}^{\max} - C_{\mathrm{D}}^{\min} = C_{\mathrm{S}}^{\max} - C_{\mathrm{S}}^{\min} . \tag{6.54} $$

Then, we may define

1. the homogeneity factors for the starting material (α_{S}) and for the neutron doped material (α_{D}), respectively,

$$ \alpha_{\mathrm{S}} = \frac{C_{\mathrm{S}}^{\min}}{C_{\mathrm{S}}^{\max}} \tag{6.55} $$

Table 6.9. Values for homogeneity factor α_D as a function of homogeneity factor α_S of the starting material and doping factor f_D, as defined in text (after [439])

α_S / f_D	0.1	0.2	0.3	0.4	0.5	0.6	0.7	0.8	0.9
1	0.1	0.2	0.3	0.4	0.5	0.6	0.7	0.8	0.9
2	0.55	0.6	0.65	0.7	0.75	0.8	0.85	0.9	0.95
5	0.82	0.84	0.86	0.88	0.9	0.92	0.94	0.96	0.98
7	0.87	0.89	0.90	0.91	0.93	0.94	0.96	0.97	0.99
10	0.91	0.92	0.93	0.94	0.95	0.96	0.97	0.98	0.99
20	0.955	0.96	0.965	0.97	0.975	0.98	0.985	0.99	0.995
50	0.98	0.98	0.99	0.99	0.99	0.99	0.99	0.996	0.998
100	0.991	0.992	0.993	0.994	0.995	0.996	0.997	0.998	0.999

and

$$\alpha_D = \frac{C_D^{min}}{C_D^{max}} \tag{6.56}$$

and

2. the doping factor

$$f_D = \frac{C_D^{max}}{C_S^{max}}. \tag{6.57}$$

From this we easily derived

$$1 - \alpha_D = \frac{1-\alpha_S}{f_D}. \tag{6.58}$$

Table 6.9 summarizes values of α_D as a function of α_S and f_D. It is seen that to obtain neutron-doped silicon with, for instance, a homogeneity factor greater than 0.9, it is necessary to use a doping factor of at least 7 when starting from "undoped" n-type material in which the homogeneity factor is typically not greater than 0.3 when taking the microcavitations (striations) into account. Examples of such neutron-doped silicon are shown in Figs. 6.32 and 6.33. Note that in terms of resistivity, which is often used for impurity characterization, a doping factor f_D means use of starting material with minimum resistivity by a factor f_D or 2.8 f_D greater than the resistivity after neutron doping for n- and p-type starting material, respectively. The difference is that the electron mobility is 2.8 times greater than the hole mobility. In concluding this section, it should be generally noted that to make neutron-doped silicon with significantly more uniform resistivity than conventionally doped material, a doping factor f_D = 5 or more should be applied.

Following Janus and Malmros [439], let us consider further the theoretical case where a cylindrical silicon crystal is surrounded by a material with

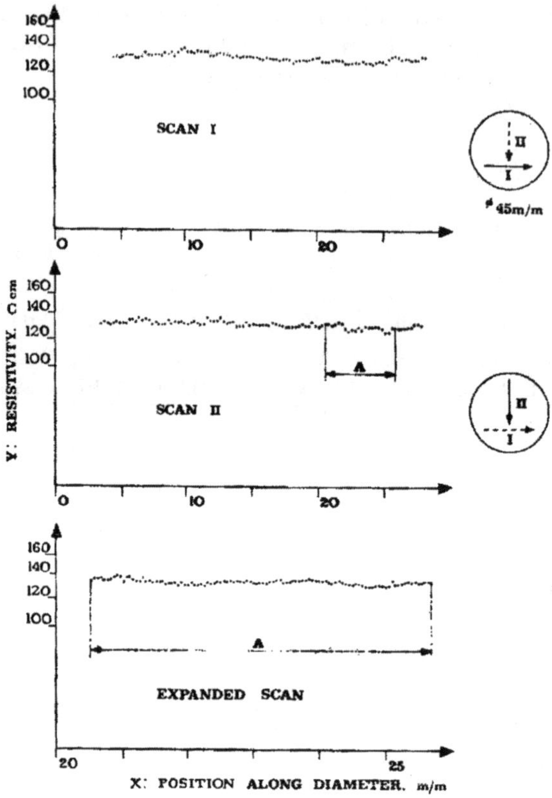

Fig. 6.32. Spreading resistance measurements of a thermal neutron irradiation doped silicon slice. The step-length on scan 1 and 2 is 250 μm, and on scan 3-step-length is 50 μm. The starting material selected was greater than 1500 Ω cm n-type. (After [439])

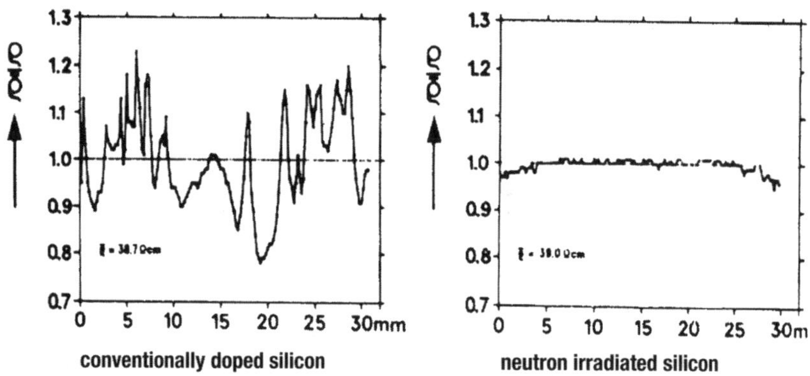

Fig. 6.33. Typical lateral microscopic resistivity distributions in conventionally doped silicon and in silicon doped by neutron irradiation (after 482])

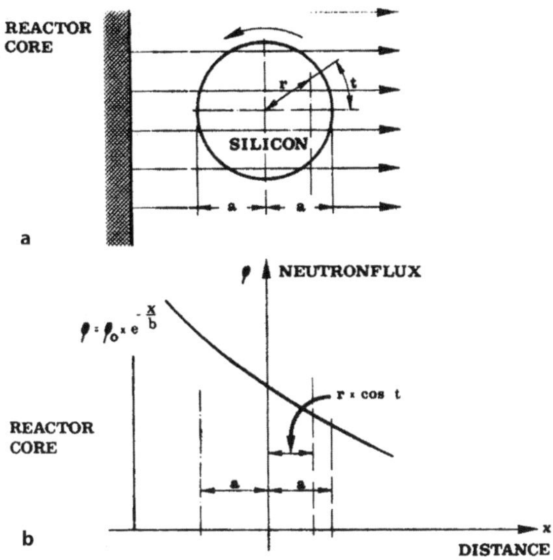

Fig. 6.34. Irradiation configuration. **(a)** Top view of the facility with the cylindrical crystal situated outside the core rotating around its cylindrical axis. Arrows indicate the overall direction of neutrons. The flux does not vary along the cylindrical axis. **(b)** The neutron flux is a function of the distance from the reactor core. (After [439])

the same neutron absorption and scattering efficiency as silicon itself (see Fig. 6.34). Let us furthermore assume a thermal neutron flux gradient along an X-axis perpendicular to the crystal axis with the neutrons coming from an external source. In this case, the neutron flux will have the form

$$\Phi = \Phi_0 \cdot \exp\left(-\frac{x}{b}\right), \tag{6.59}$$

where b, the decay length, may be obtained from the formula

$$b = (3 \cdot \sigma_{\mathrm{Si}} \cdot \sigma_{\mathrm{Si,t}} \cdot C_{\mathrm{Si}}^2)^{-0.5}. \tag{6.60}$$

$\sigma_{\mathrm{Si}} = 0.16 \cdot 10^{-24}\,\mathrm{cm}^2$ is the mean of the absorption cross sections for the three silicon isotopes, ^{28}Si, ^{29}Si, and ^{30}Si weighted with their abundances. $\sigma_{\mathrm{Si,t}} = 2.3 \cdot 10^{-24}\,\mathrm{cm}^2$ is the total cross section (absorption + scattering), and $C_{\mathrm{Si}} = 4.96 \cdot 10^{22}\,\mathrm{cm}^{-3}$ is the total number of silicon atoms in $1\,\mathrm{cm}^3$. Hence b may be calculated:

$$b_{\mathrm{silicon}} = 19\,\mathrm{cm}. \tag{6.61}$$

To improve the doping homogeneity in a cylindrical crystal, it is slowly rotated around its axis. The time average of this flux at a distance r from this axis is

$$\bar{\Phi} = \frac{1}{\pi}\int_0^{\pi} \Phi_0 \exp\left[-\frac{r}{b}\cos t\right] dt = \Phi_0 \left[1+\frac{1}{4}\left(\frac{r}{b}\right)^2 + \cdots\right]. \tag{6.62}$$

The ratio between the neutron dose at the periphery and at the axis of the crystal cylinder will then be

$$\frac{\bar{\Phi}\,(\mathrm{a})}{\bar{\Phi}\,(0)} \simeq 1 + \frac{1}{4}\left(\frac{a}{b}\right)^2, \tag{6.63}$$

where a is the crystal radius (Fig. 6.34).

As an intrinsic starting material, the irradiation-doped silicon will thus have a homogeneity factor of

$$\alpha_{\mathrm{D}} \simeq 1 - \frac{1}{4}\left(\frac{a}{b}\right)^2 \simeq 0.956 \tag{6.64}$$

for an 80-mm-diameter crystal, i.e., the absorption limiting factor for obtainable radial variations.

In the above analysis, we have neglected the effects of fast neutron moderation in silicon. By comparison, however, of irradiation in reactors with fast neutron fluxes from 10^{-4} to 1 times the thermal flux and with different flux gradients, the authors [439] have observed no influence on the resistive homogeneity due to fast neutron moderation in silicon.

In irradiated silicon crystals for semiconductor device applications, only two isotopes, ^{31}Si and ^{32}P, are important in connection with the radioactivity of neutron-doped material. For thermal neutron doses less than 10^{19} neutron/cm^2, no other elements have been detected that emit radiation. Futhermore, ^{31}Si that has a half-life of 2.62 h decays to an undetectable level in 3–5 days. For this reason, only the radioactivity of the ^{32}P isotope will be discussed. Figure 6.35 pictures ^{32}P activity as a function of final resistivity for a variety of thermal neutron flux levels typical of the nuclear test reactors in use. As shown in [439], absolute flux determination to 1% accuracy has proven obtainable, for instance, by calorimetric boron carbide monitors:

$$^{31}_{15}\mathrm{P}(n,\gamma)^{32}_{15}\mathrm{P}^{\beta^-}_{14.3\mathrm{d}} \rightarrow^{32}_{16}\mathrm{S} \tag{6.65}$$

as a secondary one with a ^{31}P concentration at each instant that is dependent on the neutron dose received and the time allowed for the $^{31}_{14}\mathrm{Si}^{\beta^-}_{2.62\,\mathrm{h}} \rightarrow^{32}_{15}\mathrm{P}$ decay.

From Fig. 6.35, it may be observed that neutron doping below 5 Ωcm can be performed only when accepting cool-down periods corresponding to the ^{32}P half-life of 14.3 days. The exempt limit for inactivity of $2 \cdot 10^{-3}$ µCi/g shown on the figure is representative of most European countries and is the value recommended by the International Atomic Energy Agency (IAEA), Vienna, Austria [511]. It should be added that careful cleaning of the silicon prior to insertion in a nuclear reactor is vital to avoid radioactive surface contamination. For the safety of personnel and end product users, a double

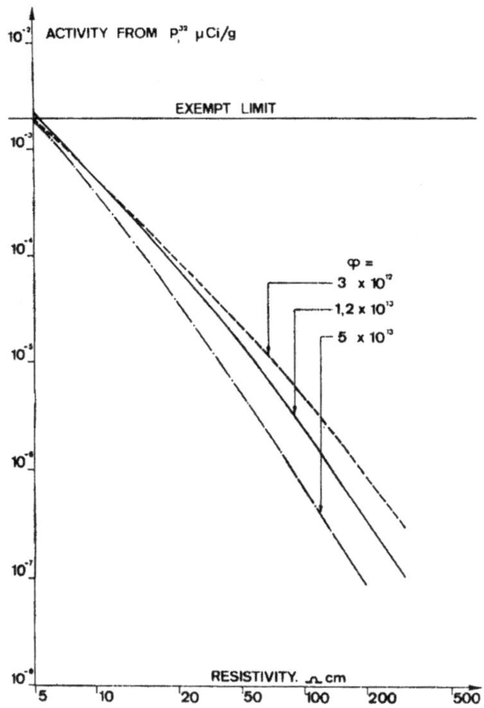

Fig. 6.35. The radioactivity of the ^{32}P isotope in silicon after 4 days of cool-down subsequent to irradiation. It may be observed that the activity as a function of the resistivity obtained depends on the neutron flux used. (After [439])

check upon shipping from the reactor sites and upon reception in the silicon plant, respectively, is carried out to ensure that only inactive material (below the exempt limit) is being further processed after neutron doping. In general, this implies shipment from the reactor not earlier than 4 days after irradiation.

The use of NTD is of particular interest to thyristor manufacturers where n-type starting material is required for the basic p-n-p structure [506, 507]. Some advantages for high power device design and performance include

1. more precise control of avalanche breakdown voltage,
2. more uniform avalanche breakdown, i.e., greater capacity to withstand overvoltages,
3. more uniform current flow in the forward direction, i.e. greater surge current capacity, and
4. narrower neutral zone and therefore narrower base and lower forward voltage drop V_f.

Summaries of some points concerning the preparation of NTD silicon for special applications on an R and D scale are given describe in papers [506, 507]. The production of large quantities of NTD silicon for power devices is described in [436]. More recently (see, e.g., [512]), the NTD technique has also been proposed for effectual doping of P in α-Si : H films (see also [147]). The results of [512] show that the NTD technique is an excellent method for doping P in α-Si : H.

Despite intensive study for many years and considerable progress, no clear understanding has emerged of one of the fundamental issues regarding the MIT in doped semiconductors and amorphous metal–semiconductor mixtures: whether and under what circumstances the Hall coefficient diverges as the transition is approached (see above for Ge). As is well known, in the localized regime, the spatial behavior of wave functions is usually described by an exponential decay length reflecting the spatial extent of the wave function (see, e.g., [513–515]). Dai et al. recently showed that the Hall coefficient of Si : P diverges at transitions, as it does in Si : B [516] and Ge : Sb [517]. The difference in the behavior of the MIT, according to these authors, may be connected with a different degree of compensation. It is also possible that the MIT is different in a persistent photoconductor, where the disorder is particularly strong and the concentration of shallow donors is varied and controlled through illumination (for details, see also [514, 515]).

6.8.3 Other Compounds

The NTD method has been used successfully in studying compound semiconductors: GaAs [430, 518–520] and GaP [521, 522]. NTD of GaAs is based on the following thermal neutron capture nuclear reactions (see also [430]):

$$^{69}\mathrm{Ga}(n,\gamma)^{70}\mathrm{Ga}^{\beta^-}_{21.1\ \mathrm{min}} \rightarrow {}^{70}\mathrm{Ge}, \tag{6.66}$$

$$^{71}\mathrm{Ga}(n,\gamma)^{72}\mathrm{Ga}^{\beta^-}_{14.1\,\mathrm{h}} \rightarrow {}^{72}\mathrm{Ge}, \tag{6.67}$$

$$^{75}\mathrm{As}(n,\gamma)^{76}\mathrm{As}^{\beta^-}_{26.3\,\mathrm{h}} \rightarrow {}^{76}\mathrm{Se}. \tag{6.68}$$

The relative abundances of the isotopes involved in the reactions and the cross sections of these reactions are such that the ratio of Se and Ge concentrations produced is

$$\frac{N_{\mathrm{Se}}}{N_{\mathrm{Ge}}} = 1.46. \tag{6.69}$$

Selenium is a typically shallow substitutional donor in GaAs with an electronic energy level a few meV from the conduction bands edge [523]. Germanium in GaAs is an amphoteric impurity which acts as a shallow donor (also a few meV from the conduction band), is situated on a Ga site, and has an acceptor level at $E_{\mathrm{V}} + 0.04\,\mathrm{eV}$ if situated on an As site [524]. Since, if electronically active, all Se atoms and some portion of the Ge atoms are expected to act as donors, NTD of GaAs is expected to dope GaAs more as the n-type. The addition of donors moves the Fermi-level (E_F) away from the valence band (E_V) to the conduction band (E_C). If a sufficiently high concentration of donors is added, E_F will move to the upper half of the band-gap, and the GaAs will be converted to the n-type. Analysis of Hall effect data as a function of temperature provides a means of measuring the donor content in irradiated GaAs samples. Young et al. were thus able to compare electrically active added donor content to the NTD-produced impurity concentrations determined from nuclear measurements. The Hall effect analysis also allows them to determine concentrations and energy levels (E) of impurities or defects in the p-type GaAs samples if the Fermi level in the material moves near E at some temperature over the range of measurements. This technique thus provides a means of identifying and measuring undercompensated acceptor content in samples. The low-temperature photoluminescence technique used in paper [430] measured donor-to-acceptor or conduction-band-acceptor luminescence. It provides an accurate determination of the position of acceptor electronic levels in GaAs, permitting positive identification of impurities or defects with known luminescence lines. Identifications of lines due to specific impurities or defects can be made using luminescence techniques, regardless of the position of the Fermi-level in a material. Little detailed information concerning an acceptor level can be obtained from the Hall effect if that acceptor is overcompensated for. However, the presence of specific acceptors can be detected by luminescence techniques, even in n-type samples. On the other hand, luminescence data do not provide the quantitative information obtainable from Hall-effect measurements.

The results of room temperature measurement of the electrical properties of eight annealed NTD GaAs samples are summarized in Table 6.10. The total NTD dose ($N_{Se} + N_{Ge}$), the carrier concentration, and carrier type (negative values of concentration indicate n-type), along with carrier mobility at room temperature are listed in Table 6.10. Note that following an NTD dose sufficient to produce $7 \cdot 10^{16}$ atoms/cm^3 initially present in the sam-

Table 6.10. Room temperature results for Hall-effect samples of GaAs annealed at 830°C/20 min (after [430])

Sample no.	NTD, (dose/cm^3)	$n(-)$ or P(+) (cm^{-3})	μ (cm^{2}V^{-1}s^{-1})
1,2	$3.8 \cdot 10^{15}$	$+\ 2.3 \cdot 10^{16}$	360
3	$8.5 \cdot 10^{15}$	$+\ 2.4 \cdot 10^{16}$	341
4	$1.7 \cdot 10^{16}$	$+\ 1.9 \cdot 10^{16}$	337
10	$2.7 \cdot 10^{16}$	$+\ 8.6 \cdot 10^{15}$	242
12	$7 \cdot 10^{16}$	$-1.6 \cdot 10^{16}$	1251
20	$1.5 \cdot 10^{17}$	$-7.7 \cdot 10^{16}$	3960
15	$2.8 \cdot 10^{17}$	$-2.3 \cdot 10^{17}$	3631
16, 18	$6.3 \cdot 10^{17}$	$-4.9 \cdot 10^{17}$	3110

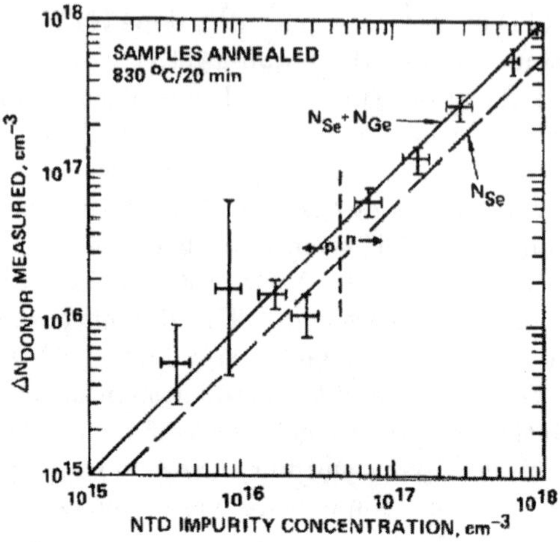

Fig. 6.36. Measured added donors vs. NTD produced impurity content (after [430])

ples, $7 \cdot 10^{16}$ donors/cm^3 would be expected to just overcompensate for the p-type material. The results presented in Table 6.10 show that p-type samples become progressively less p-type and n-type samples progressively more n-type with increasing NTD dose. Because the donor levels in GaAs are very shallow, they remain fully ionized in the temperature range of Young et al. experiments, so that the measured electron concentration is practically temperature independent (see Fig. 4 in [430]). This measured n for each sample is approximately equal to total donor minus total acceptor concentration.

Figure 6.36 shows the added electrically active donor concentration in eight NTD samples measured in [430] as a function of N_{Se} and of $(N_{\mathrm{Se}} + N_{\mathrm{Ge}})$ added by transmutation, as determined from nuclear activity measurements. The uncertainty in determining added donor content in p-type samples is large because of the complexity of analyzing material with multiple independent acceptor levels in closely compensated cases. The added donors can be much more accurately determined in the more highly doped n-type samples. The results shown in Fig. 6.36 imply that all of the selenium and a substantial fraction of the Ge atoms introduced by transmutation act as donors following annealing at 830°C/20 min. As will be shown below from photoluminescence measurements, a fraction of the Ge atoms produced by transmutation are on acceptor rather than on donor sites in GaAs samples.

Figures 6.37 and 6.38 show relative luminescence spectra for the four p-type and four n-type samples, respectively. The spectral positions indicated by arrows for carbon acceptor, Ge acceptor, and 0.07 eV acceptor correspond to donor (or band) or to acceptor luminescence lines. The most important conclusion to be drawn from a comparison of the spectra for the control and eight NTD samples is that Ge acceptors not present in the "starting material"

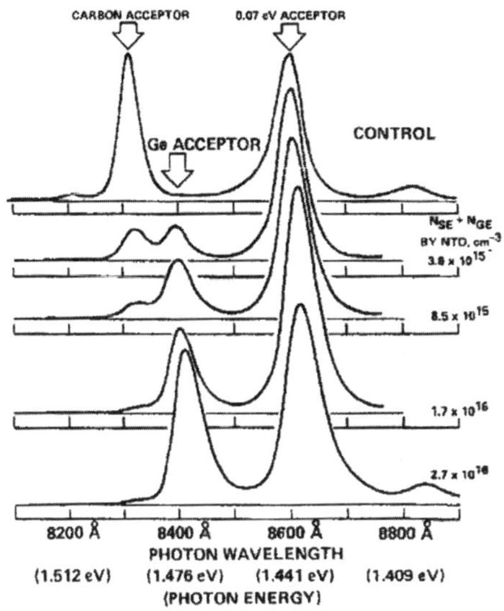

Fig. 6.37. Photoluminescence spectra for p-type NTD samples displayed with intensity of luminescence due to 0.07 eV acceptor held constant for the four spectra (after [430])

control sample are introduced by the NTD process. The increase in intensity of the Ge acceptor line with increasing dose relative to both the carbon and 0.07 eV acceptor lines indicates that Ge acceptor content increases with increasing transmutative doping. Therefore, some of the Ge atoms produced by NTD in these samples are acting as acceptors rather than donors. Photoluminescence measurement studies of the control and eight annealed NTD samples at longer wavelengths indicate another new line present only in NTD samples at about 9450 Å. The intensity of this line increases with increasing NTD dose.

The characteristic lifetimes of radioactive isotopes can be used to label and identify defect levels in semiconductors which can be detected by photoluminescence [519] and Raman-scattering spectroscopy [520]. Magerle et al. [519] show photoluminescence spectra of GaAs doped with ^{111}In that decays to ^{111}Cd. ^{111}In is isoelectronic to Ga and hence occupies Ga lattice sites in GaAs. It decays to ^{111}Cd with a lifetime $\tau_{111_{\mathrm{In}}} = 98$ h by electron capture [525]. Since the recoil energy of the Cd nucleus due to the emission of the neutrino is much smaller than the typical displacement energy in GaAs [526], ^{111}Cd atoms on Ga sites ($\mathrm{Cd_{Ga}}$) are created by the decay of ^{111}In on Ga sites ($^{111}\mathrm{In_{Ga}}$) and act there as shallow acceptors. This chemical transmutation was monitored by photoluminescence spectroscopy. Figure 6.39 shows successively taken photolumunescence spectra from the ^{111}In-doped sample. A spectrum from the undoped part is also shown. The photoluminescence spectrum of the undoped part of the sample shows features well known for

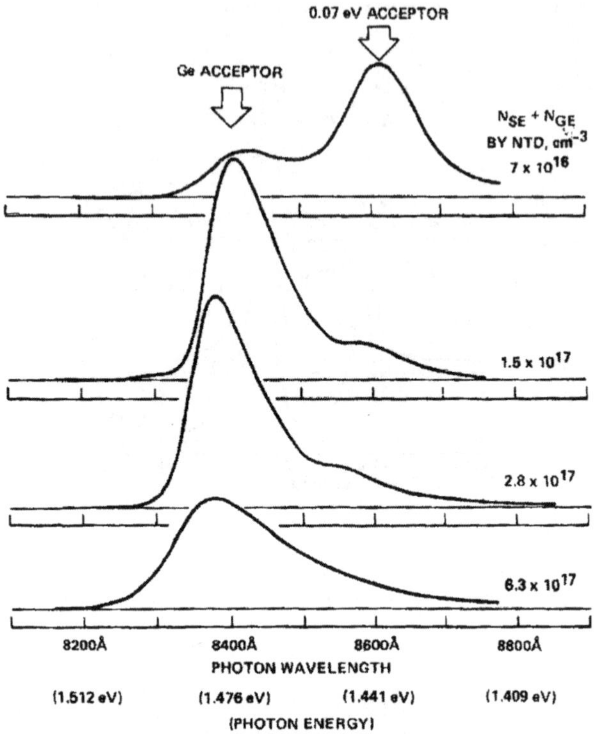

Fig. 6.38. Relative photoluminescence spectra for four n-type NTD samples. The four spectra are not normalized with respect to each other. (After [430])

undoped MBE-grown GaAs [527]. The peaks FX and AX around 819 nm are due to the recombination of free and bound excitons. The peak (e, C) at 830 nm and its LO phonon replica (e, C)-LO at 850 nm are due to recombination of electrons from the conduction band into C acceptor states. The recombination of electrons from donor states into C acceptor states appears as a small shoulder at the right-hand sides of either of these two peaks. C is a residual impurity in GaAs present in MBE-grown material with a typical concentration between 10^{14} and $10^{15}\,\mathrm{cm}^{-3}$ [527]. Magerle et al. determined the height $\frac{I_{\mathrm{Cd}}}{I_{\mathrm{C}}}$ of the (e, Cd) peak normalized to I_{C} as a function of time after doping by substracting the normalized spectrum of the undoped part from the normalized spectra of the ^{111}In-doped part. The height $\frac{I_{\mathrm{Cd}}}{I_{\mathrm{C}}}$ of the (e, Cd) peak remaining in these different spectra is displayed in the inset of Fig. 6.39. The authors fitted these data by

$$\frac{I_{\mathrm{Cd}}}{I_{\mathrm{C}}}(t) = \frac{I_{\mathrm{Cd}}}{I_{\mathrm{C}}}(t=\infty)\left(1-\mathrm{e}^{-\frac{t}{\tau}}\right) \tag{6.70}$$

and obtained a time constant $\tau = 52(17)\,\mathrm{h}$, which is not the nuclear lifetime $\tau_{^{111}\mathrm{In}} = 98\,\mathrm{h}$ of ^{111}In. Evidently $\frac{I_{\mathrm{Cd}}}{I_{\mathrm{C}}}$ is not proportional to N_{Cd}. The photo-

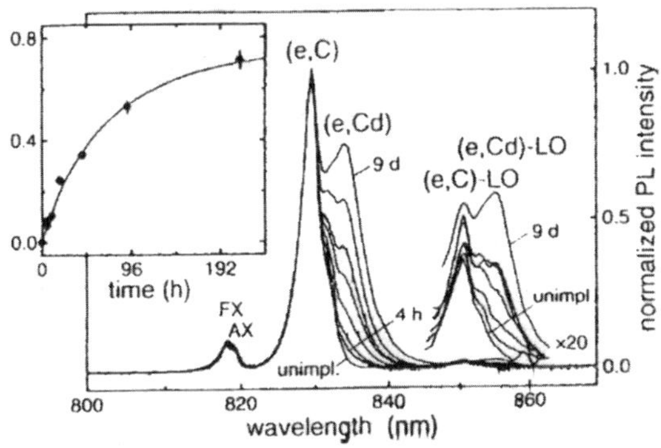

Fig. 6.39. Photoluminescence spectra of undoped and ^{111}In-doped GaAs successively taken 4 h, 7 h, 12 h, 22 h, 2 d, 4 d, and 9 d after doping. All spectra are normalized to the intensity of the (e, C) peak. In the inset, the height $\frac{I_{\mathrm{Cd}}}{I_{\mathrm{C}}}$ of the (e, Cd) peak in these spectra is shown as a function of time after doping with ^{111}In. The *solid line* is a fit to the data using (6.69). (After [519])

luminescent intensity I_{Cd} is proportional to the recombination rate of excess carriers per unit area through Cd acceptor states $\Delta n_{\mathrm{L}} B_{\mathrm{Cd}} N_{\mathrm{Cd}}$, where B_{Cd} is a recombination coefficient. The excess sheet carrier concentration in the implanted layer Δn_{L} can be expressed in terms of the total carrier lifetime in the implanted layer τ_{L} and the generation rate of excess carriers per unit area in the implanted layer $f_{\mathrm{L}} G$ by using the first of the two equilibrium conditions,

$$f_{\mathrm{L}} G = \frac{\Delta n_{\mathrm{L}}}{\tau_{\mathrm{L}}}$$

and

$$f_{\mathrm{B}} G = \frac{\Delta n_{\mathrm{B}}}{\tau_{\mathrm{B}}}. \tag{6.71}$$

The second one describes the balance between the generation rate $f_{\mathrm{B}} G$ and the recombination rate of excess carriers $\frac{\Delta n_{\mathrm{B}}}{\tau_{\mathrm{B}}}$ in the bulk. The total generation rate G is proportional to the incident photon flux, and $f_{\mathrm{L}} + f_{\mathrm{B}} = 1$. To get an expression for τ_{L}, the authors cited assumed two additional recombination processes in the implanted layer: radiative recombination via Cd acceptors and nonradiative recombination due to residual implantation damage, and write the recombination rate in the small single approximation (see, e.g., [528]) as

$$\frac{\Delta n_{\mathrm{L}}}{\tau_{\mathrm{L}}} = \frac{\Delta n_{\mathrm{L}}}{\tau_{\mathrm{B}}} + \Delta n_{\mathrm{L}} B_{\mathrm{Cd}} N_{\mathrm{Cd}} + \Delta n_{\mathrm{L}} B_{\mathrm{nr}} f_{\mathrm{nr}} N_{\mathrm{Cd}}. \tag{6.72}$$

Here $\Delta n_{\mathrm{L}} B_{\mathrm{nr}} f_{\mathrm{nr}} N_{\mathrm{Cd}}$ is the nonradiative recombination rate per unit area due to residual implantation damage, $f_{\mathrm{nr}} N_{\mathrm{Cd}}$ is the concentration of these nonradiative recombination centers, and B_{nr} is the corresponding recombination coefficient. Hence Δn_{L} and Δn_{B} can be expressed as a function of N_{Cd} and the recombination rates through all the different recombination channels, and thereby the relative photoluminescence peak intensities can be deduced. I_{C} is proportional to the sum of the (e, C) recombination rates per unit area in the implanted layer and the bulk, and within this model,

$$I_{\mathrm{C}} \propto \frac{\Delta n_{\mathrm{L}} + \Delta n_{\mathrm{B}}}{\tau_{\mathrm{C}}} = G\frac{\tau_{\mathrm{B}}}{\tau_{\mathrm{C}}}\left(\frac{f_{\mathrm{L}}}{1+\dfrac{\Phi_{\mathrm{Cd}}}{f_{\mathrm{B}}b}} + f_{\mathrm{B}}\right). \tag{6.73}$$

Here Φ_{Cd} is the dose between 10^9 and $10^{13}\,\mathrm{cm}^2$. Thereby, $\tau_{\mathrm{C}} = \frac{1}{B_{\mathrm{C}}N_{\mathrm{C}}}$ is an effective lifetime describing the recombination probability through C acceptor states and b is a constant defined below. With help of (6.63) and (6.65) (assuming that the detection efficiencies of both peaks are equal), the following relation between $\frac{I_{\mathrm{Cd}}}{I_{\mathrm{C}}}$ and Φ_{Cd} can be obtained:

$$\frac{I_{\mathrm{Cd}}}{I_{\mathrm{C}}} = \frac{\Delta n_{\mathrm{L}} B_{\mathrm{Cd}} N_{\mathrm{Cd}}}{\left(\dfrac{\Delta n_{\mathrm{L}} + \Delta n_{\mathrm{B}}}{\tau_{\mathrm{C}}}\right)} = \frac{a}{1+\dfrac{b}{\Phi_{\mathrm{Cd}}}}, \tag{6.74}$$

with

$$a = \frac{f_{\mathrm{L}}}{f_{\mathrm{B}}} \frac{B_{\mathrm{Cd}}}{(B_{\mathrm{nr}} f_{\mathrm{nr}} + B_{\mathrm{Cd}})} \frac{\tau_{\mathrm{C}}}{\tau_{\mathrm{B}}}$$

and

$$b = \frac{d}{f_{\mathrm{B}}(B_{\mathrm{nr}} f_{\mathrm{nr}} + B_{\mathrm{Cd}})\tau_{\mathrm{B}}}. \tag{6.75}$$

This model describes quantitatively the dependence of the (e, Cd) intensity of N_{Cd} and the authors cited use it to describe the increase in $\frac{I_{\mathrm{Cd}}}{I_{\mathrm{C}}}$ with time in the ^{111}In-doped sample. The authors [519] model the change of the carrier lifetime τ_{L} with time t in the ^{111}In-doped sample as

$$\frac{1}{\tau_{\mathrm{L}}} = \frac{1}{\tau_{\mathrm{B}}} + B_{\mathrm{Cd}} N_{\mathrm{In}}\left(1 - \mathrm{e}^{-\frac{t}{\tau}}\right) B_{\mathrm{nr}} f_{\mathrm{nr}}, \tag{6.76}$$

where $N_{\mathrm{In}} = \frac{\Phi_{\mathrm{In}}}{d}$ is the initial ^{111}In concentration, $\tau = \tau_{^{111}\mathrm{In}} = 98.0\,\mathrm{h}$ is the nuclear lifetime of ^{111}In, and B_{Cd}, B_{nr} and f_{nr} are the same constants as above. Thereby, we assume, following Magerle et al., that the same kinds of nonradiative recombination centers are produced by In doping as by Cd doping and that the Cd concentrations are identical to the ^{111}In concentration profile. Taking into account all of the above, we can write

$$\frac{I_{\mathrm{Cd}}}{I_{\mathrm{C}}} = \frac{a}{1 + b/\Phi_{\mathrm{In}}\left(1 - \mathrm{e}^{-\frac{t}{\tau}}\right) + \dfrac{c}{\left(\mathrm{e}^{-\frac{t}{\tau}} - 1\right)}}, \tag{6.77}$$

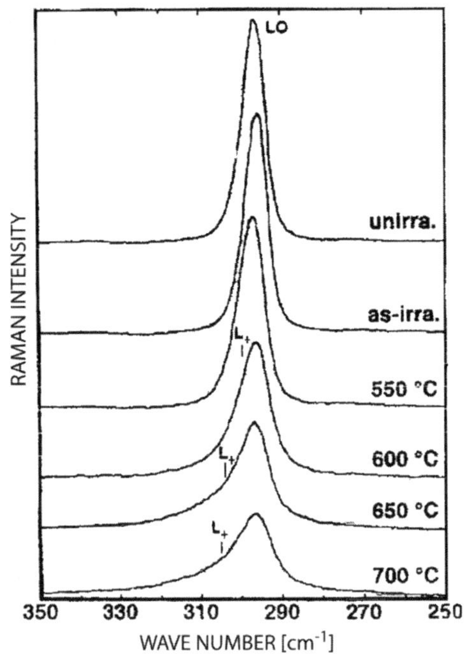

Fig. 6.40. Raman spectra at room temperature taken for the various annealing temperatures of (100)-oriented NTD GaAs irradiated with neutron doses (a thermal neutron of $1.5 \cdot 10^{18}\,\text{cm}^{-2}$ and a fast neutron of $7 \cdot 10^{17}\,\text{cm}^{-2}$). The coupling L_+ mode is observed at annealing temperatures above 600°C (see Table 6.11). (After [520])

where a and b are the same constants as above and $c = \frac{B_{\text{nr}} f_{\text{nr}}}{(B_{\text{nr}} f_{\text{nr}} + B_{\text{Cd}})}$. The c term accounts for the fact that in the ^{111}In-doped sample, the concentration of nonradiative centers is not changing with Cd concentration. Magerle et al. fitted (6.77) to the data shown in the inset of Fig. 6.39, keeping $\tau = 98.0\,\text{h}$, $a = 1.25$, and $b = 3.0 \cdot 10^{11}\,\text{cm}^{-2}$ and obtained $\Phi_{\text{In}} = 4.49 \cdot 10^{11}\,\text{cm}^{-2}$ and $c = 0.5(2)$. This fit is shown as a *solid line* and agrees perfectly with the experimental data. In concluding this section, note that this identification technique is applicable to a large variety of defect levels since suitable radioactive isotopes exist for most elements (for details, see [525]).

Coupling between the LO phonon mode and the longitudinal plasma mode in NTD semi-insulating GaAs was studied in paper [520] using Raman-scattering spectroscopy and a Fourier-transform infrared spectrometer. Raman spectra are shown in Fig. 6.40 for unirradiated, as-irradiated, and annealed samples. The remarkable feature is the low intensity and asymmetrical linewidth of the LO-phonon spectrum observed in samples that are annealed above 600°C. The behavior is not understood by considering only the LO phonon. We should pay attention to the NTD impurities, which begin to activate electrically around 600°C. In the long-wavelength limit, the valence

Table 6.11. Electron concentrations and the coupling modes of NTD GaAs (after [520])

Sample	EC (cm^{-3})	LO-phonon frequency (cm^{-1})	L_+ mode (cm^{-1})	PF (cm^{-1})
Unirradiated	$1 \sim 2 \times 10^7$	296.6		
As-irradiated	[a]	295.6		
500°C annealed	[a]	297.8		
600°C annealed	$8.2 \cdot 10^{16}$	296.0	299	96.4
650°C annealed	$2.2 \cdot 10^{17}$	296.6	304	158
700°C annealed	$2.5 \cdot 10^{17}$	296.2	305	168

[a] Since the conduction is dominated by Mott-type hopping conduction (Satoh, M., Kuriyama, K., Phys. Rev. **B40**, 3473 (1989)), the electron concentration cannot be measured by the van der Pauw method

EC: Electron concentration

PF: Plasma frequency

electrons, the polar lattice vibrations, and the conduction electrons make additive contributions to the total dielectric response function [529]:

$$\varepsilon_{\mathrm{T}}(0,\omega) = \frac{\varepsilon_\infty + (\varepsilon_0 - \varepsilon_\infty)}{\left[\left(1 - \dfrac{\omega^2}{\omega_{\mathrm{t}}^2}\right) - \dfrac{\omega_{\mathrm{p}}^2 \varepsilon_\infty}{\omega^2}\right]} . \tag{6.78}$$

The high-frequency value (L_+) of the mixed LO-phonon-plasmon modes is calculated from the roots of the dielectric constant of (6.78). The frequencies of the L_+ mode and of the longitudinal plasma mode $\omega_{\mathrm{p}} = \left(\frac{4\pi n e^2}{\varepsilon_\infty m^*}\right)^{\frac{1}{2}}$ for various annealing temperatures are listed in Table 6.11. Here n is the electron concentration, m^* the effective mass in the conduction band ($= 0.07\, m_0$), and $\varepsilon_\infty (= 11.3)$ the optical dielectric constant. The mixed LO-phonon-plasma mode appears around $300\,\mathrm{cm}^{-1}$ for electron concentration of $(0.8$–$2) \cdot 10^{17}\,\mathrm{cm}^{-3}$. The phonon strength [529] for the high-frequency mode (L_+) of the interacting plasmon-LO-phonon mode is about 0.95 for an electron concentration of $1 \cdot 10^{17}\,\mathrm{cm}^{-3}$, and that for the low-frequency mode (L_-) is below 0.1. Therefore, the asymmetrical linewidth of the Raman spectrum observed in annealed NTD GaAs arises from both the LO-phonon and L_+ modes, but the L_- mode is not observed because of very weak phonon strength. As a result, the LO-phonon intensity decreases with increasing coupling, and the L_+ mode appears beside the LO-phonon peak.

The absorption spectra at the various annealing temperatures for NTD GaAs are shown in Fig. 6.41. In unirradiated samples, an absorption around $2350\,\mathrm{cm}^{-1}$ is assigned as the antisymmetric stretching vibration from the CO_2 in the ambient atmosphere. The absorption peaks observed around

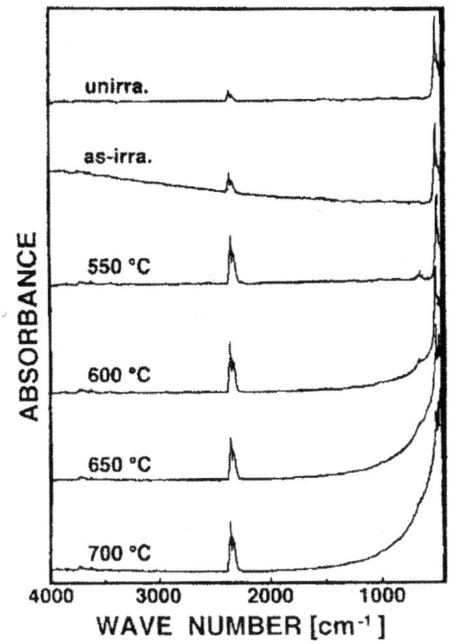

Fig. 6.41. Infrared-absorption spectra at room temperature for the various annealing temperatures of the NTD GaAs used for the Raman-scattering experiments (after [520])

500 cm^{-1} are also assigned as a two-phonon overtone scattering [530] of transverse optical phonons (TO); these were observed at 493 cm^{-1} [2TO (X)], 508 cm^{-1} [2TO (L)], and 524 cm^{-1} [2TO (Γ)], respectively. In as-irradiated samples, continuous absorption extending to the higher energy was observed, although this origin cannot be attributed to interstitial anion clusters, as discussed for neutron-irradiated GaP [531]. In samples annealed above 600°C, remarkable absorption was observed at wave numbers below 1450 cm^{-1}. The absorption increases with increasing annealing temperature (see Fig. 6.41). This behavior arises from free electron absorption due to the activation of NTD impurities, which occur at annealing temperatures above 600°C. The free-electron absorption observed is consistent with collective motion as a plasmon mode described in Raman-scattering studies.

Kuriyama et al. [522] studied the transmuted impurities Ge and S in NTD semi-insulating GaP by a photoluminescence method. In NTD GaP, Ge and S impurities are transmuted from Ga and P atoms by (n, γ) reactions, respectively. Ge in GaP is an amphoteric impurity for which both the donor and acceptor states appear to be deep. The ratio between transmuted impurities Ge and S is about 16 : 1. Unfortunately, after the transmutative reactions, the transmuted atoms are usually not in their original positions but are displaced into interstitial positions due to the recoil produced by the γ and β particles

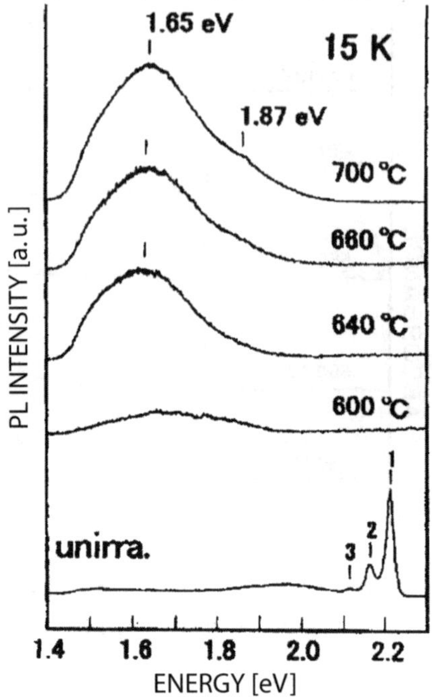

Fig. 6.42. Photoluminescence (PL) spectra at 15 K for unirradiated and NTD GaP. PL peaks 1, 2, and 3 in unirradiated GaP represent S_P–C_P DA pair recombination, its LO-phonon replica, and 2LO-phonon replica, respectively. Emission at 1.65 and 1.87 eV in NTD GaP are attributed to a Ge_{Ga}-Ge_P complex and S_P–Ge_P DA pair recombinations, respectively. (After [522])

in the nuclear reactions. In addition, the defects induced by fast neutron irradiation disturb the electrical activation of transmuted impurities. However, Frenkel type defects [521, 532] in NTD GaP were annealed out between 200 and 300°C, and P antisite (P_{Ga}) defects of $\sim 10^{18}\,cm^{-3}$ were annihilated at annealing temperatures between 600 and 650°C. Therefore, transmuted impurities, Ge and S, would be substituted at Ga and/or P lattice sites by annealing around 650°C.

Figure 6.42 shows the photoluminescence (PL) spectra of unirradiated and NTD GaP. The PL spectrum (peak 1) of unirradiated samples shows the signature of DA pair recombination involving an S donor and C acceptor [534]. Two (peaks 2 and 3) of the replicas occur at energies consistent with electronic transitions accompanied by zone-center optical phonons with energies of 50.1 (LO_Γ) and 100.2 meV ($2\,LO_\Gamma$). Sulfur, silicon, and carbon in GaP are the most common residual impurities [534]. In NTD GaP, the main transition energy observed was 1.65 eV. Since Ge in GaP is the amphoteric impurity with deep acceptor and donor levels, strong phonon cooperation will also

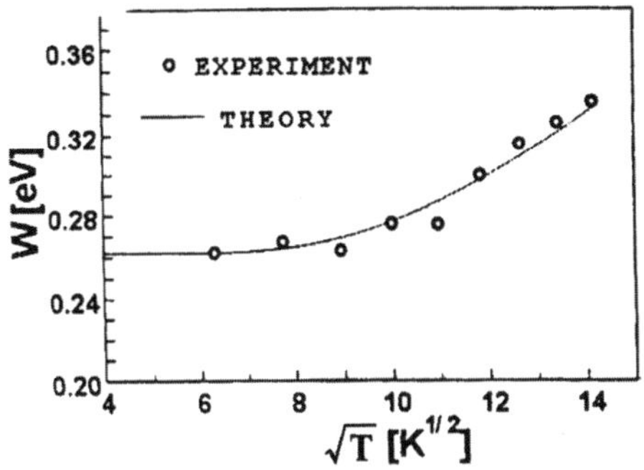

Fig. 6.43. Variation of the half-width W with the square root of the temperature T for the 1.65 eV band in NTD GaP. The theoretical curve is a plot of (6.79) with $h\nu = 0.025\,\text{eV}$. (After [522])

occur, but optical transition rates will be significant only for associates. A similar situation has been proposed for Si in GaP [533], forming a nearest neighbor Si_{Ga}-Si_P complex. Therefore, a broad emission would be expected to arise from a nearest neighbor Ge_{Ga}-Ge_P coupled strongly to the lattice. To confirm the presence of the Ge_{Ga}-Ge_P couple, the temperature dependence of the half-width, W, of the broad emission was measured. If the localized electron transitions from the excited state to the ground state of this complex center produce the characteristic luminescence, the dependence would follow the configuration-coordinate (CC) [535] model equation:

$$W = A\left[\cot\left(\frac{h\nu}{2kT}\right)\right]^{1/2}, \qquad (6.79)$$

where A is a constant whose value is equal to W as the temperature approaches 0 K and $h\nu$ is the energy of the vibrational mode of the excited state. In Fig. 6.43, (6.71) has been fitted to the experimental value for NTD GaP. To estimate W, the spectrum of the 1.65 eV band was substracted from that of the 1.87 eV band. The value of $h\nu$ used was 0.025 eV. The good fit to this equation that was found for the Ge_{Ga}-Ge_P center in NTD GaP shows the validity of applying the CC model. The results of paper [522] indicate that the NTD method is useful for introducing Ge donors, resulting from fact the Ge atoms are transmuted from Ga lattice sites in GaP. The results obtained are consistent with the presence of the Ge_{Ga}-Ge_P complex, as described earlier.

7 Optical Fiber

7.1 Optical Communication

Optical communication using fibers is a major new technology that will profoundly impact telephone systems, computer interconnections, and instrumentation (Internet). Fiber links provide several major advantages over conventional electronic communications systems. These include immunity to electromagnetic interference, thinner and lighter cables, lower transmission losses (especially for very high data rates), and potential kilometer-long link capabilities extending to the gigahertz region.

An optical waveguide is a dielectric structure that transports energy at wavelengths in the infrared or visible ranges [536, 537] of the electromagnetic spectrum. In practice, waveguides used for optical communications are highly flexible fibers composed of nearly transparent dielectric materials. The cross section of these fibers is small – comparable to the thickness of a human hair – and generally is divisible into three layers as shown in Fig. 7.1. The central re-

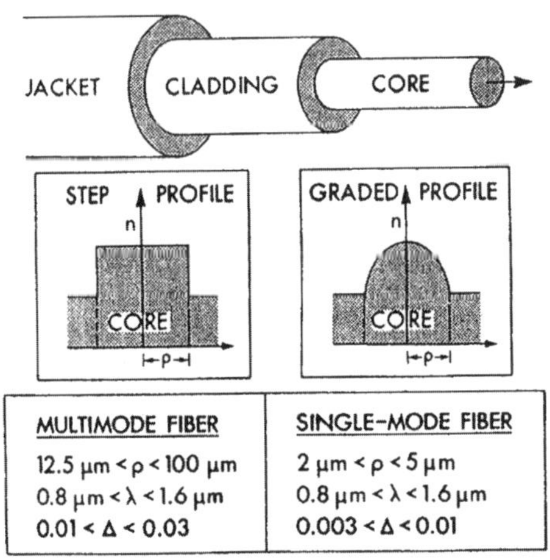

Fig. 7.1. Nomenclature, profiles, and ranges of dimensions for typical optical fibers, where ρ is the core radius, λ is the free-space wavelength of light, and $\Delta = \left(1 - n_{\mathrm{cl}}^2/n_{\mathrm{co}}^2\right)/2$. (After [538])

gion is the core, which is surrounded by cladding, which in turn is surrounded by a protective jacket. Within the core, the refractive-index profile n can be uniform or graded, whereas the cladding index is typically uniform [539]. The two situations correspond to the step-index and graded-index profiles shown in the insets in Fig. 7.1. The core index must be greater than the cladding index [540], at least in some region of the cross section, if guidance is to take place. For the majority of applications, most of the light energy propagates in the core, and only a small fraction travels in the cladding. The jacket is almost optically isolated from the core, so that we usually ignore its effect and assume unbounded cladding for simplicity in the analysis.

As usual, optical waveguides can be conveniently divided into two subclasses called multimode waveguides (with comparatively large cores) and single-mode waveguides (with comparatively small cores). The demarcation between the two is that multimode waveguides obey the condition (see, e.g., [541]) $\left(\frac{2\pi\rho}{\lambda}\right)\left(n_{\mathrm{co}}^2 - n_{\mathrm{cl}}^2\right)^{\frac{1}{2}} \gg 1$, where ρ is a linear dimension in the core, e.g., the radius of the fiber core, λ is the wavelength of light in free space, n_{co} is the maximum refractive index in the core, and n_{cl} is the uniform refractive index in the cladding.

As will be shown below, electromagnetic propagation along optical waveguides is described exactly by Maxwell's equations. However, it is well known that classical geometric optics provides an approximate description of light propagation in regions where the refractive index varies only slightly over a distance comparable to the wavelength of light. This is typical of multimode optical waveguides used for communication. Thus, the most direct and conceptually simple way to describe light propagation in multimode waveguides is by tracing rays along the core (see also [542, 543]). By using classical geometric optics, we can ignore all wave effects. In multimode waveguides, wave effects are usually negligible [538], but there are exceptional situations when such effects accumulate exponentially with the distance that light travels. Naturally in these cases, wave effects must be retained, since they can have a significant influence on long waveguides. In each such situation, we modify the classical geometric optical description by taking into account the local plane wave nature of light. The phenomenon of greatest practical interest in fibers used for long-distance communications is the spread of pulses as they propagate along the fiber. For idealized multimode fibers, pulse spreading is easily described by classical geometric optics. But since propagation in multimode guides is so complex, simple models and physical understanding are generally of much greater assistance than a precise, exact analysis [544].

7.2 Maxwell's Equations

The spatial dependence of the electric field $\boldsymbol{E}(x, y, z)$ and the magnetic field $\boldsymbol{H}(x, y, z)$ of an optical waveguide are determined by Maxwell's equations.

Further, we assume an implicit time dependence $\exp(-\mathrm{i}\omega t)$ in the field vectors, current density $\boldsymbol{J}$, and charge density σ. The dielectric constant $\varepsilon(x, y, z)$ is related to the refractive index $n(x, y, z)$ by $\varepsilon = n^2\varepsilon_0$, where ε_0 *is* the dielectric constant of free space. For the nonmagnetic materials that normally constitute an optical waveguide, the magnetic permeability μ is very nearly equal to the free-space value μ_0. Under these conditions, Maxwell's equations are expressible in the form (see, e.g., [545])

$$\nabla \times \boldsymbol{E} = \frac{-\partial \boldsymbol{B}}{\partial t} = -\mu\mu_0 \frac{\partial \boldsymbol{H}}{\partial t}, \tag{7.1}$$

$$\nabla \times \boldsymbol{H} = \frac{\partial \boldsymbol{D}}{\partial t} = \varepsilon\varepsilon_0 \frac{\partial \boldsymbol{E}}{\partial t}, \tag{7.2}$$

$$\boldsymbol{D} = \varepsilon\varepsilon_0 \boldsymbol{E}, \tag{7.3}$$

$$\boldsymbol{B} = \mu\mu_0 \boldsymbol{H}, \tag{7.4}$$

$$\nabla \cdot \boldsymbol{D} = 0 \quad \text{(no free charges)}, \tag{7.5}$$

$$\nabla \cdot \boldsymbol{B} = 0 \quad \text{(no free poles)}. \tag{7.6}$$

We will assume further that our guiding structure is in the z direction. Thus we look for solutions to problems in which the z dependence of the field is of the form

$$\boldsymbol{E} = \boldsymbol{E}_0(x, y) \exp \mathrm{i} \left(\omega t -^{z} k \cdot z\right), \tag{7.7}$$

$$\boldsymbol{H} = \boldsymbol{H}_0(x, y) \exp \mathrm{i} \left(\omega t -^{z} k \cdot z\right). \tag{7.8}$$

In this expression, we note that ω is related to ν, the frequency, by the relation $\omega = 2\pi\nu$. Furthermore, in writing the temporal and spatial variation of the field in terms of the complex exponential, it is understood, but not written, throughout this chapter that when a field is to be evaluated (e.g., for measurement), the only physical meaningful part of this complex expression is the real part. Thus, if we denote a real measurable field by ϵ and we wish to relate this to the theoretically derived field $\boldsymbol{E}$, which is given by an expression of the form of 7.7, then

$$\epsilon = \mathrm{Re}\left[\boldsymbol{E}\right]. \tag{7.9}$$

The advantage of this approach is that expressions involving the exponential of a complex quantity are more readily manipulated than the equivalent expressions involving sine and cosine. Next, we will derive expressions for the field components in planar and cylindrical geometries before studying the solutions for guided waves in planar and cylindrical waveguides (see also [544]).

7.2.1 Planar Geometry

The expression ∇x in rectangular Cartesian coordinates is

$$\nabla \times \boldsymbol{A} = \begin{vmatrix} \boldsymbol{i} & \boldsymbol{j} & \boldsymbol{k} \\ \frac{\partial}{\partial x} & \frac{\partial}{\partial y} & \frac{\partial}{\partial z} \\ {}^{x}A & {}^{y}A & {}^{z}A \end{vmatrix}, \tag{7.10}$$

where $\boldsymbol{i}$, $\boldsymbol{j}$, $\boldsymbol{k}$ are unit vectors in the x, y, z directions. Thus from equation (7.1), we obtain for the x directed component of $\nabla \times \boldsymbol{E}$

$$\left(\frac{\partial^z E}{\partial y} - \frac{\partial^y E}{\partial z} \right) = -\mu\mu_0 \frac{\partial^x H_x}{\partial t}. \tag{7.11}$$

Now, we can write similar equations for $\frac{\partial^y H}{\partial t}$, $\frac{\partial^z H}{\partial t}$, $\frac{\partial^x E}{\partial t}$, $\frac{\partial^y E}{\partial t}$, and $\frac{\partial^z E}{\partial t}$. If we substitute the field expressions of (7.7) and (7.8) in these expressions, we obtain the following results:

$$\frac{\partial^z E}{\partial y} + i^z k^y E = -i\mu\mu_0\omega^x H, \tag{7.12}$$

$$i^z k^x E + \frac{\partial^z E}{\partial x} = i\mu\mu_0\omega^y H, \tag{7.13}$$

$$\frac{\partial^y E}{\partial x} - \frac{\partial^x E}{\partial y} = -i\mu\mu_0\omega^z H, \tag{7.14}$$

$$\frac{\partial^z H}{\partial y} + i^z k^y H = -i\varepsilon\varepsilon_0\omega^x E, \tag{7.15}$$

$$-i^z k^x H - \frac{\partial^z H}{\partial x} = i\varepsilon\varepsilon_0\omega^y E, \tag{7.16}$$

$$\frac{\partial^y H}{\partial x} - \frac{\partial^x H}{\partial y} = i\varepsilon\varepsilon_0\omega^z E. \tag{7.17}$$

Analysis of planar structures will be restricted to infinite films that lie in the y-z plane. Thus, in addition to the assumption that the fields have the z dependence already postulated, we can further assume that the partial derivative with respect to y vanishes (hereafter $\frac{\partial}{\partial y} = 0$) for an infinite plane wave traveling in the z direction. With this assumption, the above equations simplify and demonstrate a fundamental relationship about the fields in such a structure. The indicated relations have these forms:

$$i^z k^y E = -i\mu\mu_0\omega^x H \quad \text{(TE group)}, \tag{7.18}$$

$$i^z k^x E + \frac{\partial^z E}{\partial x} = i\mu\mu_0\omega^y H \quad \text{(TM group)}, \tag{7.19}$$

$$\frac{\partial^y E}{\partial x} = -i\mu\mu_0\omega^z H \quad \text{(TE group)}, \tag{7.20}$$

$$i^z k^y H = i\varepsilon\varepsilon_0\,\omega^x E \quad \text{(TM group)}, \tag{7.21}$$

$$-i^z k^x H - \frac{\partial^z H}{\partial x} = i\varepsilon\varepsilon_0\omega^y E \quad \text{(TE group)}, \tag{7.22}$$

$$\frac{\partial^y H}{\partial x} = i\varepsilon\varepsilon_0\omega^z E \quad \text{(TM group)}. \tag{7.23}$$

Now, we can see that the fields have split into two separate groups, namely, ${}^yE$, ${}^xH$, and ${}^zH$ are coupled and ${}^yH$, ${}^xE$, and ${}^zE$ are also coupled. The guided waves formed by the first group are described as TE modes (transverse electric), and the latter are known as TM modes (transverse magnetic). We can now use the above relations to derive simpler expressions for the transverse field components in terms of the ${}^zE$ and ${}^zH$ components only, e.g., eliminate ${}^yH$ from (7.11), (7.13), and (7.15) to obtain a relation for ^{x}E. This yields the following:

$${}^xE = \left(\frac{-i^z k}{\varepsilon\varepsilon_0\mu\mu_0\omega^2 -^z k^2}\right)\frac{\partial^z E}{\partial x}, \tag{7.24}$$

$${}^yE = \left(\frac{-i\omega\mu\mu_0}{\varepsilon\varepsilon_0\mu\mu_0\omega^2 -^z k^2}\right)\frac{\partial^z H}{\partial x}, \tag{7.25}$$

$${}^xH = \left(\frac{-i^z k}{\varepsilon\varepsilon_0\mu\mu_0\omega^2 -^z k^2}\right)\frac{\partial^z H}{\partial x}, \tag{7.26}$$

$${}^yH = \left(\frac{-i\omega\varepsilon\varepsilon_0}{\varepsilon\varepsilon_0\mu\mu_0\omega^2 -^z k^2}\right)\frac{\partial^z E}{\partial x}, \tag{7.27}$$

Substituting these expressions into (7.18) to (7.23) yields two wave equations for propagation in the x direction:

$$\frac{\partial^{2z} E}{\partial x^2} - \left(\omega^2\varepsilon\varepsilon_0\mu\mu_0 -^z k^2\right)^z E = 0, \tag{7.28}$$

$$\frac{\partial^{2z} H}{\partial x^2} - \left(\omega^2\varepsilon\varepsilon_0\mu\mu_0 -^z k^2\right)^z H = 0. \tag{7.29}$$

These indicate that for the transverse dependence of the fields, we should seek solutions of the form $\exp(\mathrm{i}^x kx)$, where

$${}^xk^2 = \left(\omega^2\varepsilon\varepsilon_0\mu\mu_0 -^z k^2\right) = -({}^x\gamma)^2. \tag{7.30}$$

The significance of the variable ${}^x\gamma$ introduced here will become apparent later. Notice that (7.30) could have been obtained much more straightforwardly by deriving the wave equation directly from (7.31) to (7.6), setting $\frac{\partial}{\partial y} = 0$, and substituting the field equations (7.7)–(7.8). However, that route would not have yielded the detailed interrelationship between the vector components of the field that we will need for finding the conditions for guided waves. In conclusion, the relationships derived between the vector components of the fields for a planar structure lying in the y–z plane, with a wave propagating in the z direction, are summarized in (7.18) to (7.23). In addition, we have shown that fields in such a structure take the general form

$$\boldsymbol{E} = \boldsymbol{E}(x) \exp \mathrm{i}(\omega t - {}^{z}k z \pm {}^{x}k x), \tag{7.31}$$
$$\boldsymbol{H} = \boldsymbol{H}(x) \exp \mathrm{i}(\omega t - {}^{z}k z \pm {}^{x}k x), \tag{7.32}$$
$${}^{x}k^2 = \omega^2 \varepsilon \varepsilon_0 \mu \mu_0 - {}^{z}k^2, \tag{7.33}$$
$$n^2 k_0^2 = \omega^2 \varepsilon \varepsilon_0 \mu \mu_0. \tag{7.34}$$

7.2.2 Cylindrical Geometry

We now repeat the analysis of Sect. 7.2.1 but in cylindrical polar coordinates, since these are more appropriate for analyzing of optical fiber guides. The coordinates x, y and z are now replaced by r, ϕ, and z. These are related to the coordinates as follows:

$$x = r \cos \phi, \tag{7.35}$$
$$y = r \sin \phi, \tag{7.36}$$
$$z = z. \tag{7.37}$$

Since we are still concerned with a structure that is expected to guide waves in the z direction, we should postulate fields of the form

$$\boldsymbol{E} = \boldsymbol{E}(r, \phi) \exp \mathrm{i}(\omega t - {}^{z}k z), \tag{7.38}$$
$$\boldsymbol{H} = \boldsymbol{H}(r, \phi) \exp \mathrm{i}(\omega t - {}^{z}k z). \tag{7.39}$$

The relation for $\boldsymbol{\nabla} \times \boldsymbol{A}$ in polar coordinates is as follows:

$$\boldsymbol{\nabla} \times \boldsymbol{A} = \begin{vmatrix} \frac{\boldsymbol{r}}{r} & \phi & \frac{\boldsymbol{k}}{r} \\ \frac{\partial}{\partial r} & \frac{\partial}{\partial \phi} & \frac{\partial}{\partial z} \\ {}^{r}A & r^{\phi}A & {}^{z}A \end{vmatrix}, \tag{7.40}$$

where $\boldsymbol{k}$ is the unit vector in the z direction. We will now derive the expressions for the field components by using Maxwell equations (7.1)–(7.6). We obtain the set of relations equivalent to equations in Cartesian coordinates for the planar case:

$$\frac{1}{r}\left[\frac{\partial^{z} E}{\partial \phi} + i^{z}k\left(r^{\phi}E\right)\right] = -i\mu\mu_0\omega^{r}H, \tag{7.41}$$
$$i^{z}k^{r}E + \frac{\partial^{z}E}{\partial r} = i\mu\mu_0\omega^{\phi}H, \tag{7.42}$$
$$\frac{1}{r}\left[r\frac{\partial^{\phi}E}{\partial r} - \frac{\partial^{r}E}{\partial \phi}\right] = -i\mu\mu_0\omega^{z}H, \tag{7.43}$$

$$\frac{1}{r}\left[\frac{\partial^z H}{\partial \phi} + i^z k \left(r^\phi H\right)\right] = i\varepsilon\varepsilon_0 \omega^r E, \tag{7.44}$$

$$\frac{1}{r}\left[r\frac{\partial^\phi H}{\partial r} - \frac{\partial^r H}{\partial \phi}\right] = i\varepsilon\varepsilon_0 \omega^z E. \tag{7.45}$$

Solving this system of equations, we obtain expressions for the r and ϕ components only in terms of the z components:

$$^rE = \frac{-i}{T_{k^2}}\left[{}^zk\frac{\partial^z E}{\partial r} + \mu\mu_0\omega\frac{1}{r}\frac{\partial^z H}{\partial \phi}\right], \tag{7.46}$$

$$^\phi E = \frac{i}{T_{k^2}}\left[\frac{{}^zk}{r}\frac{\partial^z E}{\partial \phi} - \mu\mu_0\omega\frac{1}{r}\frac{\partial^z H}{\partial r}\right], \tag{7.47}$$

$$^rH = \frac{-i}{T_{k^2}}\left[{}^zk\frac{\partial^z H}{\partial r} - \varepsilon\varepsilon_0\omega\frac{1}{r}\frac{\partial^z E}{\partial \phi}\right], \tag{7.48}$$

$$^\phi H = \frac{-i}{T_{k^2}}\left[\frac{{}^zk}{r}\frac{\partial^z H}{\partial \phi} + \varepsilon\varepsilon_0\omega\frac{\partial^z E}{\partial r}\right], \tag{7.49}$$

$$^\mathrm{T}k^2 = \omega^2\varepsilon\varepsilon_0\mu\mu_0 -^z k^2 = n^2k_0^2 -^z k^2. \tag{7.50}$$

Here $^\mathrm{T}k$ is the total transverse component of the $\boldsymbol{k}$ in the waveguide.

7.2.3 The Electromagnetic Wave Equation

Below, we derive the standard derivation of the wave equation. Moreover, we give the form of the Laplacian operator for rectangular Cartesian and polar coordinates. If we take the curl of the first Maxwell equation, then we obtain

$$\boldsymbol{\nabla}\cdot(\boldsymbol{\nabla}\cdot\boldsymbol{E}) = -\mu\mu_0\left[\boldsymbol{\nabla}\cdot\frac{\partial \boldsymbol{H}}{\partial t}\right]. \tag{7.51}$$

Differentiating equation (7.2) with respect to time yields

$$\boldsymbol{\nabla}\cdot\frac{\partial \boldsymbol{H}}{\partial t} = \varepsilon\varepsilon_0\frac{\partial^2 \boldsymbol{E}}{\partial t^2}. \tag{7.52}$$

We then use the vector identity

$$\boldsymbol{\nabla}\cdot(\boldsymbol{\nabla}\cdot\boldsymbol{E}) = -\boldsymbol{\nabla}(\boldsymbol{\nabla}\cdot\boldsymbol{E}) - \boldsymbol{\nabla}^2\boldsymbol{E} = -\boldsymbol{\nabla}^2\boldsymbol{E} \tag{7.53}$$

since $\boldsymbol{\nabla}\cdot\boldsymbol{E} = 0$ [see also (7.5)]. Then, it follows directly by substitution that

$$\boldsymbol{\nabla}^2\boldsymbol{E} = \varepsilon\varepsilon_0\mu\mu_0\frac{\partial^2 \boldsymbol{E}}{\partial t^2} \tag{7.54}$$

and likewise

$$\boldsymbol{\nabla}^2\boldsymbol{H} = \varepsilon\varepsilon_0\mu\mu_0\frac{\partial^2 \boldsymbol{H}}{\partial t^2}. \tag{7.55}$$

Both equations have the general form

$$\boldsymbol{\nabla}^2 \boldsymbol{A} = \frac{1}{V^2}\frac{\partial^2 \boldsymbol{A}}{\partial t^2}, \tag{7.56}$$

where V is the velocity of propagation (phase velocity) of the wave in the medium. It follows, as usual, that

$$V_{\mathrm{p}} = \frac{1}{\sqrt{\varepsilon\varepsilon_0 \mu\mu_0}} \tag{7.57}$$

and that, for free space, the velocity of light, c, is given by

$$c = \frac{1}{\sqrt{\varepsilon_0 \mu_0}}. \tag{7.58}$$

For planar waveguides, described by rectangular Cartesian coordinates, or circular fibers, described by cylindrical polar coordinates, the Laplacian operator has the forms

$$\nabla^2 A = \frac{\partial^2 A}{\partial x^2} + \frac{\partial^2 A}{\partial y^2} + \frac{\partial^2 A}{\partial z^2}, \tag{7.59}$$

$$\nabla^2 A = \frac{1}{r}\frac{\partial}{\partial r}\left(r\frac{\partial A}{\partial r}\right) + \frac{1}{r^2}\frac{\partial^2 A}{\partial \phi^2} + \frac{\partial^2 A}{\partial z^2}. \tag{7.60}$$

We should stress here that the modes considered in fiber optics are exact solutions of Maxwell's equations (for details, see [538–540]).

7.3 Geometric Optics of Fibers

As noted above, the mechanism of light propagation along fibers as small as a few wavelengths in diameter can be understood almost entirely using the ray theory and well-known principles of geometric optics. It is interesting to note that, even for smaller diameter dielectric cylinders, which act as waveguides, the geometric optical theory, with some modification, helps to understand the complex mechanisms. We begin ray analysis of multimode optical waveguides with the planar, or slab waveguide, which is the simplest dielectric structure for illustrating the principles involved and has application in integrated optics [536,537]. Since we can analyze its light transmission characteristics in terms of a superposition of ray paths, it is important to appreciate fully the behavior of individual rays. We will study the trajectories of rays within planar waveguides, concentrating on those rays – bound rays – which propagate without loss of energy in a nonabsorbing waveguide, and can, therefore propagate arbitrarily across large distances. The planar, or slab, waveguide is illustrated in Fig. 7.2. It consists typically of a core layer of thickness 2ρ sandwiched between two layers that form the cladding. As

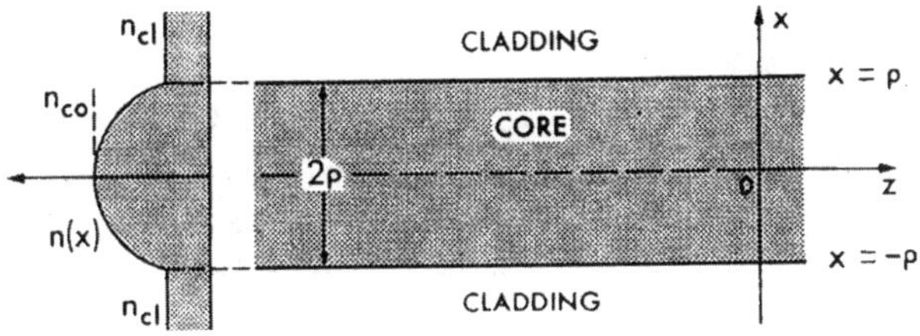

Fig. 7.2. Nomenclature and coordinates for describing planar waveguides. A representative graded profile varies over the core and is uniform over cladding, assumed unbounded. (After [538])

explained in the introduction, we assume, for simplicity, that the cladding is unbounded. The planes $x = \pm\rho$ are the core-cladding interfaces. Since the waveguide extends indefinitely in all directions orthogonal to the x-axis, the problem is two dimensional [546]. The z-axis is located along the axis of the waveguide midway between the interfaces. The refractive-index profile $n(x)$ in Fig. 7.2 can be uniform or graded across the core and assumes a uniform value n_{cl} in the cladding. The core refractive index must have some values greater than n_{cl} for the waveguide to have guidance properties. Furthermore, we assume that the profile does not vary with z, so that the waveguide is translationally invariant, or cylindrically symmetrical. The parameters defined in Fig. 7.2 can be combined with the free-space wavelength λ of the light propagating along the waveguide to form a single dimensionless parameter V, known as the waveguide parameter, or waveguide frequency. If n_{co} is the maximum value of $n(x)$, which need not concur with the on-axis value $n(0)$, then it will define

$$V = \frac{2\pi\rho}{\lambda}\left(n_{\mathrm{co}}^2 - n_{\mathrm{cl}}^2\right)^{\frac{1}{2}}. \tag{7.61}$$

The ray theory considered here is restricted to multimode waveguides, i.e., waveguides satisfying $V \gg 1$. The step-index planar waveguide, according to Fig. 7.3, has refractive-index profile defined by

$$n(x) = n_{\mathrm{co}}, -\rho < x < \rho; \quad n(x) = n_{\mathrm{cl}}, \qquad |x| > \rho, \tag{7.62}$$

where n_{co} and n_{cl} are constants and $n_{\mathrm{co}} > n_{\mathrm{cl}}$. One of the most important problems is to determine the conditions necessary for a ray to be bound, i.e., a ray propagates along a nonabsorbing waveguide without loss of power.

Propagation within the uniform core of the step-index waveguide of Fig. 7.3 is along straight lines. If a ray originates at P on one interface and makes angle θ_z with the waveguide axis, it will meet the opposite interface at Q, as shown in Fig. 7.3. The situation at Q is pictured in more detail in Fig. 7.4. Reflection in this situation is governed by Snell's law [539, 546]. Though these laws are usually expressed in terms of angles relative to the

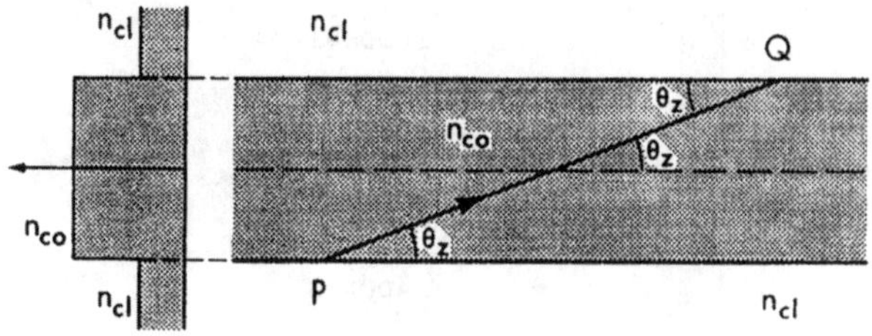

Fig. 7.3. Propagation along a straight line between interfaces in the core of a step-profile planar waveguide (after [538])

normal QN, following [538], we prefer to retain the complementary angle θ_z. Thus, in terms of complementary angles, the incident ray at Q is totally internally reflected, if $0 \preceq \theta_z \prec \theta_c$, and is partly refracted, if $\theta_c \prec \theta_z \preceq \frac{\pi}{2}$, where θ_c is the complement of the critical angle, defined by [538]

$$\theta_c = \cos^{-1}\left[\frac{n_{cl}}{n_{co}}\right] = \sin^{-1}\left[1 - \frac{n_{cl}^2}{n_{co}^2}\right]^{\frac{1}{2}}. \tag{7.63}$$

In the first case, Fig. 7.4a shows the reflected ray leaving the interface at the same angle θ_z as the incident ray; the second case (see Fig. 7.4b) shows that the ray bifurcates; part of it is reflected at angle θ_z, and part of it is transmitted into the cladding at angle θ_t to the interface, which satisfies Snell's law,

$$n_{co} \cos\theta_z = n_{cl} \cos\theta_t. \tag{7.64}$$

Only total internal reflection returns all the ray power, i.e., the energy flows along the ray, back into the core medium. A ray is reflected from the interface back into the core at angle θ_z, regardless of whether partial or total reflection occurs. If we repeat this procedure at successive reflections from the interfaces (see Fig. 7.5), then we construct zigzag paths, or trajectories. The

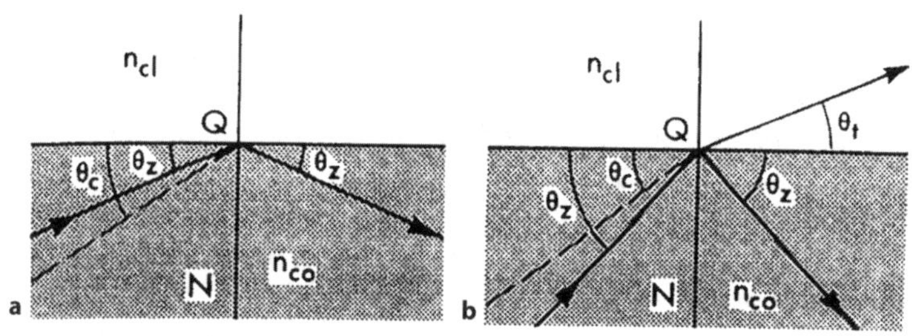

Fig. 7.4. Reflection at a planar interface, unbounded regions of refractive indexes n_{co} and n_{cl}, showing **(a)** total internal reflection and **(b)** partial reflection and refraction (after [538])

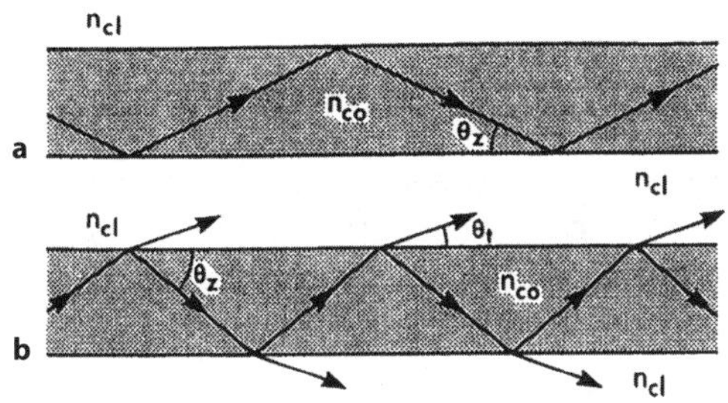

Fig. 7.5. Zigzag paths within the core of a step-profile planar waveguide for **(a)** bound rays and **(b)** refracting rays (after [538])

path depicted in Fig. 7.5a is a ray that is totally reflected at every reflection. We refer to this as a bound ray, since its path is entirely confined within the core. The path in Fig 7.5b is for a ray that is partly reflected at each reflection. We refer to this as a refracting ray. Rays may be categorized by the value of θ_z; according to [538],

$$\text{Bound rays}: \ 0 \preceq \theta_z \prec \theta_c, \tag{7.65a}$$

$$\text{Refracting rays}: \ \theta_c \preceq \theta_z \preceq \frac{\pi}{2}. \tag{7.65b}$$

Since the power of a bound ray is totally reflected back into the core at every reflection, the ray can propagate indefinitely without any loss of power. A refracting ray loses a fraction of its power at each reflection and therefore attenuates as it propagates.

As was in many textbooks (see, e.g., [538–544]), it is useful to introduce parameters that characterize ray propagation because these parameters, rather than the spatial dependence of the ray path, are important. As can be seen from Fig. 7.6, the ray trajectory is fully characterized once the angle θ_z is prescribed. We define the path length L_p between successive reflections as the distance PQ. According to a geometry of the picture,

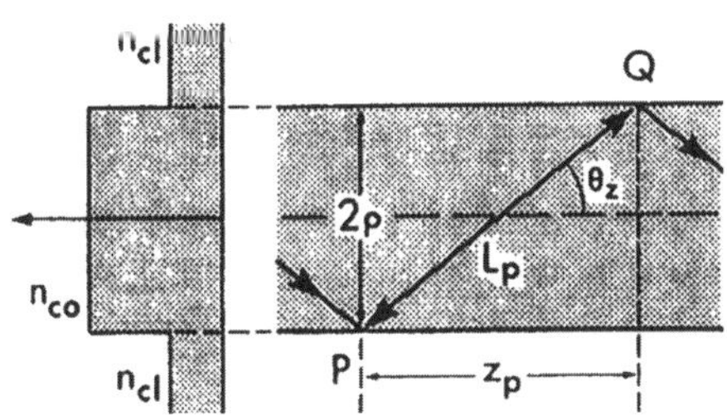

Fig. 7.6. Path length and ray half-period mz_{p} for a ray in the core of a step-profile planar waveguide (after [538])

$$L_{\mathrm{p}} = \frac{2\rho}{\sin\theta_z} = \frac{2\rho n_{\mathrm{co}}}{\left(n_{\mathrm{co}}^2 - \bar{\beta}^2\right)^{\frac{1}{2}}}, \tag{7.66}$$

where $\bar{\beta} = n_{\mathrm{co}}\cos\theta_z = n_{\mathrm{cl}}\cos\theta_{\mathrm{t}}$ is a ray invariant. For the ray transit time below, we require the optical path length L_{O}. In a homogeneous medium, this is given by the product of path length and refractive index:

$$L_{\mathrm{O}} = n_{\mathrm{co}} L_{\mathrm{p}} = \frac{2\rho n_{\mathrm{co}}}{\sin\theta_z} = \frac{2\rho n_{\mathrm{co}}^2}{\left(n_{\mathrm{co}}^2 - \bar{\beta}^2\right)^{\frac{1}{2}}}, \tag{7.67}$$

A quantity that appears frequently in attenuation problems is the ray half-period z_{p}. This is the distance between successive reflections, mesured along the waveguide axis:

$$z_{\mathrm{p}} = \frac{2\rho}{\tan\theta_z} = L_{\mathrm{p}}\cos\theta_z = \frac{2\rho\bar{\beta}}{\left(n_{\mathrm{co}}^2 - \bar{\beta}^2\right)^{\frac{1}{2}}}. \tag{7.68}$$

Closely related is the number of reflections N per unit length of waveguide, which is the reciprocal of the ray half-period. Hence

$$N = \frac{1}{z_{\mathrm{p}}} = \frac{\tan\theta_z}{2\rho}. \tag{7.69}$$

It is clear from these definitions that over an arbitrary distance z along the waveguide, the accumulated path length, optical path length, and number of reflections are given proportionally by

$$\frac{z}{z_{\mathrm{p}}} L_{\mathrm{p}}, \quad \frac{z}{z_{\mathrm{p}}} L_{\mathrm{O}}, \qquad N_z = \frac{z}{z_{\mathrm{p}}}, \tag{7.70}$$

respectively. These parameters are indicated in [538, Table I-1].

The most important quantity required to describe pulse spreading is the ray transit time t. This is the time a ray takes to propagate distance z along the waveguide, following a zigzag ray path (see also Fig. 7.5). The velocity of light v_{g} in fiber along the path is given by [546]

$$v_{\mathrm{g}} = \frac{c}{n_{\mathrm{co}}}, \tag{7.71}$$

where c is the free-space speed of light. The transit time is given by the following relation:

$$t = \frac{z}{z_{\mathrm{p}}} \frac{L_{\mathrm{p}}}{v_{\mathrm{g}}} = \frac{z L_{\mathrm{O}}}{z_{\mathrm{p}} c} = \frac{z n_{\mathrm{co}}^2}{c\bar{\beta}} = \frac{z n_{\mathrm{co}}}{c\cos\theta_z}, \tag{7.72}$$

so that the greater θ_z, the longer the transit time.

We can account for material dispersion, which occurs when the refractive index varies with the wavelength of light λ, i.e., $n_{\mathrm{co}} = n_{\mathrm{co}}(\lambda)$. This requires more sophisticated reasoning relying on treating a ray as if it were a plane wave in local regions. Ray energy propagates at the group velocity v_g, which is given by (7.71) in a dispersionless medium, but, allowing for material dispersion, it has the more general form [545]

$$v_g = c\left[n_{co}(\lambda) - \lambda\frac{dn_{co}(\lambda)}{d\lambda}\right]^{-1}. \tag{7.73}$$

It is convenient to introduce the group index n_g, described by

$$n_g = n_{co}(\lambda) - \lambda\frac{dn_{co}(\lambda)}{d\lambda}, \tag{7.74}$$

in which case, the transit time is expressible as

$$t = \frac{zn_g}{c\cos\theta_z} = \frac{zn_g n_{co}}{c\bar{\beta}}, \tag{7.75}$$

and varies with both θ_z and λ.

Earlier we established the basic concepts for ray analysis of planar waveguides. Now we extend the analysis to optical fibers, which are used for high-capacity communication over long distances. As far as ray tracing is concerned, the only difference between fibers and planar waveguides is the introduction of the third dimension. Thus, although the ray concepts are the same as earlyier, the analysis and resulting expression are generally more complicated because of the fiber geometry [543]. Nevertheless, one of the important results of fiber optics shows that ray transit times for step and clad power-law profiles circular and noncircular fibers are identical to those of the corresponding planar waveguides. If this remarkable simplification is acceptable without proof, then pulse spreading in such fibers can be studied directly. An optical fiber is illustrated in Fig. 7.7. Unless otherwise stated, the core is assumed to have a circularly symmetrical cross section of radius ρ, surrounded by cladding, which, for simplicity, is assumed as unbounded. The

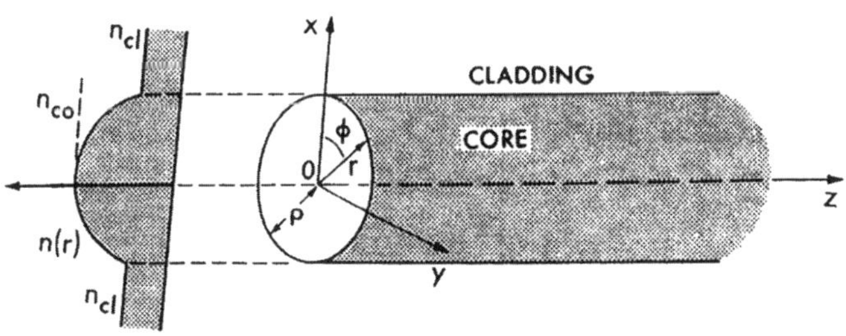

Fig. 7.7. Nomenclature for describing circular fibers. Cartesian coordinates x, y, z and cylindrical coordinates r, ϕ, z are oriented so that the z-axis lies along the fiber axis. A representative graded profile varies over the core and is uniform over the cladding, assumed unbounded. (After [538])

core–cladding interface is the cylindrical surface $r = \rho$. Over the core, the axisymmetric refractive-index profile $n(r)$ is either uniform or graded, and it takes the uniform value n_{cl} in the cladding. The dimensionless parameter V of (7.61) also applies to fibers and will be called the fiber parameter, where ρ is the core radius. The quantity $(n_{\mathrm{co}}^2 - n_{\mathrm{cl}}^2)^{\frac{1}{2}}$ is often referred to as the numerical aperture of the fiber, and a related expression $\left[n_{\mathrm{co}}^2(r) - n_{\mathrm{cl}}^2\right]^{\frac{1}{2}}$ is sometimes called the local numerical aperture (for details, see [543, 544]).

7.4 Waveguide Mode Propagation

Electromagnetic energy in a light pipe is propagated down the pipe by reflection from the walls of the structure. If the transverse dimensions are comparable to the wavelength of the light, only certain field distributions (modes) will satisfy Maxwell's equations and the boundary conditions. In this case, the light pipe is more appropriately considered a waveguide. Even in very large structures, there are so many of them, and their numbers increase as the area, that in most cases a geometric optics description is more fruitful (see above).

As is well known, the distinction between metallic and dielectric wave guides is in the reflection mechanism responsible for confining the energy. The metallic guide does so by reflection from a good conductor at the boundary. In a dielectric waveguide, this is accomplished by total internal reflection, which results when the central dielectric is made of a material of higher index of refraction than the surrounding dielectric. The two regions will henceforth be referred to as the core and cladding. In a metallic guide, there are two sets of solutions, the transverse electric and transverse magnetic modes. In the dielectric guide, all but the cylindrically symmetrical modes $\mathrm{TE}_{0\mathrm{m}}$ and $\mathrm{TM}_{0\mathrm{m}}$ are hybrid, i.e., both have electric and magnetic z components [547–549]. In general, one would expect two sets of such hybrid modes because the boundary conditions give a characteristic equation which is quadratic in the Bessel functions describing the field in the central dielectric (for details, see [539, 543, 544] and references therein).

The cylindrical dielectric waveguide consists of a core of high refractive index n_{co} and radius a surrounded by a cladding of lower refractive index n_{cl}. Let the cladding material of index n_{cl} extend to infinity. We shall use both Cartesian $(x,\ y)$ and cylindrical polar coordinates $(r,\ \phi)$. The propagation constant β of any mode of this fiber is limited within the interval $n_{\mathrm{co}} \geq \beta \geq n_{\mathrm{cl}}k$, where $k = \frac{2\pi}{\lambda}$ is the wavenumber in free space. If we define parameters

$$u = a\left(k^2 n_{\mathrm{co}}^2 - \beta\right)^{1/2}, \tag{7.76}$$

$$w = a\left(\beta - k^2 n_{\mathrm{cl}}^2\right)^{1/2}, \tag{7.77}$$

the mode field can be expressed by Bessel function $J(ur/a)$ inside the core and modified Hankel function $K(wr/a)$ outside the core [539, 548]. The quadratic summation;

$$v^2 = u^2 + w^2 \tag{7.78}$$

leads to a third parameter

$$v = ak\left(n_{\text{co}}^2 - n_{\text{cl}}^2\right)^{\frac{1}{2}}, \tag{7.79}$$

which can be considered a normalized frequency. By matching the fields at the core–cladding interface, we obtain characteristic functions $u(v)$ or $w(v)$ for every mode; the propagation constant and all other parameters of interest can be derived from these functions. For weak guidance,

$$\Delta = \frac{(n_{\text{co}} - n_{\text{cl}})}{n_{\text{cl}}} \ll 1. \tag{7.80}$$

In this case, we can construct modes whose transverse field is essentially polarized in one direction. Below, we postulate transverse field components (see also [543])

$$^{y}E = {}^{z}H = \frac{Z_0}{n_{\text{co}}} = E_l \left[\frac{J_l\left(\frac{ur}{a}\right)}{J_l(\text{u})}\right] \cos l\phi \tag{7.81a}$$

$$^{y}E = {}^{z}H = \frac{Z_0}{n_{\text{cl}}} = E_l \left[\frac{K_l\left(\frac{wr}{a}\right)}{K_l(w)}\right] \cos l\phi. \tag{7.81b}$$

Here, as in following, the "a" holds for the core and the "b" for the cladding; Z_0 is the plane wave impedance in a vacuum, and E_l the electrical field strength at the interface. Figure 7.8b–e illustrates the case $l = 1$. Since we have the freedom of choosing $\sin l\phi$ or $\cos l\phi$ in (7.90) and two orthogonal states of polarization, we can construct a set of four modes for every l as long as $l > 0$. For $l = 0$, we have only a set of two modes polarized orthogonally with respect to each other. The longitudinal components can be obtained from the equations in [550]

$$^{z}E = \frac{\text{i}Z_0}{k}\left[\frac{1}{n_{\text{co}}^2}\right]\frac{\partial^z H}{\partial y}, \tag{7.82a}$$

$$^{z}E = \frac{\text{i}Z_0}{k}\left[\frac{1}{n_{\text{cl}}^2}\right]\frac{\partial^z H}{\partial y}, \tag{7.82b}$$

and

$$^{z}H = (\frac{\text{i}}{kZ_0})(\frac{\partial^y E}{\partial x}). \tag{7.82c}$$

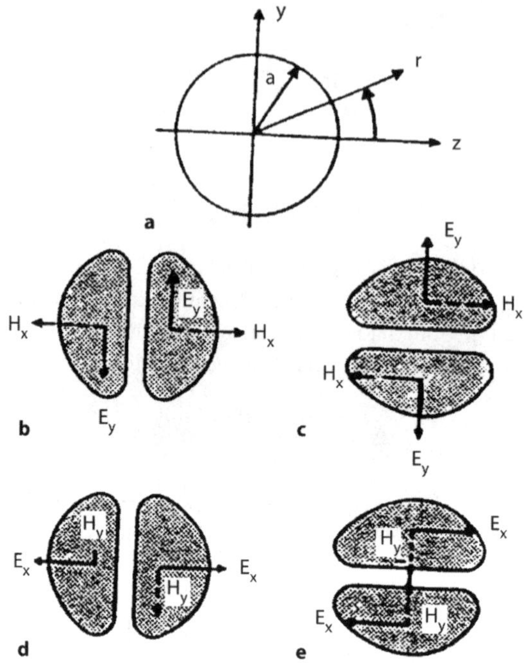

Fig. 7.8. Sketch of the fiber cross section and the four possible distributions of LP_{11} (after [549])

By introducing (7.81a, 7.81b), we have

$$
{}^{z}E = \frac{-\mathrm{i}E_l}{2ka}\left[\frac{u}{n_{\mathrm{co}}}\frac{J_{l+1}\left(\dfrac{ur}{a}\right)}{J_l(u)}\sin(l+1)\phi + \frac{u}{n_{\mathrm{co}}}\frac{J_{l-1}\left(\dfrac{ur}{a}\right)}{J_l(u)}\sin(l-1)\phi\right], \tag{7.83a}
$$

$$
{}^{z}E = \frac{-\mathrm{i}E_l}{2ka}\left[\frac{w}{n_{\mathrm{cl}}}\frac{K_{l+1}\left(\dfrac{wr}{a}\right)}{K_l(w)}\sin(l+1)\phi - \frac{w}{n_{\mathrm{cl}}}\frac{K_{l-1}\left(\dfrac{wr}{a}\right)}{K_l(w)}\sin(l-1)\phi\right], \tag{7.83b}
$$

$$^{z}H = \frac{-\mathrm{i}E_l}{2kZ_0a}\left[u\frac{J_{l+1}\left(\frac{ur}{a}\right)}{J_l(u)}\cos(l+1)\phi\right.$$

$$\left.-u\frac{J_{l-1}\left(\frac{ur}{a}\right)}{J_l(u)}\cos(l-1)\phi\right], \tag{7.83c}$$

$$^{z}H = \frac{-\mathrm{i}E_l}{2kZ_0a}\left[w\frac{K_{l+1}\left(\frac{wr}{a}\right)}{K_l(w)}\cos(l+1)\phi\right.$$

$$\left.+w\frac{K_{l-1}\left(\frac{wr}{a}\right)}{K_l(w)}\cos(l-1)\phi\right]. \tag{7.83d}$$

For small Δ, the longitudinal components (7.83a)–(7.83d) are small compared to the transverse components. The factors involved are $\frac{u}{ak}$ and $\frac{w}{ak}$ because (7.76) and (7.77) are both of the order $\Delta^{\frac{1}{2}}$. Repeated differentiation of (7.83a)–(7.83d) leads to transverse components that are not identical to the postulated field (7.71) but small of order Δ compared to it. We shall neglect these fields in the following. It is this approximation that determines the accuracy of the assumption of linearly polarized modes (see also [544]).

To match the fields at the interface let us write (7.80) in terms of cylindrical components:

$$^{\phi}E = \frac{E_l}{2}\left[\frac{J_l\left(\frac{ur}{a}\right)}{J_l(u)}\right][\cos(l+1)\phi + \cos(l-1)\phi], \tag{7.84a}$$

$$^{\phi}E = \frac{E_l}{2}\left[\frac{K_l\left(\frac{wr}{a}\right)}{J_l(w)}\right][\cos(l+1)\phi + \cos(l-1)\phi], \tag{7.84b}$$

$$^{\phi}H = -\frac{E_l}{2Z_0}\left[n_{\mathrm{co}}J_l\left(\frac{ur}{a}\right)/J_l(u)\right][\sin(l+1)\phi - \sin(l-1)\phi], \tag{7.84c}$$

$$^{\phi}H = -\frac{E_l}{2Z_0}\left[n_{\mathrm{cl}}K_l\left(\frac{wr}{a}\right)/K_l(w)\right][\sin(l+1)\phi - \sin(l-1)\phi], \tag{7.84d}$$

If we set $n_{\mathrm{co}} = n_{\mathrm{cl}}$ in (7.83)–(7.84) and use the recurrence relations for J_l and K_l, we can match all tangential field components at the interface by the one equation,

$$u\left[\frac{J_{l-1}(u)}{J_l(u)}\right] = -w\left[\frac{K_{l-1}(w)}{K_l(w)}\right]. \tag{7.85}$$

This is the characteristic equation for linearly polarized (LP) modes. Setting $w = 0$ yields the cutoff values $J_{l-1}(u) = 0$. For $l = 0$, this includes the roots of the Bessel function $J_{-1}(u) = -J_1(u)$, which we shall count so as to include

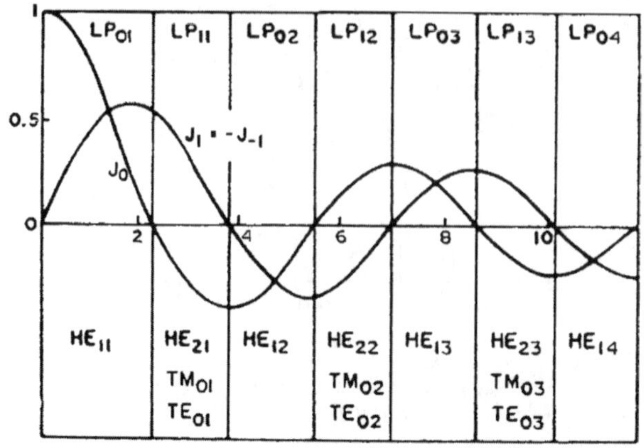

Fig. 7.9. The regions of the parameter u for modes of order $l = 0; 1$ (after [544])

$J_1(0) = 0$ as the first root. In this way Gloge [549] obtained the cutoff values indicated in Fig. 7.9 for $\mathrm{LP_{0m}}$ and $\mathrm{LP_{1m}}$. In the limit of $w \to \infty$, $J_1(0) = 0$. Thus, the solutions for u are between the zeros of $J_{l-1}(u)$ and $J_l(u)$. Every solution is associated with one set of modes designated $\mathrm{LP}_{l\mathrm{m}}$. For $l \geq 1$, each set comprises four modes.

The accuracy of the characteristic equation can be improved if we retain n_{cl} and n_{co} as different in (7.83) and (7.84). In this case, as was shown by Gloge [549], terms with $(l+1)\phi$ and $(l-1)\phi$ satisfy two different characteristic equations:

$$\left(\frac{u}{n_{\mathrm{co}}}\right)\left[\frac{J_{l\pm1}(u)}{J_l(u)}\right] = \pm\left(\frac{w}{n_{\mathrm{cl}}}\right)\left[\frac{K_{l\pm1}(w)}{K_l(w)}\right]. \tag{7.86}$$

By using the recurrence relations for J_l and K_l, Gloge showed that these two equations converge into (7.85) for $n_{\mathrm{co}} = n_{\mathrm{cl}}$. For $n_{\mathrm{cl}} \neq n_{\mathrm{co}}$, this degeneracy ceases to exist; each mode $\mathrm{LP}_{l\mathrm{m}}$ breaks up into modes with terms $(l+1)\phi$, which can be identified as $\mathrm{HE_{l+1,m}}$ or $\mathrm{TE_m}$ and $\mathrm{TM_m}$ (see also [538, 548]). A more rigorous proof of this result is given below.

As mentioned above, the problem of the dielectric cylinder with a sharp index step can be solved exactly. Using the preceding nomenclature, one can write the exact characteristic equation in the form

$$\left(Q - D - 2\Delta\left\{\left[\frac{(l\pm1)}{\omega^2}\right] \pm \left[\frac{K_l(\omega)}{K_{l\pm1}(\omega)}\right]\right\}\right)(Q - D) = Q^2\left[1 - 2\Delta\left(\frac{u^2}{v^2}\right)\right], \tag{7.87}$$

where

$$Q = (l \pm 1)\left(v^2/u^2w^2\right). \tag{7.88}$$

$$D = \left[\frac{J_l(u)}{uJ_{l\pm1}(u)}\right] \mp \left[\frac{K_l(w)}{wK_{l\pm1}(w)}\right], \tag{7.89}$$

and

$$2\Delta = \frac{\left(n_{\mathrm{co}}^2 - n_{\mathrm{cl}}^2\right)}{n_{\mathrm{co}}^2}. \tag{7.90}$$

The upper sign holds for HE_{l+1} modes and the lower sign for EH_{l-1}, TM, and TE. Equation (7.90) agrees with (7.80) for small index differences. If Δ is set to zero in (7.87), we find $D = 0$ and (7.89) then becomes the simplified characteristic (7.85). For small Δ, D is also small. Now wc simplify (7.78)) to the extent that we retain terms linear in Δ or D. This results in

$$D = \Delta\left\{Q\left(\frac{u^2}{v^2}\right) - \left[\frac{(l \pm 1)}{w^2}\right] \mp \left[\frac{K_l(w)}{wK_{l\pm1}(w)}\right]\right\} \tag{7.91}$$

and with (7.88),

$$D = \mp\Delta\left[\frac{K_l(w)}{wK_{l\pm1}(w)}\right]. \tag{7.92}$$

By introducing this into (7.89) and inserting (7.80), we find

$$\left(\frac{u}{n_{\mathrm{co}}}\right)\left[\frac{J_{l\pm1}(u)}{J_l(u)}\right] = \pm\left(\frac{w}{n_{\mathrm{cl}}}\right)\left[\frac{K_{l\pm1}(w)}{K_l(w)}\right]. \tag{7.93}$$

This is exactly the characteristic (7.86).

Evidently, a guided wave traveling along a circular guide carries energy. The respective amounts are readily calculated using the Poynting vector to estimate the energy flow [544]. The Poynting vector in the axial direction can be calculated from the cross product of the transverse fields given in (7.81). Integration over the cross section of the core and cladding leads to tabulated integrals (see also [544, 547, 551]); the results are

$$P_{\mathrm{co}} = \left[1 + \left(\frac{w^2}{u^2}\right)\left(\frac{1}{k}\right)\right]\left(\frac{\pi a^2}{2}\right)\left(\frac{Z_0}{n_{\mathrm{co}}}\right)E_l^2 \tag{7.94}$$

and

$$P_{\mathrm{cl}} = \left[\left(\frac{1}{k}\right) - 1\right]\left(\frac{\pi a^2}{2}\right)\left(\frac{Z_0}{n_{\mathrm{cl}}}\right)E_l^2 \tag{7.95}$$

for the power flow in the core and cladding, respectively. If we ignore the small difference between n_{co} and n_{cl}, the total power in a specific mode becomes

$$P = P_{\mathrm{co}} + P_{\mathrm{cl}} = \left(\frac{v^2}{u^2}\right)\left(\frac{1}{k}\right)\left(\frac{\pi a^2}{2}\right)\left(\frac{Z_0}{n_{\mathrm{co}}}\right)E_l^2 \tag{7.96}$$

Practical fibers have small heat and scattering losses that cause significant attenuation over long distances. In general, these losses are attributable to specific parts of the fiber and are proportional to the power propagating in this part. For considerations of this kind, it is convenient to use the power fractions

$$\frac{P_{\mathrm{co}}}{P} = 1 - \left(\frac{u^2}{v^2}\right)(1-k) \tag{7.97}$$

and

$$\frac{P_{\mathrm{cl}}}{P} = \left(\frac{u^2}{v^2}\right)(1-k)\,. \tag{7.98}$$

As expected, the mode power is concentrated in the core far away from the cutoff. As the cutoff is approached, the power of lower order modes ($l = 0, 1$) withdraws into the cladding, whereas modes with $l \geq 2$ maintain a fixed ratio of $(l-1)$ between the power in the core and cladding at the cutoff. The power density is related to the mode power P by

$$p(r) = k\frac{u^2}{v^2}\frac{2P}{\pi a^2}\left[\frac{J_l^2\left(\frac{ur}{a}\right)}{J_l^2(u)}\right]\cos^2 l\phi, \tag{7.99a}$$

$$p(r) = k\frac{u^2}{v^2}\frac{2P}{\pi a^2}\left[\frac{K_l^2\left(\frac{wr}{a}\right)}{K_l^2(w)}\right]\cos^2 l\phi. \tag{7.99b}$$

By averaging over ϕ at $r = a$, we obtain the mean density

$$\bar{p}(r) = k\left(\frac{u^2}{v^2}\right)\frac{P}{\pi a^2}\left[\frac{J_l^2\left(\frac{ur}{a}\right)}{J_l^2(u)}\right] \tag{7.100a}$$

$$\bar{p}(r) = k\left(\frac{u^2}{v^2}\right)\frac{P}{\pi a^2}\left[\frac{K_l^2\left(\frac{wr}{a}\right)}{K_l^2(w)}\right]. \tag{7.100b}$$

At the core–cladding interface, $r = a$ and

$$\bar{p}(a) = k\left(\frac{u^2}{v^2}\right)\frac{P}{\pi a^2}. \tag{7.101}$$

The normalized density $\frac{\pi a^2 \bar{p}(a)}{P}$ is plotted in Fig. 7.10. For modes of order $l = 0, 1$ this density approaches zero at both the cutoff and far away from it and has a maximum in between. Modes with $l \geq 2$ have $\bar{p}(a) = \left[1 - \left(\frac{1}{l}\right)\right]\frac{P}{\pi a^2}$ at the cutoff. For $r \gg \frac{a}{w}$, we can replace the K functions in (7.89) by their approximation for a large argument and obtain

$$\bar{p}(r) \approx k\left(\frac{u^2}{v^2}\right)(P/\pi a r)\exp\left[-2w(r-a)/a\right], \quad \text{for } r \gg a, \tag{7.102}$$

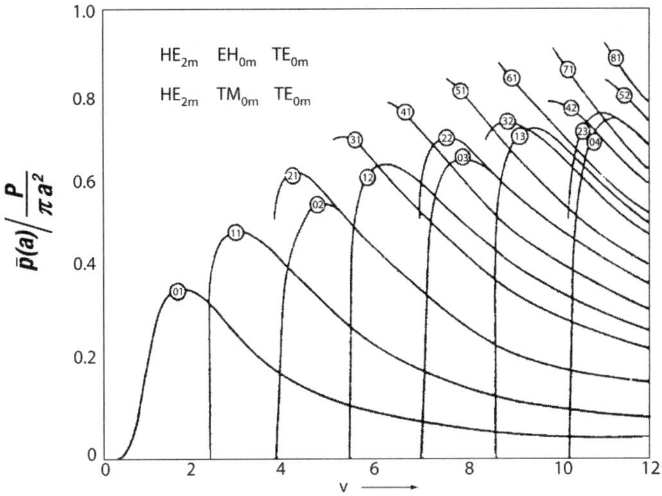

Fig. 7.10. Normalized power density at the core–cladding interface plotted vs. v (after [549])

as long as w is not too small. The power density decreases exponentially with the distance from the interface (see also [542]). It decreases sharply as the cutoff is approached and is zero at cutoff. For sufficiently small w, we may set $u = v$ and replace the K functions in (7.89) by their approximation for a small argument and obtain

$$\bar{p}\,(r) \approx k_1 \left(\frac{P}{\pi a^2}\right)\left(\frac{a}{r}\right)^1 \tag{7.103}$$

for $r > a$, $w = 0$. This function describes the cutoff power distribution in the cladding. It decreases with the distance from the axis for all but the lowest azimuthal order, whose cladding field is independent of the radius.

7.5 Pulse Spreading

Information is normally transmitted along optical fibers by sending out a sequence of pulses of light energy. However, as an individual pulse propagates, it spreads out due to the dispersive properties of the fiber. Clearly, if this spread becomes sufficiently large, the pulse will overlap adjacent pulses, leading to a decrease in information-carrying capacity because of the loss of resolution at the end of the fiber.

Below, a formalism is presented for describing the propagative characteristics of graded-index, multimode fibers. The index profiles of cylindrically symmetrical waveguides can be conveniently specified by the equation

$$n^2(r) = n_1^2 \left[1 - 2\Delta f\left(\frac{r}{a}\right)\right], \tag{7.104}$$

where $n(r)$ is the refractive index of the waveguide as a function of distance r from the axis and n_1 is the index along the axis. The profile function $f\left(\frac{r}{a}\right)$ is defined so that it is zero on the axis

$$f(0) = 0, \tag{7.105}$$

and becomes equal to unity at the core–cladding boundary located at $r = a$,

$$f\left(\frac{r}{a}\right) = 1, \quad \text{for } r \geq a. \tag{7.106}$$

The cladding index n_2 $(= n_{\text{cl}})$ is thus defined as

$$n_2 = n_1 \left(1 - 2\Delta\right)^{\frac{1}{2}}. \tag{7.107}$$

The quantity Δ provides a useful measure of the core–cladding index difference. From (7.107),

$$\Delta = \frac{\left(n_1^2 - n_2^2\right)}{2n_1^2}. \tag{7.108}$$

Each mode of the waveguide can be specified by the pair of integers μ and ν, which, respectively, specify the number of radial nodes and azimuthal nodes in the transverse electromagnetic fields of that mode. The propagative constant $\beta_{\mu\nu}$ of each mode depends explicitly on all quantities that specify the waveguide structure and on the wavelength λ of the propagating light (see also [552, 553]),

$$\beta_{\mu\nu} = \beta_{\mu\nu}\left(n_1, \Delta, a, \lambda\right). \tag{7.109}$$

The propagative constants depend, as usual, on the wavelength explicitly and implicitly through the wavelength variation of n_1 and Δ. Although the index profile $f\left(\frac{r}{a}\right)$ may also vary slightly with wavelength, such effects are not considered.

For analyzing pulse transmission, one is concerned with the group delay time per unit length for the mode μ, ν. This is given by

$$\tau_{\mu\nu} = \frac{\mathrm{d}\beta_{\mu\nu}}{\mathrm{d}\omega}. \tag{7.110}$$

If the free space propagative constant, $k = \frac{2\pi}{\lambda}$, is introduced, (7.110) can be rewritten as

$$\tau_{\mu\nu} = \frac{1}{c}\frac{\mathrm{d}\beta_{\mu\nu}}{\mathrm{d}k}. \tag{7.111}$$

If the fiber is excited by an impulse excitation and if there is no mode coupling, the impulse response, $P(t, z, \lambda)$, for spectral component λ at position z can be written as

$$P(t, z, \lambda) = \sum P_{\mu\nu}(\lambda, z)\, \delta\left[t - z\tau_{\mu\nu}(\lambda)\right], \qquad (7.112)$$

where the summation extends over all guided modes. The distribution functions $P_{\mu\nu}(\lambda, z)$ describe the power in mode μ, ν as a function of wavelength and position. At $z = 0$, the distribution function is be determined by the spatial, angular, and spectral distribution of the source, as well as by the source–fiber coupling configuration. As the impulse propagates along the waveguide, the power in each mode changes according to the attenuation in that mode. Further, it will be assumed that no mode coupling occurs.

In general, only the total power integrated over all source wavelengths is detected. Hence, the quantity of practical interest is the full impulse response

$$P(t, z) = \int_0^\infty \mathrm{d}\lambda P(t, z, \lambda). \qquad (7.113)$$

The propagation characteristics of the fiber can be described by specifying the moments $M_\mathrm{n}(z)$ of the full impulse response. These moments are defined by

$$M_\mathrm{n}(z) = \int_0^\infty \mathrm{d}t\, t^n P(t, z). \qquad (7.114)$$

In some situations, knowledge of only the first few moments is sufficient for system design. If this is the case, the required amount of pulse broadening information is reduced considerably. Equations (7.112)–(7.114) can be combined to yield

$$M_\mathrm{n}(z) = z^n \int_0^\infty \mathrm{d}\lambda \sum P_{\mu\nu}(\lambda, z)\, \tau^n_{\mu\nu}(\lambda). \qquad (7.115)$$

The predominant wavelength dependence of the distribution function $P_{\mu\nu}(\lambda, z)$ is determined by the spectral distribution $S(\lambda)$ of the source. Even for relatively broad LED sources, $S(\lambda)$ is a sharply peaked function whose rms width, at most, is a few percent of the mean source wavelength. One can thus define a new distribution function $p_{\mu\nu}(\lambda, z)$ by the expression

$$P_{\mu\nu}(\lambda, z) = (\lambda) p_{\mu\nu}(\lambda, z), \qquad (7.116)$$

where $p_{\mu\nu}(\lambda, z)$ is a slowly varying function of λ over the range where $S(\lambda)$ is nonzero.

Proceeding with the analysis, it can be assumed that $S(\lambda)$ is normalized, so that

$$\int_0^\infty \mathrm{d}\lambda S(\lambda) = 1. \qquad (7.117)$$

Consequently, the mean source wavelength λ_0 is given by

$$\lambda_0 = \int_0^\infty \mathrm{d}\lambda \lambda S(\lambda), \tag{7.118}$$

and the root mean square (rms) spectral width of the source σ_s is given by

$$\sigma_s = \left[\int_0^\infty \mathrm{d}\lambda \left(\lambda - \lambda_0\right)^2 S\left(\lambda\right)\right]^{\frac{1}{2}}. \tag{7.119}$$

The influence of the sources spectral distribution on the fiber's transmission properties can be studied by expanding the delay per unit length of the μ, νth mode, $\tau_{\mu\nu}(\lambda)$, in a Taylor series about λ_0. Substituting this series in (7.115) and using (7.116) give

$$\begin{aligned} M_n(z) = z^n &\int_0^\infty \mathrm{d}\lambda S(\lambda) \\ &\times \sum_{\mu\nu} p\left(z\right) \Big[\tau_{\mu\nu}^n\left(\lambda_0\right) + n\left(\lambda - \lambda_0\right) \tau_{\mu\nu}^{n-1}\left(\lambda_0\right) \tau_{\mu\nu}'\left(\lambda_0\right) \\ &+ n\left(\lambda - \lambda_0\right)^2 / 2\tau_{\mu\nu}^{n-1}\left(\lambda_0\right) \tau_{\mu\nu}''\left(\lambda_0\right) + \cdots \Big]. \end{aligned} \tag{7.120}$$

Treating $p_{\mu\nu}$ as independent of λ, (7.117)–(7.119) can be used to integrate (7.120) to find that

$$\begin{aligned} M_n(z) = z^n \sum p_{\mu\nu}(z) \Big(&\tau_{\mu\nu}^n\left(\lambda_0\right) + \sigma_s^2 / \left(2\lambda_0^2\right) \\ &\times \left\{n\tau_{\mu\nu}^{n-1}\left(\lambda_0\right) \lambda_0^2 \tau_{\mu\nu}''\left(\lambda_0\right) + n(n-1)\tau_{\mu\nu}^{n-2}\left(\lambda_0\right) \left[\lambda_0 \tau_{\mu\nu}'\left(\lambda_0\right)\right]^2\right\}\Big) \\ &+ 0\left(\frac{\sigma_s^3}{\lambda_0^3}\right). \end{aligned} \tag{7.121}$$

The small size of $\frac{\sigma_s}{\lambda_0}$ allows us to neglect higher order terms. The following quantities are most useful in describing the energy distribution at z. By definition, the total power arriving at z is given by

$$M_0(z) = \sum p_{\mu\nu}(z). \tag{7.122}$$

The mean delay time of the pulse $\tau(z)$ is given by

$$\tau\left(z\right) = \frac{M_1(z)}{M_0(z)}, \tag{7.123}$$

and the rms pulse width $\sigma(z)$ by

$$\sigma(z) = \left[M_2(z)/M_0(z) - \tau^2(z)\right]^{\frac{1}{2}}. \tag{7.124}$$

Combinations of higher moments further describe the power distribution, but the first three are the most important. To simplify the notation required in the following expressions, the symbol $\langle\,\rangle$ will be used to indicate the average value of a quantity with respect to the distribution $p_{\mu\nu}$. So, for example,

$$\langle A\rangle \equiv \sum p_{\mu\nu}(z)A_{\mu\nu}/M_0. \tag{7.125}$$

From (7.112)–(7.115), the full pulse delay time is

$$\tau(z) = z\left[\langle\tau\left(\lambda_0\right)\rangle + \sigma_{\mathrm{s}}^2/(2\lambda_0^2)\left\langle\lambda_0^2\tau''\left(\lambda_0\right)\right\rangle\right]. \tag{7.126}$$

To specify the fiber bandwidth for digital systems, Personick [552] has shown that one is primarily concerned with the rms width $\sigma(z)$ given by

$$\sigma(z) = (\sigma_{\text{intermodal}}^2 + \sigma_{\text{intramodal}}^2)^{\frac{1}{2}} + 0\left(\frac{\sigma_{\mathrm{s}}^3}{\lambda_0^3}\right), \tag{7.127}$$

where the definitions

$$\begin{aligned}\sigma_{\text{intramodal}}^2 &= z^2\Big\{\left\langle\tau^2\left(\lambda_0\right)\right\rangle - \left\langle\tau\left(\lambda_0\right)\right\rangle^2 \\ &\quad + \frac{\sigma_{\mathrm{s}}^2}{\lambda_0^2}\left[\lambda_0^2\tau''\left(\lambda_0\right)\tau\left(\lambda_0\right) - \left\langle\lambda_0^2\tau''\left(\lambda_0\right)\right\rangle\left\langle\tau\left(\lambda_0\right)\right\rangle\right]\Big\} \qquad (7.128)\\ \sigma_{\text{intramodal}}^2 &= z^2\frac{\sigma_{\mathrm{s}}^2}{\lambda_0^2}\left\langle\left[\lambda_0\tau'\left(\lambda_0\right)\right]^2\right\rangle \qquad (7.129)\end{aligned}$$

have been introduced. We can see that the square of the rms width has been separated into an intermodal and an intramodal component. The intermodal term (7.128) results from delay differences among the modes and vanishes only if all delay differences vanish. This term contains a dominant term and a small correction that is proportional to the square of the relative source's spectral width $\left(\frac{\sigma_s}{\lambda_0}\right)$. For the refractive index profiles considered below, this term is negligible. The intramodal term (7.119) represents an average of the pulse broadening within each mode. It becomes the only term in the dispersion of a single mode waveguide. Intramodal dispersion arises from two distinct effects, a pure material effect that corresponds to pulse broadening in bulk material and a waveguide effect. This separation can be made by writing the modal delay time in the form

$$\tau_{\mu\nu} = N_1/c + \delta\tau_{\mu\nu}, \tag{7.130}$$

where N_1 is the material group index,

$$N_1 = n_1 - \lambda\mathrm{d}n_1/\mathrm{d}\lambda, \tag{7.131}$$

and $\delta\tau_{\mu\nu}$ represents the correction to this introduced by the waveguide structure. The derivative $\tau'_{\mu\nu}$ can be written as

$$\tau'_{\mu\nu} = -\lambda n''_1 + \delta\tau'_{\mu\nu}. \tag{7.132}$$

Since the intramodal contribution to the total rms pulse width is obtained by squaring and averaging over (7.132), one can write the intramodal contributions as

$$\sigma^2_{\text{intramodal}} = z^2 \frac{\sigma_s^2}{\lambda_0^2} \left[\left(\lambda_0^2 n''_1\right)^2 - 2\lambda_0^2 n''_1 \left\langle \lambda_0 \delta\tau' \right\rangle + \left\langle \left(\lambda_0 \delta\tau'\right)^2 \right\rangle \right]. \tag{7.133}$$

Hence, the total intramodal contribution has a pure material component, a waveguide component, and a mixed component arising from the cross product. For graded-index waveguides, the derivative of the intermodal delay differences $\delta\tau'_{\mu\nu}$ *is* also small; therefore, the intramodal contribution is dominated by the pure material term,

$$\sigma_{\text{intramodal}} \approx \frac{\sigma_s}{\lambda_0} \left(\lambda_0^2 n''_1\right). \tag{7.134}$$

This quantity represents the ultimate lower limit of pulse broadening. According to [553], the meaning of $\sigma_{\text{intermodal}}$ is

$$\sigma_{\text{intermodal}} = \frac{LN_1\Delta}{2c} \frac{\alpha}{\alpha+1} \left(\frac{\alpha+2}{3\alpha+2}\right)^{\frac{1}{2}} \times \left[C_1^2 + \frac{4C_1C_2\Delta(\alpha+1)}{2\alpha+1} + \frac{4\Delta^2 C_2^2 (2\alpha+2)^2}{(5\alpha+2)(3\alpha+2)} \right]^{\frac{1}{2}}, \tag{7.135}$$

where

$$C_1 = \frac{\alpha - 2 - \varepsilon}{\alpha + 2}, \tag{7.136}$$

and

$$C_2 = \frac{3\alpha - 2 - 2\varepsilon}{2(\alpha+2)}. \tag{7.137}$$

As shown in [554], the minimum of the intermodal rms width occurs for (see also [544])

$$\alpha_c = 2 + \varepsilon - \Delta \frac{(4+\varepsilon)(3+\varepsilon)}{(5+2\varepsilon)}. \tag{7.138}$$

Here, $\varepsilon = \frac{-2n_1\lambda\Delta'}{N_1\Delta}$, $\alpha = 2 + \varepsilon$ and $N_1 = n_1$. The last term on the right of (7.138) represents a small correction of the optimal α. This correction results from a partial cancellation that occurs between the two mode dependent terms, if $\alpha - 2 - \varepsilon$ is of the order of Δ. The size of this correction changes as the distribution of power among the modes varies.

Figure 7.11A shows the fitted values of $n_1(\lambda)$ and $n_2(\lambda)$ in the range from 0.5 to 1.1 µm. Figures 7.11B and 7.11C, respectively, show the first and second derivatives of the refractive index of these two glasses. Figure 7.12 shows the values of Δ and $\lambda\Delta'$ determined from the data. It can be observed in Fig. 7.12a that, over the range of wavelength shown, Δ decreases by about 15%. From the index data in Figs. 7.11 and 7.12, $\varepsilon(\lambda)$ can be calculated, and a plot of α_c versus λ is shown in Fig. 7.13. The value of α_c minimizing the intermodal rms pulse width departs significantly from the optimal profile $\alpha = 2(1 - 6\Delta/5)$, which is predicted if the effect of material dispersion is ignored. The optimal α varies quite strongly with wavelength and decreases from about 2.5 at 500 nm to 2.2 at 1000 nm. A waveguide with a given index profile α is thus predicted that shows different pulse widths according to

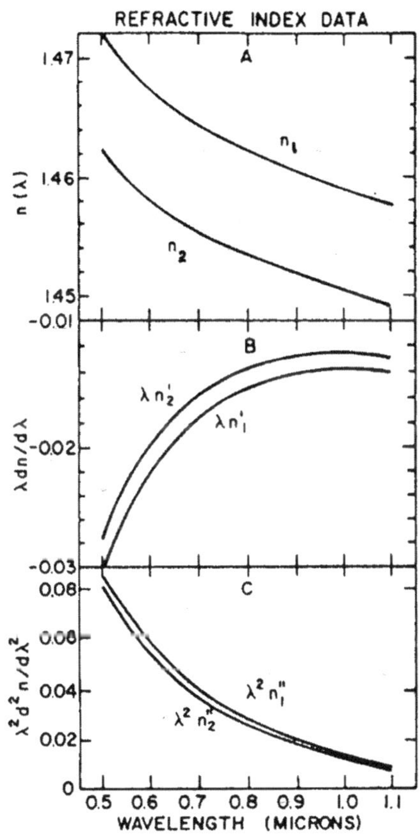

Fig. 7.11. Refractive index data for 3.4 wt% TiO_2-doped silica (n_2) and fused silica (n_1) for **(A)** Sellmeir fit to the refractive index; **(B)** $\lambda dn/d\lambda$, **(C)** $\lambda^2 d^2 n/d\lambda^2$ (after [553])

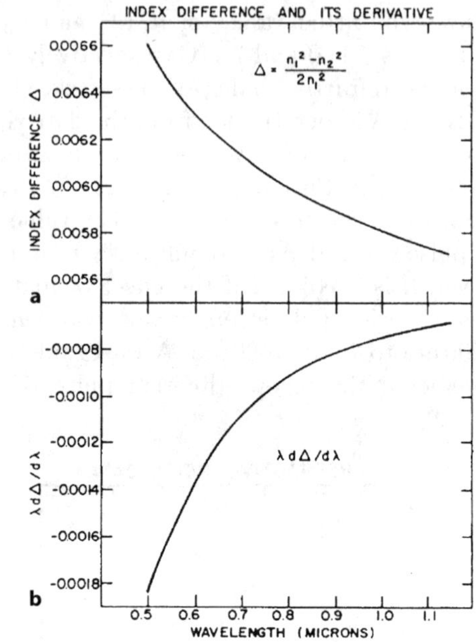

Fig. 7.12. The index difference Δ determined from the data of Fig. 7.11A is shown in **(a)**, and the derivative $\lambda \mathrm{d}\Delta/\mathrm{d}\lambda$ in **(b)**. (After [553])

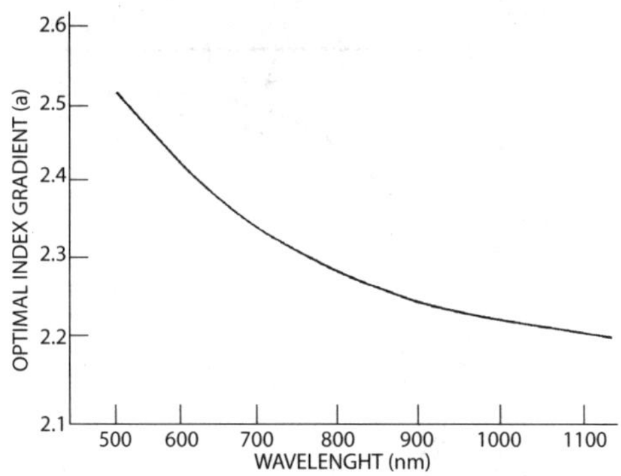

Fig. 7.13. The α value that minimizes pulse broadening as a function of wavelength (after [553])

the source wavelength used. This wavelength-dependent pulse broadening provides a tool for observing these effects and is discussed at length in [553].

According to [553], the delay time can be used to evaluate $\lambda\tau_{\mathrm{n}}'$, which is required for predicting the intramodal broadening, in the form,

$$\lambda\tau_{\mathrm{n}}' = -\lambda^2 n_1'' + N_1\Delta\left(\frac{\alpha-2-\varepsilon}{\alpha+2}\right)\left(\frac{2\alpha}{\alpha+2}\right)\left(\frac{n}{N}\right)^{\frac{\alpha}{\alpha+2}} . \tag{7.139}$$

Since $\lambda^2 n_1''$ and Δ are the same order of magnitude, both terms contribute to $\lambda n_{\mathrm{n}}'$ for large α (but with opposite signs). For α near α_{c}, the pure material term in (7.129) dominates. This reflects the fact that when the intermodal delay differences are small, the derivatives of these differences are also small. Assuming equal excitation of modes, one can perform summation over all modes required by (7.119) and find that

$$\sigma_{\text{intramodal}} = \frac{\sigma_\lambda}{\lambda}\left[\left(-\lambda^2 n_1''\right)^2 - 2\lambda^2 n_1''\left(N_1\Delta\right)\left(\frac{\alpha-2-\varepsilon}{\alpha+2}\right)\left(\frac{2\alpha}{2\alpha+2}\right)\right.$$
$$\left. + \left(N_1\Delta\right)^2\left(\frac{\alpha-2-\varepsilon}{\alpha+2}\right)^2\left(\frac{2\alpha}{3\alpha+2}\right)\right]^{\frac{1}{2}} . \qquad (7.140)$$

The total rms pulse widths for α profiles can be predicted from (7.135) and (7.140). In Fig. 7.14, this rms pulse width is shown as a function of α for three types of GaAs sources; all operate at $\lambda = 0.9\,\mu\mathrm{m}$ but have differential spectral bandwidths. The three curves correspond to an LED, an injection laser, and a distributed-feedback laser having typical rms spectral widths of

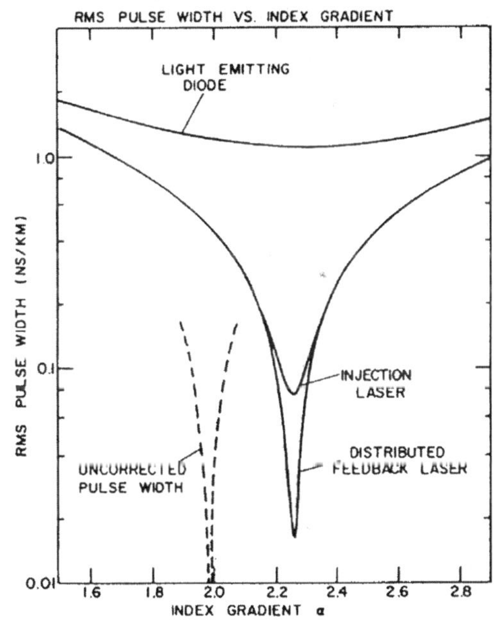

Fig. 7.14. Assuming equal power in all modes, the rms pulse width is shown as a function of α for three different sources; all operate at $0.9\,\mu\mathrm{m}$. The sources are an LED, a gallium arsenide injection laser, and a distributed feedback laser that have rms spectral widths of 150, 10, and 2 Å, respectively. The *dashed curve* shows the pulse width that would be predicted if all material dispersion effects were neglected. (After [554])

150, 10, and 2 Å, respectively. The waveguide is assumed to have $n_1(\lambda)$ and $n_2(\lambda)$ of Fig. 7.11 and to propagate equal power in all modes. For comparison, a dashed curve represents the rms width predicted if material dispersion and intramodal broadening are ignored. We shoud add that for the LED source, pulse broadening of less than 1.5 ns/km can be achieved if α is within 25% of the optimal value. For the injection laser, an α within 5% of the optimal will give pulse widths less than 0.2 ns/km, and for the distributed-feedback laser, widths of 0.05 ns/km are predicted if 1% control of α can be achieved.

In the preceding sections, it was found that the α value that minimizes intermodal pulse dispersion departs significantly from a parabolic profile and that the magnitude of this departure depends on the wavelength of the source. In Fig. 7.14, one sees that the calculated rms pulse width depends strongly on the difference between the α of the waveguide and the optimal α. In Fig. 7.15, the calculated rms width is shown as a function of source wavelength for waveguides with index profiles in the range $1.5 \leq \alpha \leq 2.9$. For the purpose of illustration, it was assumed [553] in these calculations that the rms spectral width of the source is 2 Å. As is known, for this range of α values, both intermodal and intramodal contributions must be considered. The wavelength dependence of the intermodal term depends on the specific α value. For $\alpha \prec 2.2$, the intermodal pulse width decreases as λ increases because for these $\alpha's$ the difference $|\alpha - \alpha_c|$ decreases with λ. The opposite occurs for $\alpha \succ 2.5$, and the intermodal pulse width increases with λ. For waveguides with

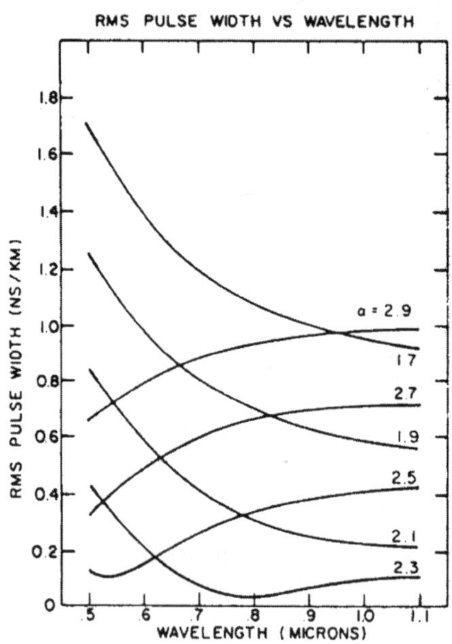

Fig. 7.15. For a source with 2 Å spectral width, the rms pulse width is shown as a function of wavelength for several different values of α. (After [553])

$2.2 \prec \alpha \prec 2.5$, the minimal intermodal broadening occurs at one wavelength in the range $500\,\text{nm} \prec \lambda \prec 1100\,\text{nm}$. As λ is varied through this range, the intermodal broadening decreases until the optimal wavelength is reached, and thereafter it increases. The intramodal contribution to the rms width is dominated by material dispersion. It is largest at the shorter wavelength and decreases quite rapidly with wavelength. For a source spectral width of 2 Å, the rms width decreases from 0.1 to 0.006 ns/km between 500 and 1100 nm. The pulse width behavior shown in Fig. 7.15 reflects the combined effect of intermodal and intramodal pulse broadening.

7.6 Materials for Optical Fibers

The choice of materials for fabricating fibers is influenced by the need to satisfy many requirements simultaneously. Obviously, the material must be formable into a fine filament, transparent, and available in two different refractive index for core and cladding, respectively. These requirements alone more or less limit the field to plastics or glasses, although a liquid has been used to form the core of a fiber drawn from a hollow tube of glass. Many plastics are excluded from further consideration because the hydrogen in their structures gives rise to very high losses and because their molecular size leads to large scattering losses. And within the infinite number of possible glasses, most are ruled out by other considerations. To appreciate this situation more fully, we need to examine in more detail the physical mechanisms involved, particularly those controlling optical loss since most optical communications systems require fibers of exceptionally low attenuation at the optical carrier frequency. Usually, less than 20 dB/km is sought [555, 556].

Glasses are formed from fused mixtures of metal oxides, sulfides, or selenides. Because they are fused mixtures rather than fixed compounds with crystal structures, their compositions are infinitely variable within certain regions of their respective phase diagrams, and large numbers of different glasses are manufactured by industry. Most of these fall into the category of oxide glasses, since these are optically transparent; the sulfide and selenide glasses are used in the infrared region of approximately 0.6 to 14 µm or more

Of the oxide glasses, by far the most common are silica (SiO_2); sodium calcium silicates, frequently used for plate and window glass; sodium borosilicates, often used for oven ware and chemical apparatus; and lead silicates, crystal glasses that have relatively high refractive indexes and thus appear "shiny." Typical starting materials for these glasses are sodium and calcium carbonates, boric oxide (B_2O_3) or boric acid, silica (sand), and lead oxide. For the typical fiber materials, new conditions arise, namely, the need for very low loss, which means very high chemical purity, so that the materials sources used are usually different from those serving large-scale industry. The structure of glasses is noticeably different from that of the solids usually encountered in the electronic industry, namely, crystals. In the latter,

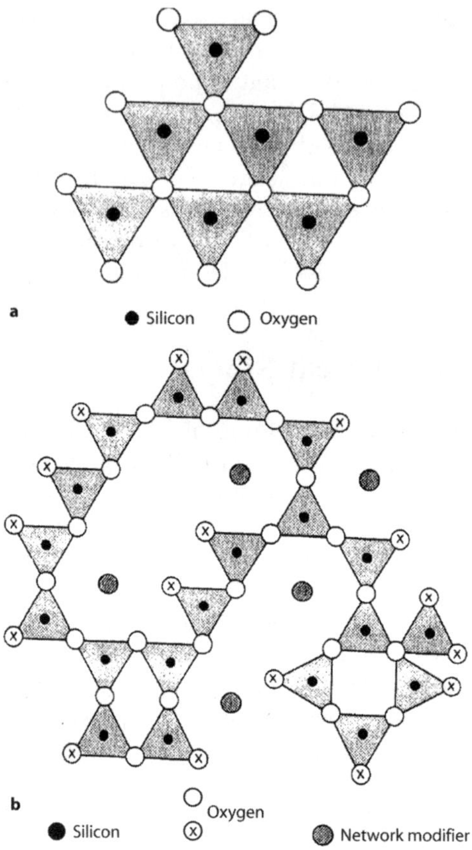

Fig. 7.16. Networks involving SiO_2 groups, shown schematically in two dimensions. **(a)** A regular SiO_2 lattice (two-dimensional quartz crystal). **(b)** The effect on the lattice of addiing of a network modifier. (after [544])

the individual atoms are well defined in space according to very precise and repeated patterns and lie in exact three-dimensional lattices. Glass, on other hand, consists of a loosely connected network formed by groups, which can be added to or modified by other components. For example, the addition of sodium (a network modifier) tends to break up the SiO_2 network, as shown in Fig. 7.16, in a sodium silicate glass. The result is less strongly bound than pure silica, and the melting temperature is thereby lowered. Since B_2O_3 is also a network former like SiO_2, a series of glasses such as sodium borates exist, paralleling sodium silicates but having very much lower melting temperatures since the B–O bond is much weaker than the Si–O bond.

Figure 7.17 shows viscosity–temperature curves for some commonly manufactured glasses. Particularly noteworthy is the fact that the viscosity of a pure silica glass is much higher than that of multicomponent glasses at the same temperature. Just as the viscosity of a glass is a function of its composition, so too are the refractive index and the thermal expansion coefficient. Since there is an infinity of glass compositions, full data are not available for all glasses; in fact, comprehensive data are restricted to a few glass groups that have been extensively studied because of large-scale commercial applications. The sodium calcium silicate (NCS) group is one such. A second glass system of great current interest is the sodium borosilicate (NBS) group of glasses. Another group of glasses that may be of interest in fiber manufacture is lead silicates since they provide large refractive index differences between core and cladding glasses. For example, a pair with indexes of 1.5 and 1.65 showing a 10% index difference can probably be obtained by using highly polarizable lead ions to dope the higher index material. Fibers with such a large index difference are characterized by a higher level of Rayleigh scattering from the lead (of the order 10 dB/km, compared to 1 dB/km for silica at 900 nm) and have not yet been produced with very low loss (i.e., under 15 or 20 dB/km). But for many short link systems applications, these disadvantages are more than offset by the large acceptance angle for such a high index difference fiber. All three groups of glasses are of interest because they can be made by nearly conventional glass melting techniques in large quantities at low cost. They are formed by mixing the appropriate powders and fusing them in a crucible to form a glass. But there is another group of materials, as already noted, all based upon pure silica with small additions of one or more other oxides.

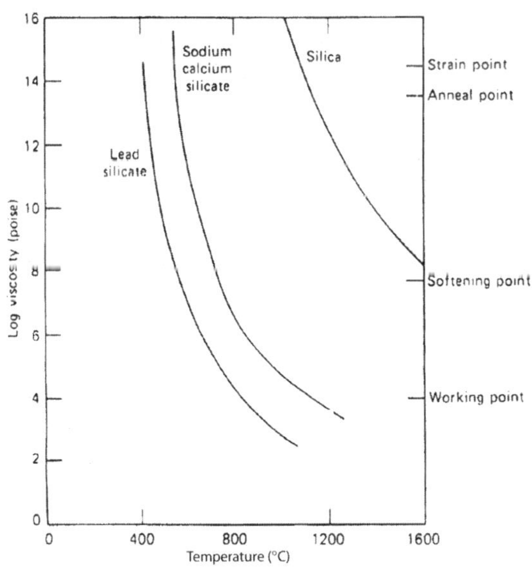

Fig. 7.17. Viscosity versus temperature for a number of glasses (after [544])

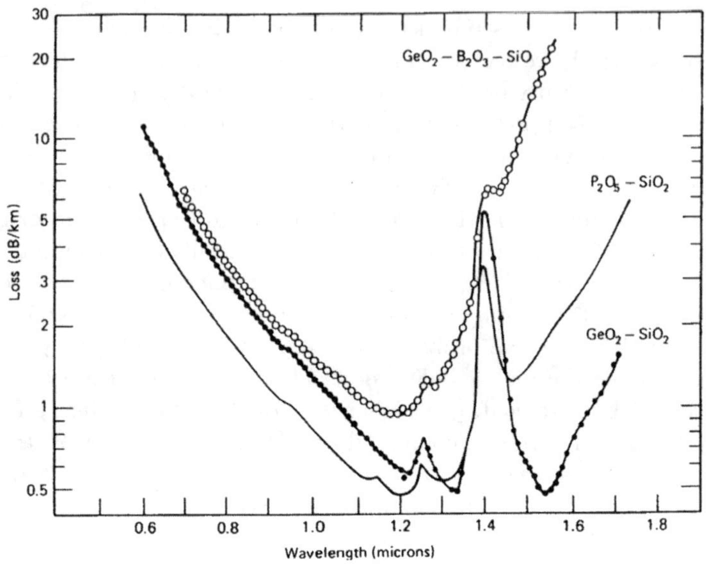

Fig. 7.18. Losses versus wavelength of some low water content fibers showing the matrix influence on infrared transmission (after [564])

7.6.1 Absorptive Losses in Glasses

Optical fibers operate in one of two wavelength bands, the GaAs device region from 800 to 900 nm or the Nd laser region at 1.06 μm, with the remote possibility of an extension to perhaps 1.2 or 1.3 μm using other semiconductor sources (see also [555,556]). This choice is dictated by good physical reasons. Wavelengths shorter than about 600 to 650 nm are ruled out because Rayleigh scattering in the glass becomes severe and impurity absorption spilling over the band edge absorption is a problem. Beyond about 1.3 μm, the first overtone of the OH stretching vibration appears at 1.4 μm, giving heavy absorption. Some spillover from multiphonon bands from the glass constituents probably adds to this, and detectors become far less efficient because the quantum energy $h\nu$ rapidly approaches kT at room temperature.

Five absorption mechanisms are of concern in a glass for use in optical fibers. The first two are associated with the basic glass constituents themselves, typically a combination of oxides of silicon, sodium, boron, calcium, germanium, and so on. The glass has a band edge absorption somewhere in the ultraviolet (UV) region of the spectrum. Such absorption is extremely intense; and although the wavelength of interest for operating a system is a considerable distance away, there has been serious debate as to whether the tail of the band edge could provide a significant loss mechanism (see also [557–562]).

Figure 7.18 shows the losses of some fibers with very low water content made with different constituents in the core by chemical vapor depo-

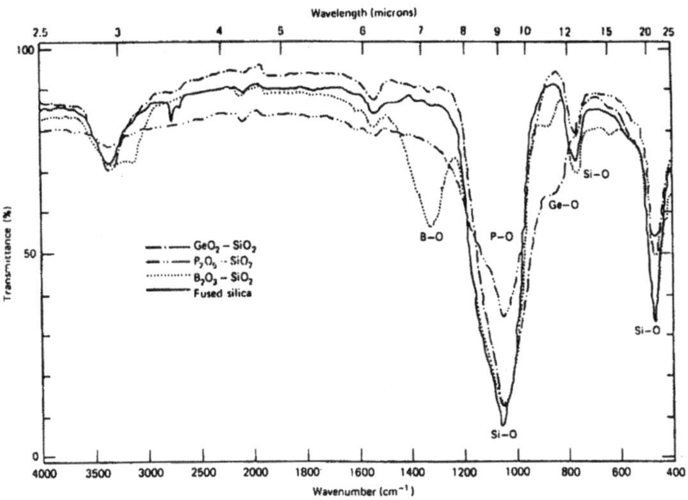

Fig. 7.19. Infrared transmission spectra of the materials associated with the fibers of Fig. 7.18, showing the various absorption bands involved (after [564])

sition [563], and in Fig. 7.19, the infrared spectra of some of these same materials, showing the relative size and position of the infrared absorptions are presented. The work suggests that GeO_2 doping for the core is the most favorable because of the longer wavelength at which GeO_2 stretching vibration occurs, and this leads to an estimate [564] for the lowest loss for this type of fiber, at about 1.5 microns, of about 0.3 dB/km, as shown in Fig. 7.20. Evidently, to obtain figures even remotely approaching these, very low water content must be achieved. The remaining absorption mechanisms are all related to impurities or defects in the glass [560] and are therefore not intrinsic. Besides that, a problem with most optical fibers is reducing the hydroxyl ion (OH) content to a sufficiently low level, usually a few parts per million (for details, see [544]).

7.6.2 Rayleigh Scattering

The phenomena of Rayleigh scattering is well known to all of us as the mechanism responsible for blue sky. It is scattering of light from microirregularities in the dielectric medium through which an electromagnetic wave is propagating. The physical scale of the irregularities is of the order of one-tenth wavelength or less, so that each irregularity acts as a point source for scattered radiation. The resulting relation pattern from the induced dipole is doughnut shaped; it is uniform in the plane perpendicular to the dipole and varies as $\sin\phi$ in the plane containing the dipole, where ϕ is the angle between the observation direction and the dipole axis. In the sky, Rayleigh scattering arises from minute density fluctuation in the atmospheric gas caused by the

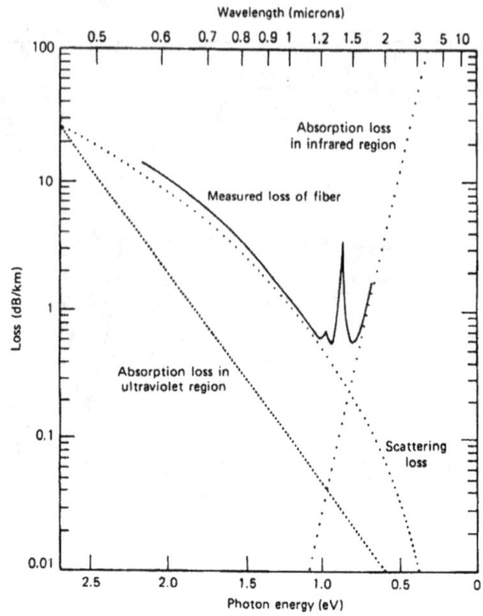

Fig. 7.20. Predicted losses arising from both ultraviolet edge absorption and infrared absorption for GeO_2–SiO_2 fiber (after [563])

constant thermal fluctuation of the medium, whereas the $\sin\phi$ component makes it highly polarized at ground level. In glass, Rayleigh scattering can arise from two separate effects, density and composition fluctuations.

We have already seen that glasses are disordered structures, loosely connected in a largely random sequence. Evidently, in such a structure, there are local regions in which the average density is higher than in other regions. The dense and less dense regions can be traced back to thermally driven fluctuations in density arising from Brownian movement of the liquid glass before it solidified. The magnitude of the fluctuation is thus expected to be related to the melting temperature – the higher the temperature, the greater the density fluctuation. One might, therefore, expect that a high melting material such as silica would show a higher Rayleigh scattering loss than a lead glass of lower melting temperature. But this is not so for another reason.

The magnitude of the density fluctuation scattering is given by the following expression [565]:

$$\alpha_{\mathrm{scat.}\rho} = \frac{8\pi^3}{3\lambda^4}\left(n^8 p^2\right)\left(kT_{\mathrm{f}}\right)\beta_{\mathrm{T}}. \tag{7.141}$$

Here p is the photoelastic coefficient for the glass; T_{f} is the fictive temperature, defined as the temperature at which it becomes possible for the glass to reach a state of thermal equilibrium and is closely related to the anneal temperature; k is Boltzmann's constant; and β_{T} is the isothermal compressibility. Certainly, the energy kT_{f} is the driving energy for the fluctuation that gives rise to the loss mechanism described by $\alpha_{\mathrm{scat.}\rho}$.

The expression for scattering due to composition is more complex and takes the form [565],

$$\alpha_{\mathrm{scat.,c}} = \frac{32\pi^3 n^2}{3\lambda^4 \rho N_{\mathrm{A}}} \sum_{j=1}^{m} \left[\left(\frac{\partial n}{\partial x_j} \right)_{T_{\mathrm{f}}, x_i \neq x_j} + \left(\frac{\partial n}{\partial \rho} \right)_{T_{\mathrm{f}}, x_i} \left(\frac{\partial \rho}{\partial x_j} \right)_{P, T_{\mathrm{f}}, x_i \neq x_j} \right]^2 M_j x_j . \tag{7.142}$$

In this formula, M_j and x_j are, respectively, the molecular weight and the weight fraction of the jth modifier, and N_{A} is Avogadro's number. The partial derivatives of the refractive index and the density, ρ, may be determined from experimental data, so that the summation can be performed. Schroeder et al. [566] derived an expression for the mean composition fluctuation in glass that takes the form

$$\langle \Delta C^2 \rangle_{T_{\mathrm{f}}} = \frac{\rho_0}{C_0} \left(\frac{kT_{\mathrm{f}}}{V} \right) \left(\frac{\partial \mu}{\partial C} \right)^{-1}_{C_0, T_{\mathrm{f}}} , \tag{7.143}$$

where C is the mole fraction of one of the constituents in the composition C_0, V is the molar volume, and μ is the chemical potential difference between the

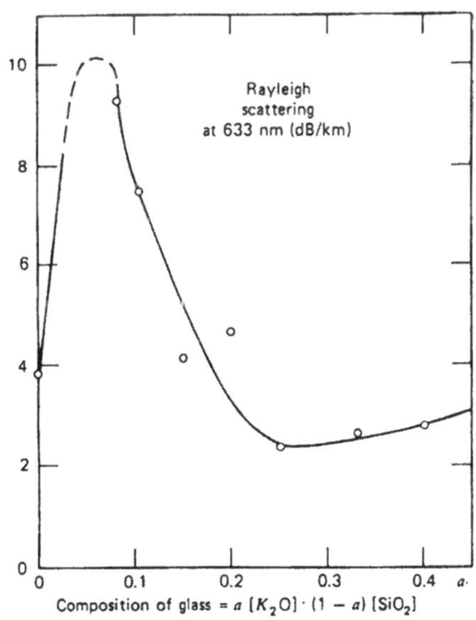

Fig. 7.21. Rayleigh scattering loss at 633 nm versus composition for a sodium silicate glass (after [566])

Table 7.1. Rayleigh scattering losses for a number of fiberglasses (quoted for 900 nm) (after [544])

Glass	Loss (dB/km)
Silica	∼ 1.2
Sodium borosilicate [a]	∼ 1.5
Phosphosilicate [b]	∼ 1.6
Germania silicate [c]	∼ 1.2
Selfoc [d]	∼ 2.0

[a] Ref. [544]
[b] Keck, D.B., Maurer, R.D., Schultz, P.C., Appl. Phys. Lett. **22**, 307 (1973) (0.7% index difference from silica)
[c] Kawachi, M., Kawara, A., Miyashita, T., Electron. Lett. **13**, 442 (1977) (0.18% index difference from silica)
[d] Morey, G.W., The Properties of Glass, 2nd ed. (Reinhold, New York,1954)

major and minor constituents of the (binary)glass. Evidently, the composition fluctuation is minimized if $\frac{\partial \mu}{\partial C}$ is large and the fictive temperature is small.

Since the refractive index of a glass is a sensitive function of composition, frequently, the composition fluctuation swamps the density fluctuation component. Thus, a low melting glass may well have a much higher scatter loss than silica, and in general this is bound to be the case. However, some particular glass compositions have been found that lead to exceptionally low total Rayleigh scattering losses. Potassium silicates [566] are one such group which show a minimum scattering loss at one composition, as shown in Fig. 7.21. In Table 7.1 are listed the Rayleigh scattering at the 900 nm wavelength for a number of glasses that are of interest for optical fiber systems.

7.7 Fiber Preparation

In this section, we briefly consider the methods by which different materials can be formed into fibers with desired properties. Several techniques have been described for melting glasses that are specifically intended for the preparation of optical fibers. Each is based on using powders that are premixed, heated in a crucible until they fuse, and then agitated to produce a homogeneous mix. The crucible may be heated through blackbody radiative coupling from the walls of an electrically heated furnace; it may be generated by the coupling of radio-frequency (rf) radiation to the crucible when the crucible is made of a conducting material such as platinum; or, alternatively, rf energy may be coupled directly to the melt glass, provided that is preheated to a temperature at which it begins to conduct. In the last case, the crucible remains relatively cold, thus helping to reduce crucible contamination. In all cases, it is necessary to use an enclosure for the melting which excludes contamination from the laboratory environment. Typically, silica liners are used to provide such isolation.

In an electric furnace, energy is transferred from the furnace wall to the crucible by blackbody radiation. A great deal of early melting work was done using platinum crucibles contained within silica enclosures for clean lines, but the heat was applied through rf energy coupled from an encircling coil, as shown in Fig. 7.22. This technique has the attraction that there is no refractory insulating anywhere near the melt [568]. However, precise temperature control is generally more difficult, the thermal time constant of the apparatus is much shorter, and the apparatus is far more expensive. It also requires a conducting crucible, ruling out the use of silica.

The production of fiber using a double-crucible apparatus is conceptually extremely simple [569, 570]. The basic apparatus is illustrated in Fig. 7.23. Two concentric crucibles are held with their axes vertical. Each crucible has a central circular nozzle in its base, the inner is carefully aligned concentrically with the outer and perhaps 1 cm above it. The inner crucible is filled with

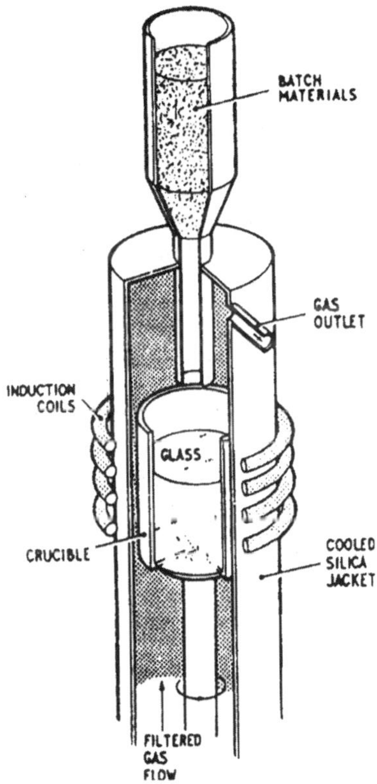

Fig. 7.22. Radio-frequency heating of a platinum crucible for glass melting in a clean enclosure (after [567])

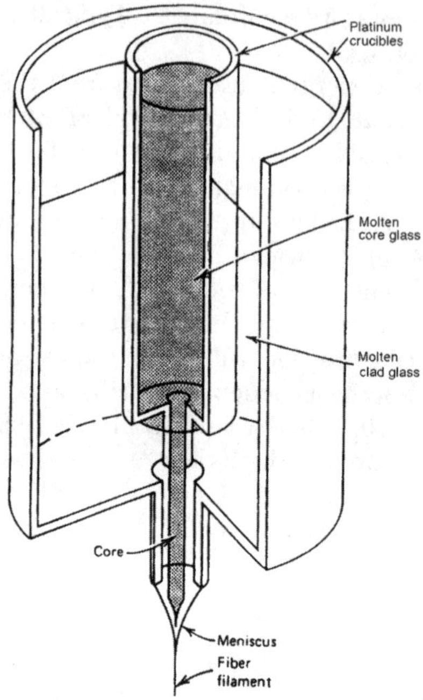

Fig. 7.23. Apparatus for double crucible fiber pulling (after [567])

core composition glass, and the outer with cladding composition glass. When the glasses are molten, the core glass flows through the inner nozzle into the cladding stream and is subsequently carried out of the cladding outer nozzle surrounded by cladding glass, so that a composite glass flow is produced. The molten glass exudes into the space outside the outer crucible, and a filament is pulled from the exudant to form a cored glass fiber. In practice, the production of low loss fiber by this process requires a great deal of careful attention to detail at every stage of the process: glass preparation, cleaning of crucibles, assembly of apparatus, loading of glass, atmosphere control, stabilization of temperature, and so on.

We should note that increasing interest centers upon the deposition of more than one dopant for a given fiber and also upon depositing multiple layers; the concentration in each layer is carefully monitored and varied in a predetermined manner. In the former category, a typical deposition is of B_2O_3–SiO_2 to form cladding and GeO_2–SiO_2 to form the core. Such a pair could be chosen so that the viscosities of the two layers are more equally matched, thus facilitating the collapse stage. Also, it is found that the presence of some of the dopants allows greatly increased gas flow and reaction rates, a factor of considerable commercial importance in implementing a process.

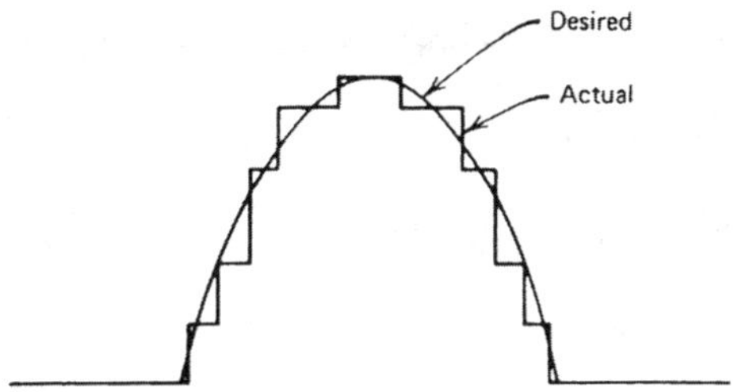

Fig. 7.24. A schematic picture of the index distribution of a graded core fiber produced by layer deposition (after [544])

The presence of layers arising from the deposition process is sometimes clearly evident in the preforms and fibers. This appears to be particularly true of CVD (chemical vapor deposition) fiber and less so of outside deposition poor fiber. Interferometric analysis of the index profiles of preforms shows an "oscillating" index profile following a mean curve of the form desired, as illustrated schematically in Fig. 7.24. To conclude this section, we mention a new class of fibers – the holey fibers, which use another technology (for details, see [571] and references therein)

7.8 Isotopes in Fibers

As a first example of possible major applications of isotopic engineering, we consider isotopic fiber optics and isotopic optoelectronics at large (see also [23, 349, 423]). It is known that for typical solids, the lattice constant variations of isotopically different samples are usually within the limits (see Sect. 3)

$$\frac{\Delta d}{d} \sim 10^{-3} - 10^{-4}. \tag{7.144}$$

Let us define an isotopic fiber as a structure in which the core and cladding have the same chemical content but a different isotopic composition. The boundary between different isotopic regions forms an isotopic interface. The difference in the refractive index on both sides of the isotopic interface could lead to the possibility of total internal reflection of light and, consequently, could provide an alternative route to the confinement of light. For a quantitative estimate, let us consider a boundary between SiO_2 (the main component of silica) where the body sides are identical chemically and structurally but have a different isotopic content, e.g., $^{28}Si^{16}O_2$ and $^{30}Si^{18}O_2$, respectively (Fig. 7.25). In the first approximation, the refractive index n is proportional to the number of light scatterers in a unit volume. From the Clausius–Mosotti relation for refractive index, one can deduce the following proportion (at $\Delta n \ll n$):

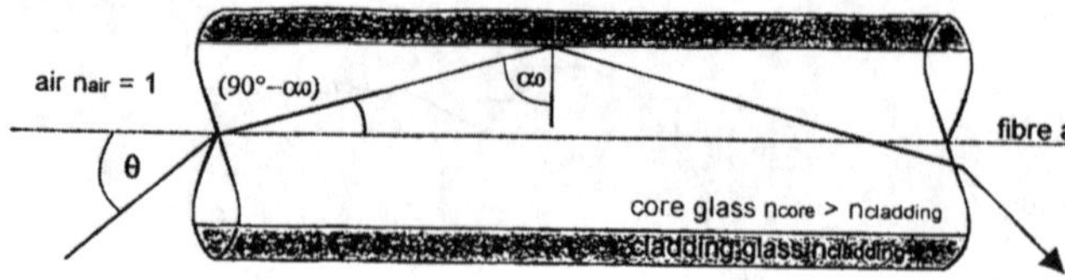

Fig. 7.25. Light guidance by total internal reflection in an optical waveguide

$$\frac{\Delta n}{n} \simeq 3c \frac{\Delta d}{d}, \tag{7.145}$$

where c is a dimensionless adjustment factor of the order of unity. Substituting (7.144) in (7.145),

$$\frac{\Delta n}{n} \sim 3 \cdot 10^{-3} - 10^{-4}. \tag{7.146}$$

Using Snell's law of light refraction, we obtain the following expression for the ray bending angle Φ when light travels through a refractive boundary:

$$\theta \simeq \arcsin \left(\frac{n_1}{n_2} \sin \alpha_0 \right), \tag{7.147}$$

where α is the angle between the falling ray and the direction normal to the interface. For a sliding ray ($\alpha \simeq 90°$), which is the control case for light confinement in fibers, combining (7.146) and (7.147) leads to an estimated $\theta \sim 1.5 - 4.5°$.

Thus, in isotopic fibers in which the core and cladding are made of different isotopes, the half-angle of the acceptance cone could be up to several degrees. The resulting lattice mismatch and strains at the isotopic boundaries are correspondingly one part per few thousand [23] and, therefore, can be tolerated. Further advancement of this "isotopic option" could open the way for an essentially monolithic optical chip with built-in isotopic channels inside a fully integrated and chemically uniform structure.

Besides that, we should mention the fact that composition (different isotopes) fluctuations are subject to the restoring force of the total free energy of the glass system which also seeks to minimize itself. Using isotopically pure materials for the core and cladding, we should have significantly less Rayleigh scattering (see, e.g., [27]).

8 Laser Materials

8.1 Some General Remarks

As is well known, the word laser is an acronym for "light amplification by stimulated emission of radiation," a phrase which covers most, though not all, of the key physical processes inside a laser. Unfortunately, that concise definition may not be very enlightening to the nonspecialist who wants to use a laser but has less concern about the internal physics than the external characteristics. A general knowledge of laser physics is as helpful to the laser user as a general understanding of semiconductor physics is to the circuit designer. From a practical standpoint, a laser can be considered a source of a narrow beam of monochromatic, coherent light in the visible, infrared, or UV parts of the spectrum. The power in a continuous beam can range from a fraction of a milliwatt to around 20 kilowatts (kW) in commercial lasers, and up to more than a megawatt in special military lasers. Pulsed lasers can deliver much higher peak power during a pulse, although the average power levels (including intervals while the laser is off and on) are comparable to those of continuous lasers.

The range of laser devices is broad. The laser medium, or material emitting the laser beam, can be a gas, liquid, crystalline solid, or semiconductor crystal and can range in size from a grain of salt to filling the inside of a moderately sized building. Not every laser produces a narrow beam of monochromatic, coherent light. A typical laser beam has a divergence angle of around a milliradian, meaning that it will spread to one meter in diameter after traveling a kilometer. This figure can vary widely depending on the type of laser and the optics used with it, but in any case it concentrates the output power onto a small area. Semiconductor diode lasers, for example, produce beams that spread out over an angle of 20° to 40° hardly a pencil-thin beam. Liquid dye lasers emit at a range of broad or narrow wavelengths depending on the optics used with them. Other types emit at a number of spectral lines, producing light that is neither truly monochromatic nor coherent. Practically speaking, lasers contain three key elements. One is the laser medium itself, which generates the laser light. The second is the power supply, which delivers energy to the laser medium in the form needed to excite it to emit light. The third is the optical cavity or resonator, which concentrates the light to stimulate the emission of laser radiation. All three elements can take various forms, and although they are not always immediately evident in all types of lasers, their functions are essential. Like most other light sources, lasers are inefficient in converting input energy into light; efficiencies can range from less than 0.01% to around 20% [572–575].

8.2 Absorption and Induced Emission

The idea of stimulated emission of radiation, as is well known, originated with Albert Einstein [576]. Until that time, physicists had believed that a photon could interact with an atom in only two ways: it could be absorbed and raise the atom to a higher energy level or be emitted as the atom dropped to a lower energy level. Einstein proposed a third possibility – that a photon with energy corresponding to that of an energy-level transition could stimulate an atom in the upper level to drop to the lower level, in the process stimulating the emission of another photon with the same energy as the first. In the normal world, stimulated emission is unlikely because at thermodynamic equilibrium, more atoms are in lower energy levels than in higher ones. Thus a photon is much more likely to encounter an atom in a lower level and be absorbed than to encounter one in a higher level and stimulate emission.

Below, we consider a cavity whose walls are at temperature T, containing radiation and an ensemble of atoms. Let each atom be represented by a two-level quantum mechanical system with an energy level separation of $\hbar\omega_0$. In thermal equilibrium, the energy density per unit angular frequency range at ω_0 is given by [577]

$$\rho_{\omega_0} = \frac{\hbar\omega_0^3}{\pi^2 c^3} \frac{n^3}{\mathrm{e}^{\frac{\hbar\omega_0}{kT}} - 1} = \frac{\omega_0^2}{\pi^2 c^3} \frac{\hbar\omega_0 n^3}{\mathrm{e}^{\frac{\hbar\omega_0}{kT}} - 1}. \tag{8.1}$$

Then,

$$\rho_{\omega_0}\left(\mathrm{e}^{\frac{\hbar\omega_0}{kT}} - 1\right) = \frac{\hbar\omega_0^3 n^3}{\pi^2 c^3}. \tag{8.2}$$

In addition, because of detailed balance,

$$AN_2^e + B_{21}\rho_{\omega_0}N_2^e = B_{12}\rho_{\omega_0}N_1^e, \tag{8.3}$$

where N_1^e and N_2^e are the equilibrium populations of atoms in the lower and upper levels, respectively; $B_{21}\rho_{\omega_0}$ and $B_{12}\rho_{\omega_0}$ are the probabilities per unit time for induced downward and upward transitions, respectively, and A is the Einstein coefficient.

Now, we can write

$$A\frac{N_2^e}{N_1^e} + B_{21}\rho_{\omega_0}\frac{N_2^e}{N_1^e} = B_{12}\rho_{\omega_0} \tag{8.4}$$

and

$$Ae^{-\frac{\hbar\omega_0}{kT}} + B_{21}\rho_{\omega_0}e^{-\frac{\hbar\omega_0}{kT}} = B_{12}\rho_{\omega_0} \tag{8.5}$$

or

$$\rho_{\omega_0}\left(B_{12}e^{-\frac{\hbar\omega_0}{kT}} - B_{21}\right) = A. \tag{8.6}$$

Set

$$B_{12} = B_{21} = B, \tag{8.7}$$

$$\rho_{\omega_0}\left(e^{-\frac{\hbar\omega_0}{kT}} - 1\right)B = A, \tag{8.8}$$

$$\frac{A}{B} = \rho_{\omega_0}\left(e^{-\frac{\hbar\omega_0}{kT}} - 1\right) = \frac{\hbar\omega_0^3 n^3}{\pi^2 c^3}, \tag{8.9}$$

$$B = \frac{\pi^2 c^3}{\hbar\omega_0^3 n^3}A = \frac{\pi^2 c^3}{n^3\hbar\omega_0^3}\frac{4n\,\omega_0^3\,|\mu|^2}{3\hbar c^3} = \frac{4\pi^2}{3\hbar^2 n^2}|\mu|^2 \tag{8.10}$$

or

$$A = \frac{4n\,\omega_0^3}{3\hbar c^3}|\mu|^2\,(\mathrm{s}^{-1}),$$

$$B = \frac{4\pi^2}{3\hbar^2 n^2}|\mu|^2\,(\mathrm{cm}^3\mathrm{erg}^{-1}\mathrm{s}^{-2}), \tag{8.11}$$

$$\frac{A}{B} = \frac{\hbar\omega_0^3 n^3}{\pi^2 c^3}(\mathrm{erg\,s\,cm}^{-3}). \tag{8.12}$$

Let us call $w(\omega)\mathrm{d}\omega$ the probability per unit time that an atom undergoes an induced transition by absorbing or emitting a photon with angular frequency $(\omega, \omega + \mathrm{d}\omega)$:

$$w(\omega)\,\mathrm{d}\omega = Bg(\omega)\,\rho_\omega\mathrm{d}\omega = \frac{4\pi^2}{3\hbar^2 n^2}|\mu|^2\rho_\omega g(\omega)\,\mathrm{d}\omega = \frac{4\pi^2}{3nc\hbar^2}|\mu|^2 I(\omega)\,g(\omega)\,\mathrm{d}\omega, \tag{8.13}$$

where

$$I(\omega)\,\mathrm{d}\omega = \rho_\omega\frac{c}{n}\mathrm{d}\omega \tag{8.14}$$

$$= \text{intensity of radiation with angular frequency in } (\omega, \omega + \mathrm{d}\omega).$$

Now, we can write

$$\int w(\omega)\,\mathrm{d}\omega = \frac{4\pi^2}{3\hbar^2 n^2}|\mu|^2\rho_\omega = \frac{4\pi^2}{3\hbar^2 n\,c}|\mu|^2 I(\omega_0). \tag{8.15}$$

Further let us assume that a plane wave goes through a certain medium in the x direction. Let the medium consist of atoms that have two possible energy levels, and let $N_1(N_2)$ be the concentration of atoms in the lower (higher) energy level (see also [573])

$$\mathrm{d}I(\omega) = -w(\omega)(N_1 - N_2)\hbar\omega \mathrm{d}x \left(\frac{\mathrm{erg}}{\mathrm{cm}^2}\right) \tag{8.16}$$

But from (8.5),

$$w(\omega) = \frac{4\pi^2}{3n\, c\hbar^2} |\mu|^2 I(\omega) g(\omega). \tag{8.17}$$

Then,

$$\begin{aligned} \mathrm{d}I(\omega) &= -\left[\frac{4\pi^2}{3n\, c\hbar^2} |\mu|^2 I(\omega) g(\omega)\right](N_1 - N_2)\hbar\omega \mathrm{d}x \\ &= -\left[\frac{4\pi^2 (N_1 - N_2)}{3nc\hbar^2} |\mu|^2 I(\omega) g(\omega)\right] I(\omega) \mathrm{d}x \\ &= -\,\alpha(\omega) I(\omega) \mathrm{d}x, \end{aligned} \tag{8.18}$$

where

$$\alpha(\omega) = \frac{4\pi^2 (N_1 - N_2)}{3n\, c\hbar^2} |\mu|^2 \omega g(\omega) = \frac{n\hbar\omega}{c} B(\omega)(N_1 - N_2) \tag{8.19}$$

is the absorption coefficient.

The solution of (8.18) is

$$I(\omega; x) = I(\omega; x = 0)\,\mathrm{e}^{-\alpha x}. \tag{8.20}$$

We define the absorption cross section of the radiative transition as follows (for details, see also [574]):

$$\sigma(\omega) = \frac{\alpha(\omega)}{N_1 - N_2} = \frac{4\pi^2}{3n\, c\hbar^2} |\mu|^2 \omega g(\omega) \quad (\mathrm{cm}^2). \tag{8.21}$$

Note the following:

$$\int \alpha(\omega) \mathrm{d}\omega = \frac{4\pi^2 (N_1 - N_2)}{3n\, c\hbar^2} |\mu|^2 \omega_0 = \frac{n\hbar\omega_0}{c} B(N_1 - N_2) \quad (\mathrm{cm}^{-1}\,\mathrm{s}^{-1}). \tag{8.22}$$

8.3 Semiconductor Lasers

8.3.1 Heterojunction Lasers

Semiconductor lasers, like other lasers, have population inversions that lead to stimulated emission of photons. A semiconductor laser is different from an another lasers primarily because the energy levels in semiconductors must be treated as continuous distributions of levels rather than as discrete levels [578]. We shall assume that the densities of states in the conduction and valence bands of the semiconductor are known functions of energy and that the occupations of these levels are characterized by quasi-Fermi levels [503]. Then the probability that the state of energy E in the conduction band is occupied by an electron is

$$f_{\mathrm{c}}(E) = \frac{1}{\{1 + \exp\left[(E - F_{\mathrm{c}})/kT\right]\}}, \tag{8.23}$$

where F_{c} is the quasi-Fermi level for the conduction band, k is Boltzmann's constant, and T is absolute temperature. A corresponding expression applies in the valence band, with quasi-Fermi level F_{v}. For a system in thermal equilibrium, the quasi-Fermi levels are equal to each other and become the Fermi level E_{F}. In an excited system, $F_{\mathrm{c}} \succ F_{\mathrm{v}}$, and we can use the separation of the quasi-Fermi levels as a measure of the excitation. The use of quasi-Fermi levels greatly simplifies the treatment of systems with many energy levels, or with continuous distributions of levels, because one quantity represents the occupation probability of many levels. The concept of quasi-Fermi level in an excited system is valid, provided that the scattering rate of carriers within a band is rapid compared to the recombination rate between bands, i.e., if the carriers within the conduction band and within the valence band rapidly establish a quasi-equilibrium among themselves, although the conduction band and the valence band are not in equilibrium with each other. For semiconductors with substantial numbers of free carriers, carrier–carrier scattering leads to the establishment of quasi-equilibrium (see, e.g., [579, 580]).

The original semiconductor lasers were p–n junctions prepared by diffusion of acceptor impurities into n-type GaAs, and this is still one of the most common structures. Semiconductors with $\boldsymbol{k}$ conserving transitions at the energy gap are strongly favored for lasing [581], but some impurity levels can lead to stimulated emission in indirect-gap semiconductors, e.g., [582]. All p–n junction lasers are excited by passing current through the p–n junction, and the excitation rate is characterized by the current density. When a forward current flows, electrons are injected into the p-type material, and holes are injected into the n-type material, the latter to a much smaller extent partly because of lower hole mobility. In heterojunctions, potential barriers play a major role in the injection of carriers [583]. The excess of electron and hole concentrations over their equilibrium values creates a local population inversion and leads to stimulated emission of photons at sufficiently high excitation levels. The layer near the p–n junction where this occurs is called the active region or active layer of the device. Figure 8.1 shows that the effective thickness of the active layer in graded junction lasers increases as the current density increases. This leads to smaller quasi-Fermi level separations and to less efficient use of the excitation for lasing. Heterostructure lasers (see also [583]) contain built-in potential barriers for electrons that tend to

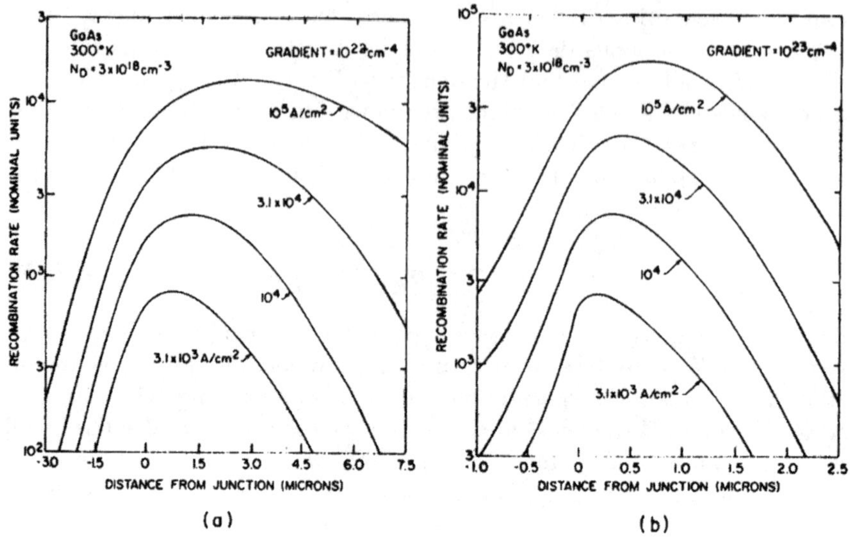

Fig. 8.1. Spatial distribution of the recombination rate for graded GaAs p–n junction lasers with the indicated forward current densities at 300 K. The donor concentration is $3\cdot10^{18}$ cm^{-3}, and the acceptor gradients at the junction are **(a)** 10^{22} cm^{-4} and **(b)** 10^{23} cm^{-4}. The ordinate is the nominal current density defined in (8.31), in A/cm^2, equivalent to $6.2\cdot10^{22}$ recombination/cm^3 s. Note the increase in effective active layer thickness as the current density increases, particularly in **(a)**. (After [584])

confine them to regions of fixed width. Thus, the excitation can be used most effectively.

A second class of excitation methods involves excitation with photons [585] or with an electron beam [584]. For optical excitation, the active layer thickness will be of the order of $\frac{1}{\alpha}$, where α is the absorption coefficient of the incident photons. For electron beam excitation, the active layer thickness will be of the order of the penetration depth of the electrons, which is a function of their energy (see, e.g., [586]). In both cases, diffusion of carriers will add a distance of the order of the diffusion length to the thickness given. If sufficiently thin samples are used, the excitation state may be relatively uniform through the sample, provided surface recombination is unimportant. The excitation rate for an externally excited structure can be converted to an equivalent current density. For photons, the absorbed photon flux is multiplied by the electronic charge, provided that each absorbed photon gives rise to an electron–hole pair. It should be added that values of about $3E_g$ are necessary to create one e–h pair [585]. This means that the main part of the incident pump energy is transformed into heat. This is one of the disadvantages of e-beam pumping.

The value of the gain coefficient g_{th} at the lasing threshold is given by a very simple calculation (see also [584]). If the laser cavity is bounded by parallel surfaces with reflectivities R_1 and R_2, separated by a distance L, then the amplification of the photon intensity on a round trip through the device is $R_1 R_2 \exp\left[2\left(g-\alpha\right)L\right]$, where α is the effective absorption coefficient, e.g., for free-carrier absorption. The lasing condition is that this amplification factor be unity (see also [210]). Thus,

$$g_{\mathrm{th}} = \alpha + \left(\frac{1}{L}\right)\log\left(\frac{1}{R}\right), \tag{8.24}$$

where $R = \left(R_1 R_2\right)^{\frac{1}{2}}$. The current density for which lasing occurs, i.e., the lowest current density, for which g reaches g_{th} at any photon energy, is the threshold current density J_{th}, one of the principal characteristics of a semiconductor laser.

The rate at which photons are emitted per unit volume can be written as (see, e.g., [587])

$$r\left(E\right)\mathrm{d}E = \left[r_{\mathrm{spon}}\left(E\right) + N r_{\mathrm{stim}}\left(E\right)\right]\mathrm{d}E, \tag{8.25}$$

where N is the average number of photons per mode, given for thermal equilibrium by

$$N\left(E\right) = \frac{1}{\exp\left(\frac{E}{kT}\right) - 1}. \tag{8.26}$$

The term r_{spon} in (8.25) gives the rate of spontaneous downward transitions of the electronic system, and r_{stim} is the difference between the stimulated rates of downward and upward transitions.

The spontaneous and stimulated functions can be written

$$r_{\mathrm{spon}}\left(E\right) = a\int \rho_{\mathrm{c}}\left(E_{\mathrm{u}}\right)\rho_{\mathrm{v}}\left(E_{\mathrm{l}}\right)\left|M\right|^2 f_{\mathrm{u}}\left(1-f_{\mathrm{l}}\right)\mathrm{d}E_{\mathrm{u}}, \tag{8.27a}$$

$$r_{\mathrm{stim}}\left(E\right) = a\int \rho_{\mathrm{c}}\left(E_{\mathrm{u}}\right)\rho_{\mathrm{v}}\left(E_{\mathrm{l}}\right)\left|M\right|^2 \left(f_{\mathrm{u}}-f_{\mathrm{l}}\right)\mathrm{d}E_{\mathrm{u}}, \tag{8.27b}$$

where

$$a = 4Ne^2E/m^2\hbar^2c^3, \tag{8.27c}$$

where f_{u} and f_{l} are the probabilities that the upper and lower states involved in the transition are occupied, N is the index of refraction, and $E_{\mathrm{l}} = E - E_{\mathrm{g}} - E_{\mathrm{u}}$ with the convention for measuring energies from nominal band edges separated by the nominal energy gap E_{g}. The squared matrix element $\left|M\right|^2$ is averaged over all polarizations of the radiation and is a function

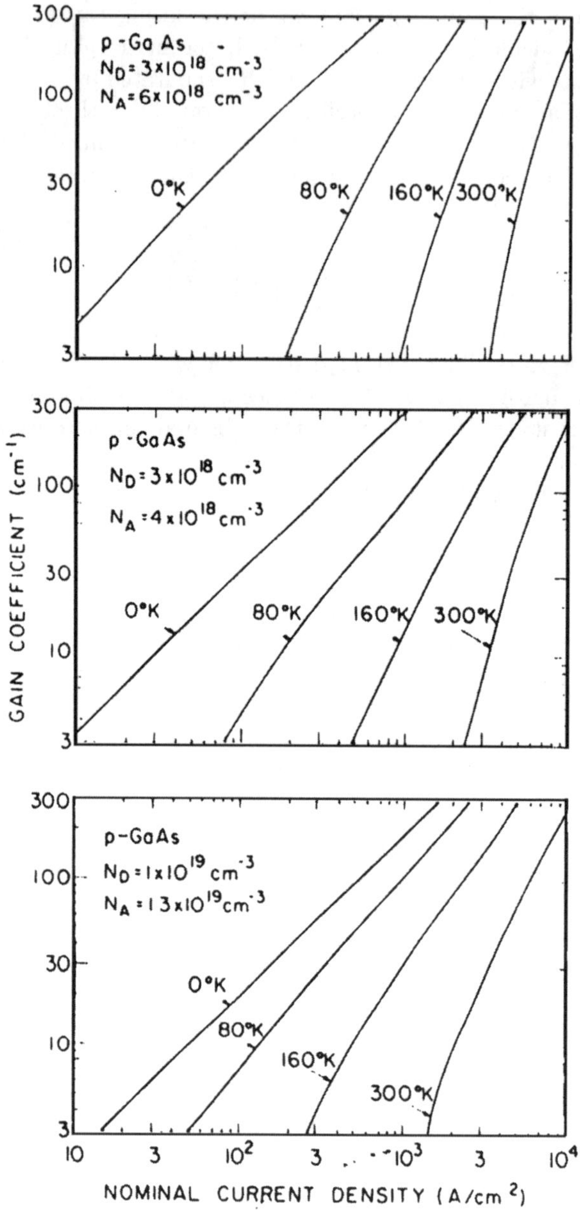

Fig. 8.2. Maximum local gain coefficient [the value of $g_{\text{local}}(E)$ at the photon energy with maximum gain] vs. nominal current density, defined in (8.31), for various temperatures and for **(a)** $N_D = 3 \cdot 10^{18}$, $N_A = 6 \cdot 10^{18}\,\text{cm}^{-3}$; **(b)** $N_D = 3 \cdot 10^{18}$, $N_A = 4 \cdot 10^{18}\,\text{cm}^{-3}$; **(c)** $N_D = 10^{19}$, $N_A = 1.3 \cdot 10^{19}\,\text{cm}^{-3}$. The ion screening temperature is 30,000 K. (After [584])

of both E_u and E_1. Equations (8.27a)–(8.27c) ignore Coulomb interaction between electron and hole (see below), which should be relatively weak when large carrier concentrations are present. Lasing in materials for which exciton effects are important has been treated by Haug [588].

When the populations are characterized by quasi-Fermi levels as in (8.13), the relation between r_{stim} and r_{spon} is

$$r_{\text{stim}}(E) = r_{\text{spon}}\left\{1 - \exp\left[\frac{(E - \Delta F)}{kT}\right]\right\}, \tag{8.28}$$

where $\Delta F = F_c - F_v$ is the difference in the quasi-Fermi levels for the conduction band and the valence band. The relation between the local gain coefficient g_{local} and r_{stim} is

$$g_{\text{local}}(E) = -\alpha(E) = \left(\frac{\pi c \hbar^{\frac{3}{2}}}{NE}\right)^2 r_{\text{stim}}(E), \tag{8.29}$$

where the minus sign arises because we have defined r_{stim} and g_{local} as positive when radiation is emitted, whereas the absorption coefficient α is positive when radiation is absorbed. Combining (8.28) and (8.29), we obtain

$$r_{\text{spon}}(E) = \left(\frac{NE}{\pi c \hbar^{\frac{3}{2}}}\right)^2 \alpha\left(E / \left\{\exp\left[(E - \Delta F)\right] / kT\right) - 1\right\}. \tag{8.30}$$

In thermal equilibrium, when $\Delta F = 0$, this reduces to the well-known result of Van Roosbroeck and Shokley [589]. For semiconductor lasers, the term ΔF is extremely important. Equation (8.28) shows that there can be gain only for photon energies $E \prec \Delta F$.

The local gain coefficient is a function of photon energy, but for most laser cases we are interested primarily in the peak value $g_{\max}$. Figure 8.2 is a plot of $g_{\max}$ versus excitation rate, which is characterized for this purpose by the nominal current density

$$J_{\text{nom}} = 10^{-4} e \int r_{\text{spon}}(E)\, \mathrm{d}E, \tag{8.31}$$

the current density required to maintain the actual excitation rate in a homogeneous layer 1 μm thick. The curves of Fig. 8.2 are given for several sets of values of donor and acceptor concentrations and for several temperatures from absolute zero to 300 K. Note that increasing the compensation, i.e., increasing $N_D + N_A$ for a given $|N_D + N_A|$, raises the nominal current density required to reach a specified gain at low temperature but lowers it at room temperature (cf. Fig. 8.2a–c). Figure 8.3 shows the nominal current density required to reach a gain of 50 cm^{-1} for each case in Fig. 8.2. It is an indication of the temperature dependence of the threshold current density for typical semiconductor junction lasers but makes no allowance for the effect of the

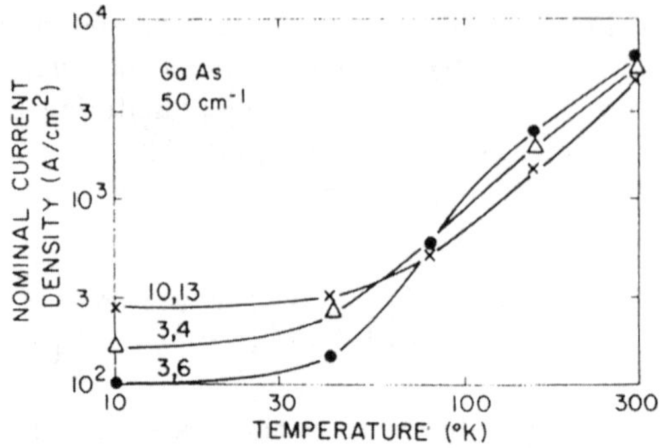

Fig. 8.3. Temperature dependence of the nominal current density required to reach a gain coefficient $g_{\max} = 50\,\mathrm{cm}^{-1}$, taken from the results of Fig. 8.2. The curves are labeled by the values of donor and acceptor concentration, respectively, in units of $10^{18}\,\mathrm{cm}^{-3}$. The temperature dependence of the threshold current density is also a function of the temperature dependence of the losses and of the effective active layer thickness. (After [584])

structure or for losses such as free carrier absorption. A similar analysis has also been made by Hwang [590], whose curves of gain versus current density are similar to those given in Fig. 8.3. Both calculations agree well with experiment. The photon energy at which lasing occurs, found by Hwang [590] is in good agreement with experimental results (for details, see [584, 585, 588]).

8.3.2 Study of Excitons Lasing

With increasing excitation intensity, frequently, laser action is observed in excitonic luminescence. However, the direct recombination of an exciton can never give rise to laser action because the coupled exciton–photon system corresponds in the resonant approximation to two linearly coupled harmonic oscillators. The equations of motion of this system do not contain the nonlinearity that is necessary for laser action. The participation of a third field is required to introduce the possibility of laser action [588], i.e., the laser action in exciton systems has to be a parametric process in which a pump field, a signal field, and an idler field participate.

Below, we describe a scheme for lasing action involving excitons in a pure crystal. In Refs. [591,592], a general theory of various spontaneous and stimulated exciton recombination processes has been developed. In our analysis, we assume a two-band model for the crystal. Atoms at lattice sites interact with each other and an electromagnetic field. The analysis is addressed to tight-binding excitons in the first step. According to [592], the weak-binding case of Wannier–Mott excitons follows a similar description but is more involved. In next section, we follow very closely the results of Liu and Liboff [592].

The Hamiltonian of the entire system of crystal and electromagnetic field is [592]

$$H = H_{\mathrm{xtal}} + H_{\mathrm{em}} + H_{\mathrm{int}}. \tag{8.32}$$

The Hamiltonian of a crystal can be written in the Wannier representation,

$$H_{\mathrm{xtal}} = \sum_{\rho} E_{\mathrm{e}} c_{\rho}^{\dagger} c_{\rho} + \sum_{\rho} E_{\mathrm{h}} d_{\rho}^{\dagger} d_{\rho} + \hbar \sum_{\rho_1,\rho_2} W\left(\rho_1 - \rho_2\right) c_{\rho_1}^{\dagger} c_{\rho_1} d_{\rho_2}^{\dagger} d_{\rho_2}. \tag{8.33}$$

In this expression, operators c_ρ and d_ρ refer, respectively, to electrons and holes at site ρ. Their respective energies are E_{e} and E_{h}. The indicated operators obey the anticommutation relations:

$$\begin{aligned} \left\{c_{\rho}, c_{\rho'}^{\dagger}\right\} &= \left\{d_{\rho}, d_{\rho'}^{\dagger}\right\} = \delta_{\rho,\rho'}, \\ \left\{c_{\rho}, d_{\rho'}\right\} &= \left\{c_{\rho}, d_{\rho'}^{\dagger}\right\} = \left\{c_{\rho'}^{\dagger}, d_{\rho'}\right\} = \left\{c_{\rho}^{\dagger}, d_{\rho'}^{\dagger}\right\} = 0. \end{aligned} \tag{8.34}$$

The function in (8.23) represents the interaction between electrons and holes and has the integral representation

$$\hbar W\left(\rho_1 - \rho_2\right) = \int\int d\boldsymbol{r} d\boldsymbol{r}' w_{\mathrm{c}}^{*}\left(\boldsymbol{r} - \boldsymbol{\rho}_2\right) w_{\mathrm{m}}^{*}\left(\boldsymbol{r}' - \boldsymbol{\rho}_2\right) \frac{e^2}{|\boldsymbol{r} - \boldsymbol{r}'|} \\ w_{\mathrm{c}}\left(\boldsymbol{r}' - \boldsymbol{\rho}_2\right) w_{\mathrm{v}}\left(\boldsymbol{r} - \boldsymbol{\rho}_1\right), \tag{8.35}$$

where $w_{\mathrm{c}}\left(\boldsymbol{r} - \boldsymbol{\rho}\right)$ and $w_{\mathrm{v}}\left(\boldsymbol{r} - \boldsymbol{\rho}\right)$ are Wannier functions relevant to the conduction and valence bands, respectively.

The Hamiltonian H_{xtal} may be diagonalized as follows (see also [593]). For this purpose, we define the operators

$$B_{\rho}^{\dagger} = c_{\rho}^{\dagger} d_{\rho}^{\dagger}, \tag{8.36a}$$

$$B_{\rho} = d_{\rho} c_{\rho}. \tag{8.36b}$$

The operator $B_{\rho}^{\dagger}$ creates an electron–hole pair, whereas B_{ρ} annihilates the pair. The operator

$$\mu_{\rho} = \frac{1}{2}\left(c_{\rho}^{\dagger} c_{\rho} + d_{\rho}^{\dagger} d_{\rho}\right) \tag{8.37}$$

indicates or not the atom at the site ρ is excited. Its eigenvalues are (1,0). Combining the latter three equations gives

$$\left[B_{\rho'}, B_{\rho}^{\dagger}\right] = \delta_{\rho,\rho'}\left(1 - 2\mu_{\rho}\right). \tag{8.38}$$

We further define the operator

$$b_{\mathrm{q}}^{\dagger} = \frac{1}{\sqrt{N}} \sum_{\rho} \mathrm{e}^{\mathrm{i}q\rho} B_{\rho}^{\dagger}, \tag{8.39}$$

which represents the creation of an exciton in the propagating state $\boldsymbol{q}$. The total number of atoms in the crystal is N. The commutation relation for exciton operators is [588]

$$\left[b_{\mathrm{q}}, b_{\mathrm{q}'}^{\dagger}\right] = \delta_{\mathrm{q},\mathrm{q}'} - \frac{2}{N}\sum_{\rho} \mathrm{e}^{\mathrm{i}\left(q-q'\right)}\mu_{\rho}. \tag{8.40}$$

In this manner, the important results that exciton operators are boson operators only when the number of excited atoms in a crystal is small compared to the total number of atoms N. In this event, the second term of (8.30) can be neglected, giving $\left[b_{\mathrm{q}}, b_{\mathrm{q}^{\dagger}}^{\dagger}\right] = \delta_{\mathrm{q},\mathrm{q}'}$. The transformation (8.29) then diagonalizes H_{xtal} as

$$H_{\mathrm{xtal}} = \sum_{q} E_{\mathrm{q}} b_{\mathrm{q}}^{\dagger} b_{\mathrm{q}}, \tag{8.41a}$$

where

$$E_{\mathrm{q}} = E_{\mathrm{e}} + E_{\mathrm{h}} + \epsilon_{\mathrm{q}}, \tag{8.41b}$$

and

$$\epsilon_{\mathrm{q}} = \hbar \sum_{\rho} W(\rho)\, \mathrm{e}^{-\mathrm{i}q\rho}. \tag{8.41c}$$

The diagonalization (8.41a) demonstrates that excitons are collective excitations of the entire crystal.

In the limit of intense lasing, authors cited [592] replaced the exciton and photon amplitudes with their classical counterparts (see also [594–596]):

$$b_{\mathrm{q}}^{\dagger} = \beta_{\mathrm{q}} \exp\left[\mathrm{i}\left(\omega_{\mathrm{q}}^{0} + \Omega_1 + i\eta\right)t\right], \tag{8.42}$$

$$a_j^{\dagger} = a_j\left[i\left(\omega_j^{0} + \Omega_2 + i\eta\right)t\right], \tag{8.43}$$

where $\hbar\omega_{\mathrm{q}}^{0} = E_{\mathrm{q}}$. The operator $a_j^{\dagger}$ creates a photon of frequency ω_j, and γ is the decay rate of this mode. The coefficient η is introduced to describe the decay of excitons. The frequencies Ω_1 and Ω_2 remain arbitrary. Equations for $a_j^{\dagger}$ and β_{q} have the following expressions:

$$\left(i\Omega_1 - \eta\right)\beta_{\mathrm{q}} = -\frac{i}{\sqrt{N}}\sum_{\rho} G_{\rho}\sigma_{\rho} a_j^{\dagger} \mathrm{e}^{\mathrm{i}q\rho}, \tag{8.44}$$

$$\left(i\Omega_2 - \gamma\right)a_j^{\dagger} = \frac{i}{\sqrt{N}}\sum_{\rho}\sum_{q} G_{\rho}^{*}\beta_{\mathrm{q}} \mathrm{e}^{-\mathrm{i}q\rho}. \tag{8.45}$$

Substituting (8.44) in (8.45), and taking into account the resonance condition,

$$\omega_{\mathrm{q}}^{0} + \Omega_1 = \omega_j^0 + \Omega_2, \tag{8.46}$$

and equating the real and imaginary parts of the resulting equation gives

$$\gamma = \frac{\eta}{\eta^2 + \Omega_1^2} \sum_\rho |G_\rho|^2 \sigma_\rho, \tag{8.47}$$

$$\Omega_2 = -\frac{\Omega_1}{\eta^2 + \Omega_1^2} \sum_\rho |G_\rho|^2 \sigma_\rho. \tag{8.48}$$

It follows that

$$\frac{\Omega_1}{\Omega_2} = -\frac{\eta}{\gamma}. \tag{8.49}$$

Through the difference frequency

$$\Delta \equiv \Omega_1 - \Omega_2, \tag{8.50}$$

(8.39) yields

$$\Omega_1 = \frac{\Delta}{1+\frac{\gamma}{\eta}}, \qquad \Omega_2 = -\frac{\Delta}{1+\frac{\gamma}{\eta}}. \tag{8.51}$$

Substituting the exponential forms (8.42) and (8.43) in the next equation,

$$\frac{\mathrm{d}}{\mathrm{d}t}\sigma_\rho = \frac{1}{\tau}\left(s_0 - \sigma_\rho\right) + \frac{2i}{\sqrt{N}}\left(\sum_q G_\rho b_{\mathrm{q}} a_j^\dagger \mathrm{e}^{\mathrm{i}q\rho} - \sum_q G_\rho^* b_{\mathrm{q}}^\dagger a_j \mathrm{e}^{-\mathrm{i}q\rho}\right), \tag{8.52}$$

in the steady state,

$$0 = \frac{1}{\tau}\left(s_0 - \sigma_\rho\right) + \frac{2i}{\sqrt{N}}\left[\sum_q G_\rho b_{\mathrm{q}}^* a_j^* \mathrm{e}^{\mathrm{i}q\rho} - \sum_q G_\rho^* b_{\mathrm{q}} a_j \mathrm{e}^{-\mathrm{i}q\rho}\right]. \tag{8.53}$$

Inserting (8.44) into the latter equation gives

$$\sigma_\rho = s_0 - \frac{4\tau}{\eta\left[1 + \Delta^2/\left(\eta+\gamma\right)^2\right]} |G_\rho|^2 \langle n\rangle \sigma_\rho, \tag{8.54}$$

where s_0 is an asymptotic inversion value. Here, we have written

$$n = \langle a_j^* a_j\rangle \tag{8.55}$$

for the average number of photons in the jth mode.

Solving (8.54) for the local inversion, we obtain

$$\sigma_\rho = s_0 \left[1 + \frac{8\pi\tau \langle n \rangle e^2 d_{\mathrm{f}}^2}{\hbar\omega_j^0 m^2 V \eta (1+\Delta^2)/(\eta+\gamma)^2} \sin^2 \boldsymbol{k}_j \boldsymbol{\rho} \right]^{-1} . \tag{8.56}$$

For the average inversion,

$$\langle \sigma \rangle = \frac{1}{N} \sum_\rho \sigma_\rho , \tag{8.57}$$

we find

$$\langle \sigma \rangle = \frac{s_0}{N} \sum_\rho \left[1 + \frac{8\pi\tau \langle n \rangle e^2 d_{\mathrm{f}}^2}{\hbar\omega_j^0 m^2 V \eta (1+\Delta^2)/(\eta+\gamma)^2} \sin^2 \boldsymbol{k}_j \boldsymbol{\rho} \right]^{-1} . \tag{8.58}$$

In the continuum limit, (8.58) becomes

$$\langle \sigma \rangle = \frac{s_0}{V} \int \mathrm{d}r \left[\frac{1}{1 + \left(\frac{4\tau n \gamma}{\rho S} \right) \sin^2 \boldsymbol{k} \boldsymbol{r}} \right] , \tag{8.59}$$

where $N = \rho V$ and $\langle n \rangle = nV$. Thus, ρ is the atomic density and n is the photon density. Furthermore,

$$S = \frac{\hbar\omega_j^0 m^2 \gamma \eta}{2\pi e^2 \rho \delta_{\mathrm{f}}^2} \left[1 + \frac{\Delta^2}{(\eta+\gamma)^2} \right] . \tag{8.60}$$

Substituting the form for γ given by (8.47) in the last equation, together with the continuum form of the next equation,

$$G_\rho = -\frac{e}{m} \sqrt{\frac{2\pi}{\hbar\omega_j V}} \sin(\boldsymbol{k}_j \boldsymbol{\rho})\, e_j \delta_j \tag{8.61}$$

gives the following relation:

$$S = \frac{s_0}{V} \int \mathrm{d}r \frac{\sin^2 \boldsymbol{k} \boldsymbol{r}}{1 + \left(\frac{4\tau n \gamma}{\rho S} \right) \sin^2 \boldsymbol{k} \boldsymbol{r}} , \tag{8.62}$$

and with $\int \mathrm{d}r V$; (8.59) and (8.61) may be manipulated to yield

$$n = \frac{\rho s_0}{4\pi\gamma} \left(1 - \frac{\langle \sigma \rangle}{s_0} \right) . \tag{8.63}$$

Finally, integrating (8.59), we obtain

$$\frac{\langle \sigma \rangle}{s_0} = \frac{1}{\sqrt{1 + 4\tau n\gamma/\rho S}}. \tag{8.64}$$

With (8.53), this expression becomes

$$\langle \sigma \rangle = \frac{S}{2}\left(1 + \sqrt{1 + \frac{4s_0}{S}}\right). \tag{8.65}$$

For laser action to be realized, one must have the photon density (at frequency ω_j), $n > 0$. From (8.63), this condition is satisfied providing $s_0 > \langle \sigma \rangle$, or from (8.1), $s_0 > S$, which with (8.63) gives

$$s_0 > \frac{\hbar\omega_j^0 m^2 \gamma\eta}{\pi e^2 \rho \delta_\mathrm{f}^2}\left[1 + \frac{\Delta^2}{(\eta + \gamma)^2}\right]. \tag{8.66}$$

At resonance, $\Delta = 0$ and the right-hand side of the preceding expression is minimized. In this limit, (8.62) reduces to the same form given by the Schawlow–Townes criterion [597] for randomly distributed atoms. However, in (8.62), the matrix element $\delta_\mathrm{f} \equiv \langle v\rho\,|p|\,c\rho\rangle$ is appropriate for excitons, whereas the matrix element in the Schawlow–Townes expression is relevant for a single atom.

For Wannier–Mott excitons, the criterion (8.62) is maintained with δ_f replaced by

$$\delta_{\mathrm{W-M}} \equiv \sum_{\rho'} F(\rho') \int \mathrm{d}r w_\mathrm{v}^* (r - \rho)\, p w_\mathrm{c}(r - \rho - \rho'). \tag{8.67}$$

In this expression $F(\rho)$ is the wave function of a Wannier–Mott exciton [167]. In a preceding investigation, Liu and Liboff considered rigid lattices. If lattice vibrations are present, exciton–phonon interactions cause exciton diffusion [167], which may diminish coherence for lasing. However, lasing will still occur [588], providing the relaxation time for exciton diffusion is longer than the exciton relaxation time, γ, in (8.42). This is the case for sufficiently weak excito–phonon interaction (for details, see below).

As was shown before optical transitions in pure III–V compounds, which can be used for laser action, are band–band transitions. In II–VI compounds (as well as LiH [23] and etc.), the recombination process of electrons and holes via exciton states is more favorable than the band–band transition [169, 183, 185]. During the last four decades, laser action has been obtained in II–VI compounds by electron beam bombardment [598–600] and by optical excitation [169, 601, 602]. The laser transitions involve the A_1-nLO phonons, where $n = 1, 2$. Gain measurements [600, 602] and simultaneous measurements of the emission intensities of the A_1 line (direct A–exciton recombination [216]) and the A_1-LO line also confirm the statement that in

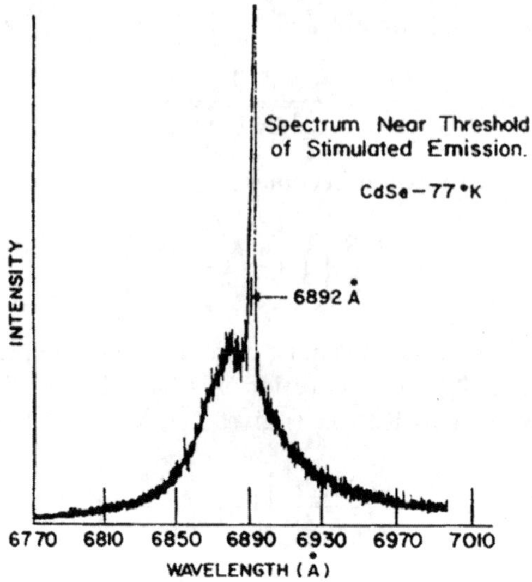

Fig. 8.4. The onset of stimulated emission in CdSe at 477 K (after [599])

CdS the A_1-LO (A_1-2LO) line starts to lase at sufficiently high pump rates (see also Fig. 8.4 (for details, see [599]). In the papers cited, Haug [588] calculated the temperature dependence of the maximum gain frequency at the threshold (see Fig. 8.5). The result is simple in the low-temperature limit:

$$\Delta_{\text{max}} \to \frac{3}{2}kT, \tag{8.68a}$$

and also in the high-temperature limit:

$$\Delta_{\text{max}} \to (3\kappa kT/B)^{\frac{2}{5}} \exp(-2\hbar\nu/5\,kT, \tag{8.68b}$$

where $2\kappa = 1.25 \cdot 10^{12}\,\text{s}^{-1}$, corresponding to losses of $100\,\text{cm}^{-1}$, $B = 1.55 \cdot 10^{35}\,\text{erg}^{-3/2}\text{s}^{-1}$ for CdS crystals. These limiting results have also been given by Mashkevich et al. [603]. The typical experimental gain $I_{\text{thr}}(T)$ dependence, obtained in paper [602] is presented in Fig. 8.6. There are shown some spectra of stimulated emission at different temperatures. The authors of [602] indicated some contradiction of their experimental results with the theoretical description.

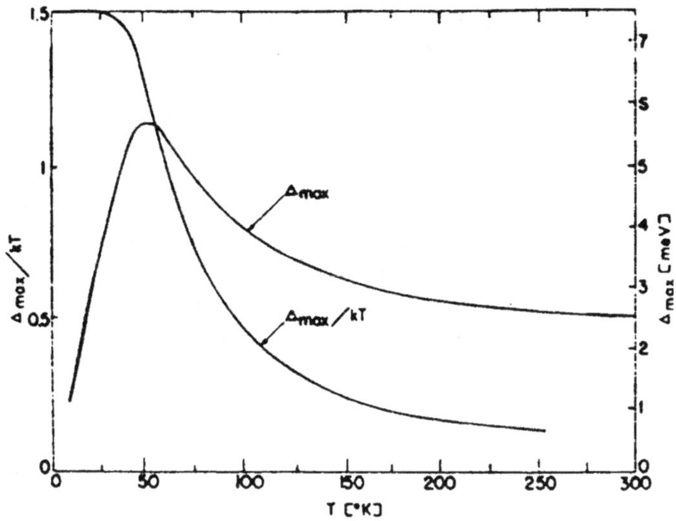

Fig. 8.5. Maximum gain frequency Δ_{max} vs. temperature (after [588])

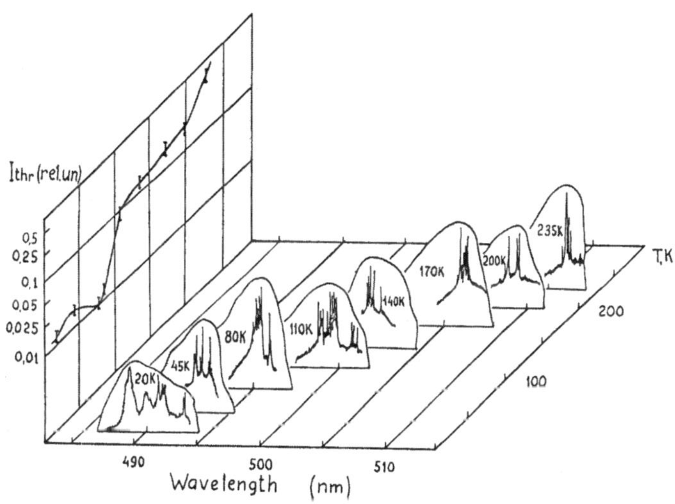

Fig. 8.6. The dependence $I_{thr}(T)$ and some examples of (above-threshold) lasing spectra (in the range of A_1 – 1LO; A_1 – 2LO phonons) at different temperatures (after [602])

8.4 Nonlinear Properties of Excitons in Isotopically Mixed Crystals

Another application of isotopically pure and isotopically mixed crystals that is discussed here is related to the possibility of using an isotopically mixed medium (e.g., LiH_xD_{1-x} or ${}^{12}C_x\,{}^{13}C_{1-x}$) as an oscillator of coherent radiation in the ultraviolet spectral range [204,604]. To achieve this, the use of indirect electron transitions involving, say, LO phonons was planned [210,588]. As shown above, using indirect electron transitions involving phonons to degenerate coherent radiation in semiconductors was originally proposed by Basov et al. (see [598] and reference therein). Kulevsky and Prokhorov [601] were the first to observe stimulated radiation using emission lines of LO phonon repetitions in CdS crystals on two-photon excitation (see also [605]). The detection of LO phonon replicas of free-exciton luminescence in wide-gap insulators attracted considerable attention to these crystals (see e.g., Plekhanov [204] and Plekhanov [109]). At the same time, it allowed one to pose a question about the possibility of obtaining stimulated emission in the UV (VUV) region ($4-6$ eV) of the spectrum, where no solid-state sources for coherent radiation exist yet. In the first place, this related to the emitters working on the transitions of the intrinsic electronic excitation (exciton). The last one provides a high energetic yield of coherent emission per unit volume of the substance. The results obtained on solidified xenon (Basov et al. [606]) and argon (Schwenter et al. [607]) under electron beam excitation with following excimer molecule emission are an exception.

In this section, we discuss the results of the investigation of the influence of excitation light density on the resonant secondary emission spectra of free excitons in wide-gap insulator LiH_xD_{1-x} ($LiH_{1-x}F_x$) crystals. Cubic LiH crystals are typical wide-gap, ionic insulators with $E_g = 4.992$ eV [23] and relatively weak exciton–phonon interaction, however: $\frac{E_B}{\hbar\omega_{LO}} = 0.29$ where E_B and $\hbar\omega_{LO}$ are exciton binding energy and longitudinal optical phonon energy, respectively. Besides, it might be pointed out that the analogous relations for CdS, diamond, and NaI are 0.73, 0.45, and 12.7, respectively (Plekhanov [608]). Figure 8.7 depicts, as an example, the exciton luminescent spectrum of pure (LiH and LiD) and mixed (LiH_xD_{1-x}) crystals at a low temperature. Analogous results for ${}^{12}C_x\,{}^{13}C_{1-x}$ mixed diamond crystals are shown in Fig. 8.8. A common feature of all three spectra depicted in Fig. 8.7 is the phononless emission line of free excitons and its 1LO and 2LO phonon replicas. An increase in the density of the exciting light causes a burst of radiation energy in the long-wave wing of the emission of 1LO and 2LO repetitions (see Fig. 8.9) at a rate higher for the 1LO repetion line [210]. The detailed dependence of the luminescent intensity and the shape of the 2LO phonon replica line are presented in Figs. 8.10 and 8.11, respectively. Further investigations have shown [610] that with the increase in excitation light intensity at the beginning, a certain narrowing can be observed, followed by widening of the line of the 2LO phonon replica with the simultaneous ap-

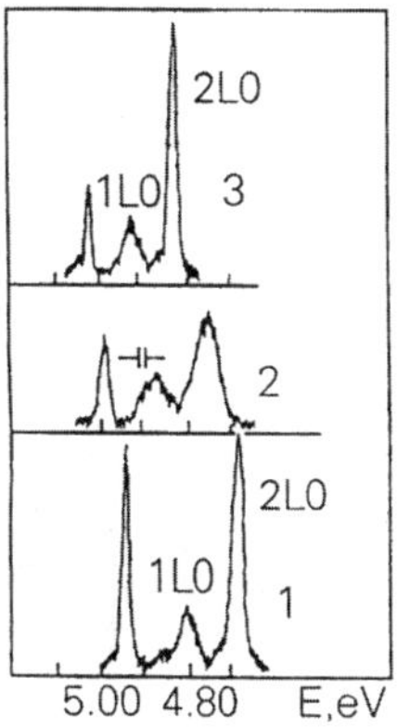

Fig. 8.7. Photoluminescent spectra of free excitons at 4.2 K. 1: LiH; 2: LiH_xD_{1-x}; and 3: LiD crystals (after [109])

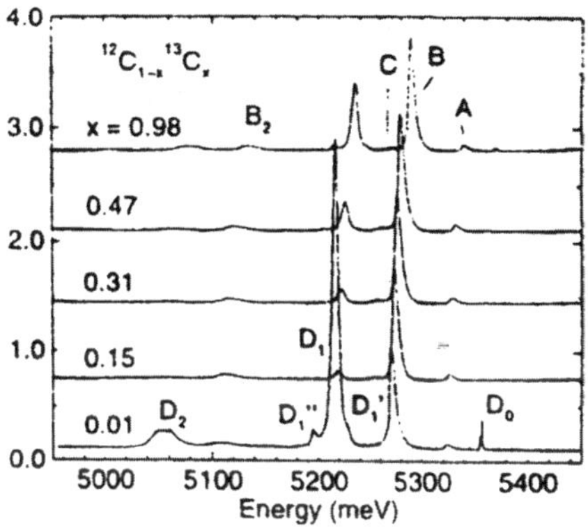

Fig. 8.8. Cathode-luminescent spectra of isotopically modified diamond at 36 K. Intrinsic photoassisted recombination peaks are labeled in the top spectrum; those from boron-bound excitons in that at the bottom (after [609])

pearance of a characteristic, probably, mode structure. From Fig. 8.10, it can be seen that the coupling between long-wavelength luminescent intensity and excitation light intensity is not only linear, but, in fact, of a threshold character as in other crystals [578, 605]. The proximity of the exciton parameters of LiH and CdS (ZnO) crystals allowed us to carry out the interpretation of the density effects in LiH in analogy with these semiconducting compounds. Coming from this in paper [610], it was shown that for the observed experimental picture of LiH crystals the exciton–phonon mechanism of light generation [588] is enough for exciton density of about $10^{15}\,cm^{-3}$. This is a reasonable value, if the high quality of the resonator mirror – the crystal cleavage "in situ" and the relatively large exciton radius ($r = 40$ Å [78]) are taken into account. This light mechanism generation must also be promoting

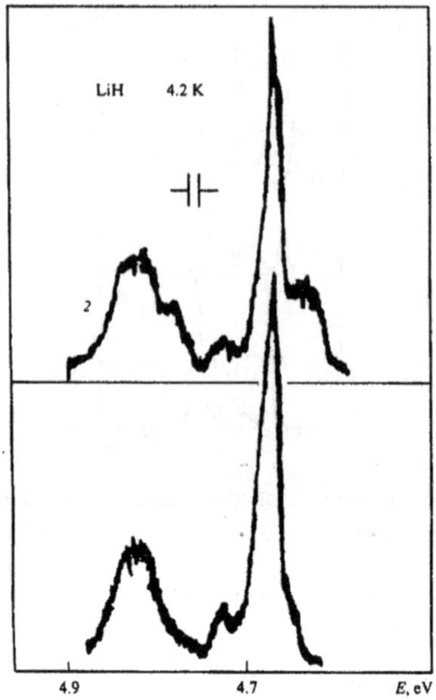

Fig. 8.9. Luminescenct spectra of free excitons in LiH crystals in the region of the emission lines of 1LO and 2LO phonon repetitions at 4.2 K for low (1) and high (2) density of excitations of 4.99 photons. The scales of the curves are different. (After [610])

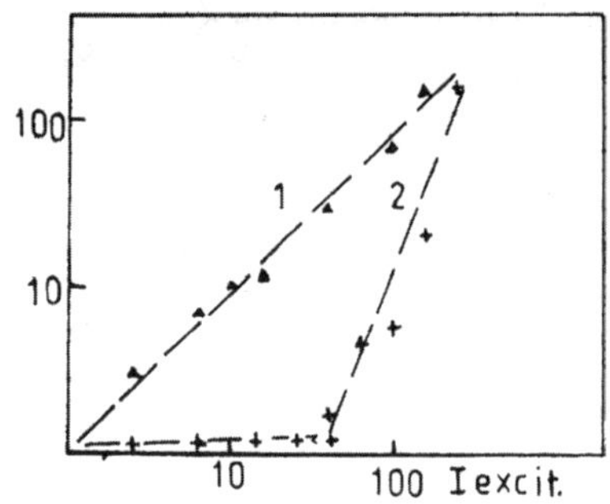

Fig. 8.10. The dependence of the intensity in the maximum (1) and on the long-wavelength side (2) of the 2LO replica emission line of LiH crystals on the excitation light intensity (after [610])

a large value of LO phonon energy ($\hbar\omega_{\rm LO}$140 meV). Owing to this, radiative transition is being realized in the spectral region with a small value of the absorption coefficient and thus with small losses in the resonator (for details, also see [608]).

In this section, we briefly analyze the shift of free exciton luminescence on the crystal lattice dependence. In the first step, F doped of LiH crystals are

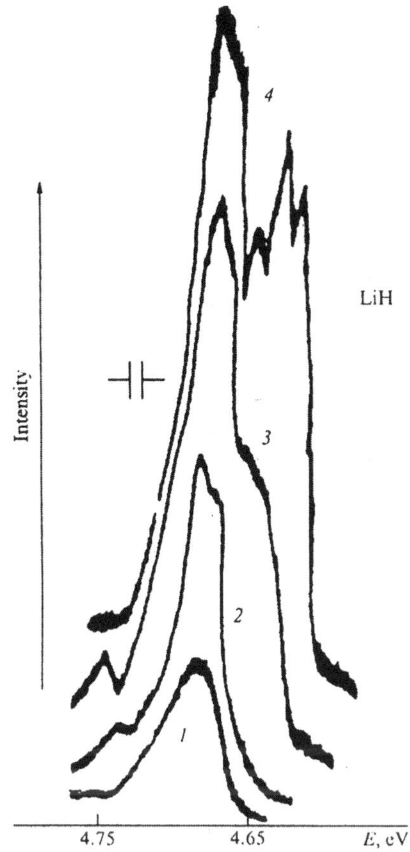

Fig. 8.11. The dependence of the shape of 2LO replica line on the excitation intensity (I_0) light: 1 – $0.05I_0$; 2 – $0.09I_0$; 3 – $0.35I_0$; 4 – I_0 (after [610])

considered. The reflectance spectra of the crystals investigated with clean surfaces (cleaved in LHeT) had a distinctly expressed excitonic structure. Typical reflection spectra of $LiH_{1-x}F_x$ with a mirror surface are depicted in Fig. 8.12 [611]. The crystal is cleaved in superfluid helium in a helium cryostat bath. In Fig. 8.12 for comparison, the reflection spectrum of pure crystal LiH (curve 1) is shown. All spectra possess the identical long-wavelength exciton structure: we can see 1 s and 2 s exciton states. It is clearly seen, that the F addition in LiH crystals leads, as naturally expected, to a short-wavelength shift of the spectrum as a whole. With the growth of x, the energetic interval is increased between 1s and 2 s exciton states and at the same time, their short-wavelength shift is different and has a saturated character [210]. At the maximum value of $x \leq 2.5\%$, the exciton Rydberg (E_B), obtained on hydrogen-like formulas is equal to 75 ± 3 meV ($E_B = 40$ meV for LiH). For $LiD_{1-x}F_x$ crystals at $x \leq 1.6\%$, the exciton binding energy is equal

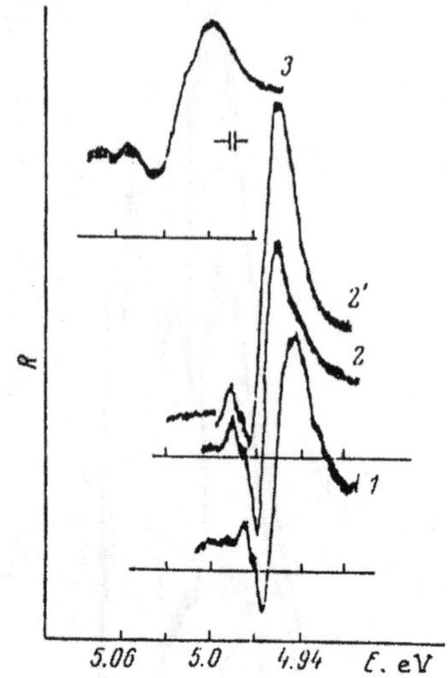

Fig. 8.12. Reflection spectra of LiH (1); $LiH_{1-x}F_x$ ($x = 0.06\%$) (2); and ($x = 1.6\%$) (3) at 78 and 4.2 K (2') (after [611])

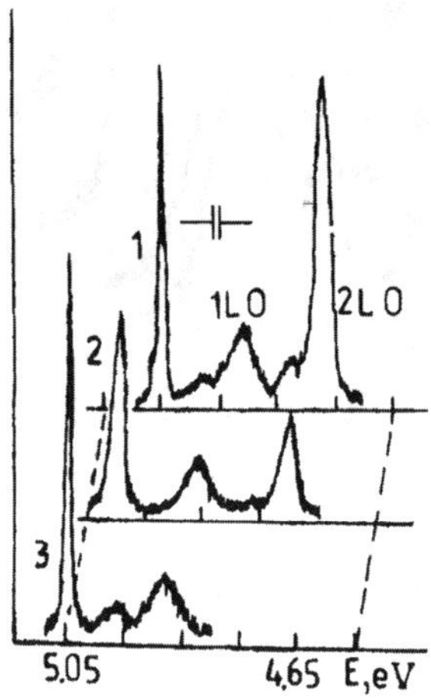

Fig. 8.13. Free-exciton luminescent spectra of LiH (1), $LiH_{0.984}F_{0.016}$(2), and $LiD_{0.992}F_{0.008}$(3) crystals cleaved in liquid helium (after [614])

$E_B = 57 \pm 2$ meV. Supposing the linear dependence, E_B on the x concentration in $LiH_{1-x}F_x$ crystals, we obtain $E_B = 62 \pm 25$ meV for $x = 1.6\%$ and the experimental value of this parameter is equal to 67 ± 3m meV. Such a large error from theoretical extrapolation is connected with large errors when determining E_B for LiF crystals [612, 613]. However, despite the identical structure of all free-exciton luminescent spectra, it is necessary to note a rather big variation of the luminescent intensity of crystals from different

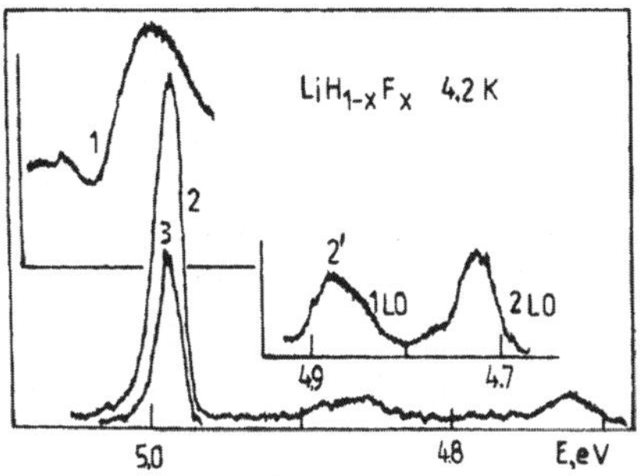

Fig. 8.14. Reflection (1) and luminescence (2, 2') spectra at 4.2 K and 260 K (3) of $LiH_{1-x}F_x$ mixed crystals cleaved in liquid helium (after [614])

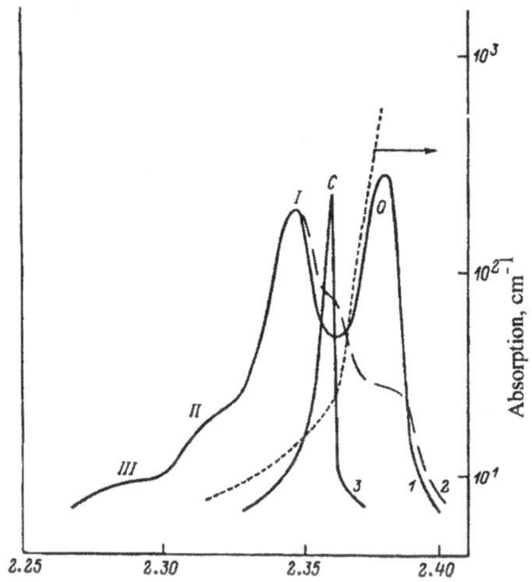

Fig. 8.15. Emission spectra of $CdS_{0.9}Se_{0.1}$ at different levels of excitation: 10^2(1), 10^4(2), and 10^6 (3) Wt/cm^2. 0: zero-phonon line; I–III: LO-phonon replicas, and *dashed line* is the absorption spectrum, $T = 2K$ (after [615])

batches in the experiment. Therefore, crystals possessing the maximum value of free-exciton luminescent quantum yield were chosen for measurements of the density effects.

The luminescent spectra of virgin and mixed crystals are similar and consist of a narrow, zero-phonon line and its more wider LO replicas [210, 610]. As well as in the reflection spectra (see above), doping LiH crystals with deuterium or fluorine causes a short-wavelength shift of the luminescent spectrum as a whole (Figs. 8.7 and 8.13). Increasing in the deuterium concentrations leads to widening of the luminescence line (see also [78]). An increase in the fluorine concentration causes (Figs. 8.13 and 8.14), except for the spectrum shift, a sharp ignition of the zero phonon line intensity in comparison with the lines of LO–replica intensity (see also [611]). Except for the indicated effects, fluorine activation of the LiH (LiD) crystals shifts the temperature quenching of free-exciton luminescence to a higher region. As one example, Fig. 8.14 depicte the reflection and luminescence spectra of $LiH_{1-x}F_x$ crystals in a wide temperature range [109]. It can be seen that the free-exciton zero-phonon emission line in these crystals is reliably registered practically up to room temperature. Taking into account the mixed crystal lattice potential relief (compare Fig. 8.15), the possibility to obtain laser generation on the zero-phonon line emission that was already obtained in paper [615] cannot be absolutely excluded.

In concluding of this section, we should underline that if the observable mode structure is really caused by laser generation, it may be smoothly tuned in the region of energies 4.5±5.1 eV owing to the smooth transition of the line emission energy in LiH_xD_{1-x} (LiH_xF_{1-x}; LiD_xF_{1-x}) mixed crystals [210] as well as in the range 5.35 – 5.10 eV in $^{12}C_x\,^{13}C_{1-x}$ mixed crystals (see also Fig. 20 in [23].

9 Other Unexplored Applications of Isotopic Engineering

9.1 Isotopic Information Storage

Current rapid progress in the technology of high-density optical storage makes the mere announcing of any other imaginable alternatives a rather unthankful task. The obvious query "who needs it and what for?" has, nevertheless, served very little purpose in the past and should not be used to veto the discussion of orthodox technological possibilities. One such possibility, the technology of isotopic information storage (IIS), is discussed in this section.

Isotopic information storage may consist of assigning the information "zero" or "one" to mono isotopic microislands (or even to a single atoms) within a bulk crystalline (or thin film) structure. This technique could lead to very high density ROM-type (read-only memory or permanent storage) information storage, probably up to 10^{20} bits per cm^3. The details are discussed by Berezin et al. [349, 616, 617]: here, we note only that the use of tri-isotopic systems (e.g., ^{28}Si, ^{29}Si, ^{30}Si) rather than di-isotopic (e.g., ^{12}C, ^{13}C) could naturally lead to direct three-dimensional color imaging without the need for complicated redigitizing (it is known that any visible color can be simulated by a properly weighted combination of three prime colors, but not of two).

Indeed, let us assume that the characteristic size of one information-bearing isotopic unit (several atoms) is 100 Å. Then 1 cm^3 of crystalline structure, e.g., diamond, is able to store roughly $(10^8)^3/100 = 10^{22}$ bits of information [617]. This capacity greatly exceeds that needed to store the information content of all literature ever published ($\cong 10^{17}$ bits), including all newspapers.

The main potential advantage of isotopically mixed crystals lies in the fact that the information is incorporated in the chemically homogeneous matrix. There are no chemically different impurities (as in optical storage with color centers) or grain boundaries between islands of different magnetization (as in common magnetic storage). The information in isotopically mixed crystals exists as a part of the regular crystal lattice. Therefore, the stored information in isotopically mixed crystals is protected by the rigidity of the crystal itself. There are no "weak points" in the structure (impurities, domain wells, lattice strain, etc.) which can lead to information loss due to bond strains, enhanced diffusion, remagnetization, and so on. Differences in bond lengths between different isotopes (e.g., $^{28}Si - ^{29}Si$ or $^{29}Si - ^{30}Si$; H–D and so on) are due to the anharmonicity of zero-point vibrations (see, e.g., [28]). This is not enough for the development of any noticeable lattice strains although these differences are sufficiently large to be distinguishably detected in IIS reading).

The mechanism potentially available in IIS for writing-in the information may be divided into two general categories. The first category refers to all techniques that can direct externally a particular atom to a specified position on the surface of a growing crystal structure. Any beam technique with the ability of focusing on a 1 Å scale could, in principle, be feasible for such a purpose. The second category relates to all "internally operated" possibilities, i.e., delivering the required isotope as a part of the molecule and depositing it in a particular position through some chemical process (e.g., exchange reaction, chemisorption, etc.). This group of possibilities is similar to the DNA–RNA mechanism in actual biological information transfer in living systems. Some chemically simple crystals have, nevertheless, very complex lattice structures. One known example is elementary boron [618, 619], which can crystallize in a beta-rhombohedral structure with 105 atoms in a unit cell with 15 crystallographically nonequivalent positions. Moreover, various atoms have three different coordination numbers: 91 atoms have the coordination number 6; 12 atoms, 8 and 2 "special" atoms have 9 nearest neighbors [619]. This peculiarity of crystalline boron is rather surprising in view of the fact that it is an elementary (monoatomic) crystal. Similar possibilities exist for lattices with two elements, e.g., the silicides of manganese are known to form very complex structures [620]. It is, therefore, legitimate to consider the ability of such structures with complex crystal lattice to provide the basis for compact information storage within the frameworks of the model of an alternative (nonorganic) genetic code. In principle, isotopic combinations could provide the basis for storing information, even in "simple" crystals (e.g., in carbon or silicon-based structures) not only in crystals with complex unit cells. It is even possible to raise the question of what (if any) effects could be connected with isotopic permutations in "regular" biology (e.g., ^{12}C and ^{13}C combinations in various fragments of DNA) Nevertheless, in crystals with large and sophisticatedly constructed unit cells, the already "preexisting" significant level of structural complexity makes them, apparently, more preferable candidates for the evolutionary game of isotopic information-bearing arrays (for details, see [619]).

The possible key to 3-D access could, in principle, be provided by any method which is able to probe the nuclear mass and/or magnetic moment of a single atom at a particular lattice site below the surface. Without detailed elaboration the following possibilities present themselves:

(1) Localized crystal vibrations (see also, Sect. 6), which generally contain information on the vibrational frequencies of an individual atom's (due to the square root of the mass) dependence on vibrational frequencies could be a spectroscopically pronounced effect.

(2) Recoil phenomena (e.g., Rutherford backscattering).
(3) Nuclear magnetic resonance (NMR).
(4) Spin-sensitive neutron scattering.

Of course, there has to be a great deal of refinement in these or other techniques before they can actually be used in isotopically mixed crystals.

9.2 Isotopic Structuring for Fundamental Studies

Isotopic substitution has made, it possible to produce objects of research that earlier were simply inaccessible (exception of the LiH–LiD system). The use of such objects allows investigating isotopic effects in lattice dynamics (elastic, thermal, and vibrational properties (see reviews [27, 423, 428]) and also the influence of such effects on electronic states (renormalization of the band-to-band transition energy E_g, exciton binding energy E_B, and the size of longitudinal–transverse splitting Δ_{LT} [78, 109]).

Furthermore, it is widely known that the melting and boiling points of ordinary water and heavy water (D_2O) differ by a few degrees centigrade. For elements heavier than hydrogen, the isotopic differences in melting points (ΔT) of elemental and complex solids are generally smaller but also detectable. It is quite surprising, however, that there are almost no reports of direct measurement of these differences in the literature.

Another noticeable fact is that sometimes the isotopic effect shows a drastic "self-amplification," e.g., isotopic replacements of Ba and Ti in $BaTiO_3$ (both are heavy elements) can shift the phase transition temperatures by as much as 200 K [621]. The reason(s) for such selective anomalies are not yet clearly established. This was considered early in this book [23].

9.3 Other Possibilities

Here, we briefly list a few additional possibilities of isotopic structuring; see also [27, 423, 428].

(1) Isotope engineering could be based on exploiting the differences in thermal conductivity (see above) between isotopically pure and isotopically mixed solids for purposes such as phonon focusing, precise thermometry based on isotopically-gradiented structures, and so on.
(2) The use of isotopically structured Ni films for neutron interference filters has been reported by Antonov et al. [622].
(3) Isotopically structured light devices. This could slightly shift spectral characteristics and lead to some changes in the kinetics of energy transfer, modify the lifetimes, recombination rate, and so on.

(4) Since the speed of sound is proportional to $\sqrt{M}$, variations in isotopically structured acoustoelectronic devices (transducers, surface acoustic wave devices, etc.) could be significant, especially in achieving phase differences over relatively short isotopically distinguished paths (see also [423]).
(5) The possibility to get, at rather low pressure, the transition of metal–insulator with metallic conductivity in the zone genetically related to hydrogen in LiH crystals (see also [423] and references therein).
(6) The use of an isotopic boundary for Mössbauer filtration of synchrotron radiation, since this makes it possible to get rid of the background noise caused by the interaction between synchrotron radiation and the electrons in matter [623].
(7) Isotope-based quantum computers (see, e.g., [423, 624, 625]).

We have outlined several, mostly untested possibilities arising from exploiting differences among various stable isotopes and purposeful isotopic structuring. These examples of the potential capabilities of isotopic engineering are by no means an exhaustive list.

10 Conclusion

In this book, we have presented briefly the results of experimental and theoretical studies of the objects of research that earlier were simply inaccessible (except for LiH_xD_{1-x} crystals). The use of such objects allows the investigation of isotopic effects in lattice dynamics (elastic, thermal and vibrational properties) and also the influence of such effects on electronic states via electron–phonon coupling (renormalization of the band-to-band transition energy E_g, exciton binding energy E_B, and the size of longitudinal–transverse splitting Δ_{LT}). The first three chapters of the book briefly describe the influence of isotopic effects on lattice dynamics as well as the energy spectrum renormalization of electronic excitation in isotopically-mixed semiconducting and insulating crystals. The thermal conductivity enhancement in isotopically enriched materials amounts (C; Ge; Si) to almost 60% at room temperature and is close to a factor of 6 at the thermal conductivity maximum around 20 K (Si-case). The change in the lattice constant is $\Delta a/a \sim 10^{-3}-10^{-4}$, and the change δc_{ik} in elastic constants amounts to several percent. In addition, crystals of different isotopic compositions possess different Debye temperatures. This difference between a LiH crystal and its deuteride exceeds 100 K. The difference between Debye temperatures for diamond crystals is of the same order of magnitude. Very pronounced and general effects of isotopic substitution are observed in phonon spectra. The Raman lines in isotopically mixed crystals are not only shifted (the shift of LO phonon lines exceeds 100 cm^{-1}) but are also broadened. This broadening is related to the isotopic disorder of a crystal lattice. It is shown in this book that the degree of change in the scattering potential is different for different isotopically mixed crystals. In semiconducting crystals (C, Ge, Si, α-Sn, etc.), phonon scattering is weak, which allows one to successfully apply the coherent potential approximation (CPA) for describing the shift and broadening of scattering lines in Raman spectra. In LiH, the change in the scattering potential is so strong that it results in phonon localization, which is directly observed in experiments.

Substituting a light isotope with a heavy one increases the interband transition energy E_g (excluding Cu salts) and the binding energy of the Wannier–Mott exciton E_B as well as the magnitude of longitudinal–transverse splitting Δ_{LT}. The nonlinear variation of these quantites with isotopic concentration is due to isotopic disordering of the crystal lattice and is consistent with the concentration dependence of the line half-width in exciton reflection and luminescent spectra. A comparative study of the temperature and isotopic shift of the edge of fundamental absorption for a large number of different semiconducting and insulating crystals indicates that the main (but not the only) contribution to this shift comes from zero oscillations whose magnitude may be quite considerable and comparable with the energy of LO phonons. The theoretical description of the experimentally observed dependence of the binding energy of the Wannier–Mott exciton E_B on a nuclear mass requires simultaneous consideration of the exchange of LO phonons between the electron and hole in the exciton and the separate interactions of carriers with LO phonons. The experimental dependence $E_B \sim f(x)$ for LiH_xD_{1-x} crystals fits in well enough with the calculation according to the model of a large-radius exciton in a disordered medium; hence, it follows that the fluctuation smearing of the band edges is caused by isotopic disordering of the crystal lattice.

Detailed analysis of self-diffusion in isotopically pure and heterostructures was done in Chap. 5. This chapter was organized around general principles that are applicable to all materials. There, we briefly discussed the very popular SIMS technique. As is well known, self-diffusion is the migration of constituent atoms (isotopes) in materials. The knowledge obtained in self-diffusion studies is pivotal for understanding many important mass transport process including impurity diffusion in solids for use in semiconductor devices.

The new reactor technology – neutron transmutative doping (NTD) of semiconductors was described in Chap. 6. Capture of thermal neutrons by isotopic nuclei followed by nuclear decay produces new elements, resulting in a large number of possibilities for isotopic selective doping of solids. There are different facilities presented that use this reactor technology. The feasibility of constructing reactors dedicated to the production of NTD silicon, germanium and other compounds was analyzed in terms of technical and economic viability, and the practicality of such a proposal was examined. The importance of this technology for studies of semiconductor doping (materials for different devices) as well as metal–insulator transitions and neutral impurity scattering process is underlined. The introduction of particle irradiation into processing of semiconductor materials and devices creates a new need for additional understanding of atomic-displacement-produced defects in semiconductors. The measurement of decay rates of induced radioactivity and the system of clearance and certification to allow the solids to be internationally transported as “Exempt Material”, as defined in IAEA Regulations, is dealt with.

The use of the isotopes in the theory and technology of optical fibers was considered in Chap. 7. This chapter is addressed to readers who wish to learn about fiber communications systems and, particularly, about the properties of optical fibers. Very briefly in this chapter we describe the Maxwell equations as well as wave electromagnetic equations. In this chapter, we also describe the properties of optical fibers and also the materials for optical fiber and fiber technology. The influence of isotopes on the properties of optical fibers was also shown.

Chapter 9 was devoted to the application of the isotopic effect in laser physics. There is a short description of the theory and practice of semiconductor lasers. The discovery of the linear luminescence of free excitons observed over a wide temperature range has placed lithium hydride [23], as well as crystals of diamond [604] in line as prospective sources of coherent radiation in the UV spectral range. LiH isotope tuning of exciton emission has also been shown.

The last chapter of this book is devoted to a description of other unexplored applications of isotopic engineering. In the first place, we considered materials for information storage in modern personal computers as well as in biology. There are large possibilities for isotope-based quantum computers. We should add here that the strength of the hyperfine interaction is proportional to the probability density of the electron wave function in a nucleus. In semiconductors, the electron wave function extends over large distances through the crystal lattice. Two nuclear spins can consequently interact with the same electron, leading to electron-mediated or indirect nuclear spin coupling. Because an electron is sensitive to externally applied electric fields, the hyperfine interaction and electron-mediated nuclear spin interaction can be controlled by voltages applied to metallic gates in a semiconductor device, enabling the external manipulation of nuclear spin dynamics that is necessary for quantum computation in quantum computers (for details, see [625]).

The wide possibilities of isotopic engineering discussed in this book hold the greatest promise for application in solid-state and quantum electronics, biology, human memory, optoelectronics, different electronic devices, electronic and quantum computers, and many other modern and new technologies that are even now difficult to imagine. And we should repeat that the main aim of this book is to familiarize readers with the present and some future applications in isotopic science and engineering.

References

1. Born, M., Huang, K., *Dynamical Theory of Crystal Lattices* (Clarendon Press, Oxford, 1968)
2. Born, M., Oppenheimer, R., Ann. Phys. **84**, 457 (1927)
3. Hardy, J.R., Caro, A.M., *The Lattice Dynamics and Statics of Alkali Halide Crystals* (Plenum Press, New York, 1979)
4. Bötger, H., *Principles of the Theory of Lattice Dynamics* (Weinheim-Verlag, 1983)
5. Reissland, J.A., *The Physics of Phonons* (Wiley and Sons, London–Toronto, 1973)
6. Brüesch, P., *Phonons: Theory and Experiments*; Lattice Dynamics and Models of Interatomic Forces, Springer Series in Solid State Sciences, Vol. **34** (Springer-Verlag, Berlin–Heidelberg–New York, 1982)
7. Maradudin, A.A., Montroll, E.W., Weiss, G.H., Ipatova, I.P., *Theory of Lattice Dynamics in the Harmonic Approximation*, 2nd ed. (Academic Press, New York, 1971)
8. Barker, A.S., Sievers, A.J., Rev. Mod. Phys. **47**, Suppl. 2, p. 515 (1975)
9. Taylor, D.N., Dynamics of Impurities in Crystals, in Horton, G.K., Maradudin, A.A. (eds.) *Dynamical Properties of Solids* (North-Holland, Amsterdam, 1975, Vol. 2, Chap. 5)
10. Economou, E.N., *Green's Function in Quantum Physics* (Springer-Verlag, Heidelberg–New York, 1983)
11. Lifshitz, I.M., *Physics of Real Crystals and Disordered Systems* (Science, Moscow, 1987) (in Russian)
12. Musgrave, M.J.P., *Crystal Acoustics* (Hakden Day, London, 1970)
13. Nye, J.F., *Physical Properties of Crystals* (Oxford University Press, Oxford, 1957)
14. Neighbours, J.R., Schacher, G.E., J. Appl. Phys. **38**, 5366 (1967)
15. Gluyas, M., Hunter, R., James, B.W., J. Phys. C: Solid State Phys. **8**, 271 (1975)
16. Watherman, P.C., Phys. Rev. **113**, 1240 (1959)
17. James, B.W., Kherandish, H., J. Phys. C: Solid State Phys. **15**, 6321 (1982)
18. Hayes, W., Loudon, R., *Scattering of Light by Crystals* (Wiley, New York, 1978)
19. Guinan, M.W., Cline, C.F., J. Nonmetals **1**, 11 (1972)
20. Gerlich, D., Smith, S.S., J. Phys. Chem. Solids **35**, 1587 (1974)
21. Haussuhl, S., Skorczyk, I., Zs. Krist. **130**, 340 (1969)
22. Laplaze, D., Boissier, M., Vaher, R.: Solid State Commun. **19**, 445 (1976)
23. Plekhanov, V.G., *Isotope Effects in Solid State Physics* (Academic Press, New York–London, 2001)

24. Leibfried, G., Gittertheorie der mechanischen und termischen Eigenschaften der Kristalle, in Fluge, S. (ed.) *Encyclopedia of Physics* (Springer-Verlag, Berlin, 1955, Vol. 7, Pt. 1, p. 309)
25. Plekhanov, V.G., Opt. Spectrosk. **82**, 95 (1997)
26. Wilson, W.D., Johnson, R.A., Phys. Rev. B**1**, 3510 (1970)
27. V.G. Plekhanov, Materials Science and Eng. R: Reports **35**, 139 (2001); Physics -Uspekhi (Moscow) **46** 689 (2003).
28. Leibfried, G., Ludwig, W., *Theory of Anharmonic Effects*, in Seitz, F., Turnbull, D. (eds.), *Crystals* (Academic Press, New York, 1961, Vol. 12)
29. Yates, B., Wostenholm, G.H., Bingham, J.L., J. Phys. C: Solid State Phys. **7**, 1769 (1974)
30. Grimsditch, M.H., Ramdas, A.K., Phys. Rev. B **11**, 3139 (1975)
31. Sondercock, J., Festkörperprobleme, Vol. 15 of Advances in Solid State Physics (Pergamon Press, Braunschweig, 1975, p. 183)
32. Ramdas, A.K., Rodriguez, S., Grimsditch, M., Anthoni, T.R., Banholzer, W.F., Phys. Rev. Lett. **71**, 189 (1993)
33. Vogelgesang, R., Ramdas, A.K., Rodriguez, S., Anthony, T.R., Phys. Rev. B **54**, 3989 (1996)
34. Zoubolis, E.S., Ramdas, A.K., Grimsditch, M., Phys. Rev. B**57**, 2889 (1998)
35. Hurley, D.C., Gilmore R.S., Banholzer, W.F., J. Appl. Phys. **76**, 7726 (1994)
36. Musgrave, M.J.P., Pople, J.A., Proc. R. Soc. (London) **A268**, 474 (1962)
37. Keating, P.N., Phys. Rev. **145**, 637 (1966)
38. Martin, R.M., Phys. Rev. B**1**, 4005 (1970)
39. Lax, M., Lattice Dynamics of Ionic Crystals, in Wallis, R.F. (ed.), *Lattice Dynamics* (Pergamon Press, New York, 1965, p. 583)
40. Holloway, H., Hass, K.C., Tamor, M.A., Anthony, T.R., Banholzer, W.F., Phys. Rev. B **44**, 7123 (1991)
41. Mykolajewycz, R., Kalnajs, J., Smakula, A., J. Appl. Phys. **35**, 1773 (1964)
42. McSkimin, H.J., Andreatch, P., J. Appl. Phys. **43**, 2944 (1972)
43. Muinov, M., Kanda, H., Stishov, S.M., Phys. Rev. B**50**, 13860 (1994); Stishov, S.M., Uspekhi Fiz. Nauk **171**, 299 (2001) (in Russian)
44. Cerdeira, F., Cardona, M., Phys. Rev. B **5**, 1440 (1972)
45. Bilz, H., Kress, W.,*Phonon Dispersion Relations in Insulators* (Springer-Verlag, Berlin, 1979)
46. Bilz, H., Strauch, D., Wehner, R.K., Vibrational Infrared and Raman Spectra of Non - Metals, in Genzel, L. (eds.) *Handbuch der Physik* (Springer-Verlag, Berlin, 1984, Vol. 25/2d)
47. Zhernov, A.P., Inyushkin, A.V., Uspekhi Fiz. Nauk (Moscow) **171**, 827 (2001) (in Russian)
48. Krivoglaz, M.A., *Theory of Scattering X-Rays and Thermal Neutrons by Real Crystals* (Science, Moscow, 1967) (in Russian)
49. Dolling, G., Neutron Spectroscopy and Lattice Dynamics, in Horton, G.K., Maradudin, A.A. (eds.), *Dynamical Properties of Solids* (North-Holland, Amsterdam, 1974); Dorner, B., *Inelastic Neutron Scattering in Lattice Dynamics* (Springer Tracts in Modern Physics, Springer-Verlag, Berlin, 1982, Vol. 93)
50. Zemlianov, M.G., Browman, E.G., Chernoplekov, N.A., Shitikov, N.Kl., Inelastic neutron Scattering in LiH - LiD Systems, in *Inelastic Scattering of Neutrons* (International Atomic Energy Agency, Vienna, 1965, Vol. 2, p. 431)
51. Verble, J.L., Warren, J.L., J.L. Yarnell, J.L., Phys. Rev. **168**, 980 (1968)

52. Plekhanov, V.G., Sov. Phys. Opt. Spectrosk. **62**, 768 (1987)
53. Warren, J.L., Yarnell, J.L., Dolling, G., Phys. Rev. **158**, 805 (1967)
54. Dolling, G., Woods, A.D.V., Neutron Scattering in Solids, in Egelstaff, P.A. (ed.), *Thermal Neutron Scattering* (Academic Press, New York, 1965)
55. Nilsson, G., Nelin, G., Phys. Rev. B **3**, 364 (1971)
56. Tubino, R., Piseri, L., Zerbi, G., J. Chem. Phys. **56**, 1022 (1972)
57. Baroni, S., Gianozzi, P., Testa, A., Phys. Rev. Lett. **58**, 1861 (1987)
58. Dolling, G., Cowley, R.A., Proc. Phys. Soc. **88**, 463 (1966)
59. Cardona, M., Introduction, in *Light Scattering of Solids* Topics in Applied Physics, **8** (1975)
60. Lax, M., Burstein, E., Phys. Rev. **97**, 39 (1955)
61. Loudon, R., Adv. Phys. **13**, 423 (1964)
62. Birman, J.L., *Handbuch für Physik* (Springer-Verlag, Berlin–New York, 1974, Vol. 25/26)
63. Jaswal, S.S., Wolfram, G., Sharma, T.R., J. Phys. Chem. Solids **35**, 571 (1974)
64. Laplaze, D., J. Phys. C: Solid State Phys. **10**, 3499 (1977)
65. Pavone, P., Karch, K., Schutt, O., Windl, W., Phys. Rev. B **48**, 3156 (1993)
66. Windl, W., Pavone, P., Karch, K., Schutt, O., Strauch, D., Gianozzi, P., Baroni, S., Phys. Rev. B **48**, 3164 (1993)
67. Baroni, S., de Gironcoli, S., Dal Corso, A., Gianozzi, P., Rev. Mod. Phys. **73**, 515 (2001)
68. Dean, P., Rev. Mod. Phys. **44**, 127 (1972)
69. Bell, R., Rep. Progr. Phys. **35**, 1321 (1972)
70. I.F. Chang and S.S. Mitra, Adv. Phys. **20**, 360 (1971).
71. Elliott, R.J., Krumhansl, J., Leath, P., Rev. Mod. Phys. **46**, 465 (1974); Elliott, R.J., Taylor, D.W., Proc. R. Soc. London **296**, 161 (1967)
72. Taylor, D.W., Phonon Response Theory and the Infrared and Raman Experiments, in Elliott, R.J., Ipatova, I.P. (eds.), *Optical Properties of Mixed Crystals* (North-Holland, Amsterdam, 1988, Chap. 3, p. 35)
73. Lifshitz, I.M., Nuovo Cimento, Suppl. **3**, 716 (1956)
74. Lifshitz, I.M., Kosevich, A.M., Rep. Progr. Phys. **29**, 217 (1966)
75. Plekhanov, V.G., Veltri, V.A., Fiz. Tverd. Tela, St.-Petersburg, **33**, 2384 (1991) (in Russian)
76. Jaswal, S.S., Sharma, T.P., Wolfram, G., Solid State Commun. **11**, 1159 (1972)
77. Anderson, A., Luty, F., Phys. Rev. B**28**, 3415 (1983); Tyutyunnik, V.I., Tyutyunnik, O.I., Phys. Stat. Solidi (b) **162**, 597 (1990)
78. Plekhanov, V.G., Physics Uspekhi (Moscow) **40**, 553 (1997)
79. Anderson, P.W., Phys. Rev. **109**, 1442 (1955).
80. Belitz, D., Kirkpatrick, T.R., Rev. Mod. Phys. **66**, 261 (1994)
81. Plekhanov, V.G., Phys. Rev. B **51**, 8874 (1995)
82. Behera, S.M., Tripathi, R.S., J. Phys. C: Solid State Phys. **7**, 4452 (1972)
83. Montgomery, D.J., Yeung, K.F., J. Chem. Phys. **37**, 1056 (1962)
84. Nordheim, L., Ann. Phys. (Leipzig) **9**, 641 (1931); Pant, M.M., Joshi, S.K., Phys. Rev. **184**, 635 (1969)
85. Ipatova, I.P., Universal Parameters in Mixed Crystals, in [234], p.1
86. Taylor, D.V., Phys. Rev. **156**, 1017 (1967)
87. Ehrenreich, H., Schwartz, W., The Electronic Structure of Alloys, in Ehrenreich, H., Seitz, F., Turnbull, D. (eds.) *Solid State Physics* (Academic Press, New York, 1976, Vol. 31)

88. Agekyan, V.F., Asnin, V.M., Kryukov, A.M., Fiz. Tverd. Tela, St.-Petersburg **31**, 101 (1989) (in Russian)
89. Fuchs, H.D., Grein, C.H., Thomsen, C., Cardona, M., Hansen, W.L., Haller, E.E., Itoh, K., Phys. Rev. B**43**, 4835 (1991)
90. Fuchs, H.D., Grein, C.H., Cardona, M., Hansen, W.L., Itoh, K., Haller, E.E., Solid State Commun. **82**, 225 (1992); Fuchs, H.D., Etchegoin, P., Cardona, M., Phys. Rev. Lett. **70**, 1715 (1993)
91. Etchegoin, P., Fuchs, D., Weber, J., Cardona, M., Itoh, K., Haller, E.E., Phys. Rev. B **48**, 12661 (1993)
92. Cardona, M., Etchegoin, P., Fuchs, H.D., Molina-Matra, P., J. Phys.: Condens. Matter **5**, A61 (1993); Debernardi, A., Cardona, M., Phys. Rev. B**54**, 11305 (1996)
93. Zhang, J.M., Giehler, M., Göbel, A., Ruf, T., Cardona, M., Phys. Rev. B **57**, 1348 (1998)
94. Hanzawa, H., Umemura, N., Nisida, Y., Kanda, H., Phys. Rev. B **54**, 3793 (1996)
95. Chu, C.J., D'Evelyn, M.P., Hauge, R.H., Margrave, G.L., J. Mater. Res. **5**, 2405 (1990)
96. Knight, D.S., White, W.B., J. Mater. Res. **4**, 385 (1989)
97. Hass, K.C., Tamor, M.A., Anthony, T.R., Banholzer, W.F., Phys. Rev. B **45**, 7171 (1992)
98. Chrenko, R.M., J. Appl. Phys. **63**, 5873 (1988)
99. Solin, S.A., Ramdas, A.K., Phys. Rev. B **1**, 1687 (1970)
100. Montgomery, D.J., Hardy, J.R., Lattice Vibrational Impurity Modes in LIh:D and LiD:H Crystals, in Wallis, R.F. (ed.), *Lattice Dynamics* (Pergamon Press, New York, 1965, p. 643)
101. Tripathi, R.S., Behera, J., J. Phys. C: Solid State Phys. **7**, 4440 (1974)
102. Elliott, R.J., Leath, P.L., High Concentration - Mixed Crystals and Alloys, in Horton, G.K., Maradudin, A.A. (eds.), *Dynamical Properties of Solids* (North-Holland, Amsterdam, 1975, Vol. 2, p. 387)
103. Menendez, J., Page, J.B., Guha, S., Philos. Mag. **70**, 651 (1994)
104. Nelin, G., Nilsson, G., Phys. Rev. B **5**, 3151 (1972)
105. Göbel, A., Wang, D.T., Cardona, M., Kulda, J., Itoh, K., Haller, E.E., Phys. Rev. B**58**, 10510 (1998)
106. Cardona, M., Grein, C.H., Fuchs, H.D., Zollner, S., J. Non-Cryst. Solids **141**, 257 (1992); Cardona, M., Semiconductor Crystals with Tailor-Made Isotopic Composition, in Helbig, R. (ed.), *Advances in Solid State Physics* (Vieweg, Braunschweig, Weisbaden, 1994, Vol. **34**, p. 35)
107. Spitzer, J., Etchegoin, P., Cardona, M., Anthony, T.R., Banholzer, W.F., Solid State Commun. **88**, 509 (1993)
108. Efros, A.L., Raikh, M.E., Effect of Composition Disorder on the Electronic Properties of Semiconducting Crystals, in Elliott, R.J., Ipatova, I.P. (eds.), *Optical Properties of Mixed Crystals* (North-Holland, Amsterdam, 1988, Chap. 5 p. 133)
109. Plekhanov, V.G., Rep. Progr. Phys. **61**, 1045 (1998)
110. Hass, K.C., Tamor, M.A., Anthony, T.R., Banholzer, W.F., Phys. Rev. B **44**, 12046 (1991)
111. Wang, C.Z., Chang, C.T., Ho, K.M., Solid State Commun. **76**, 483 (1990)
112. Schwoerer-Bohning, M.A., Macrander, A.T., Arms, D.A., Phys. Rev. Lett. **80**, 5572 (1998)

113. Cardona, M., Semiconductor Crystals with Tailor-Made Isotopic Composition, in Helbig, R. (ed.), *Advances in Solid State Physics* (Vieweg, Braunschweig, Wiesbaden, 1994, **34**, p. 35)
114. Fuchs, H.D., Etchegoin, P., Cardona, M., Phys. Rev. Lett. **70**, 1715 (1993)
115. Wang, D.T., Gobel, A., Zepenhagen, J., Cardona, M., Phys. Rev. B **56**, 13167 (1997)
116. Plekhanov, V.G., Altukhov, V.I., J. Raman Spectrosc. **16**, 358 (1985)
117. Plekhanov, V.G., J. Raman Spectrosc. **32**, 631 (2001)
118. Ashcroft, N.M., Mermin, N.D., *Solid State Physics* (Holt, Reinhart and Winston, New York, 1976)
119. Peierls, R.E., *Quantum Theory of Solids* (Clarendon Press, Oxford, 1955)
120. Klemens, P.G., *Thermal conductivity and lattice vibrational modes*, in Seitz, F., Turnbull, D. (eds.) *Solid State Physics* (Academic Press, 1958, Vol. 7, p. 1)
121. Berman, R., *Thermal Conduction in Solids* (Clarendon Press, Oxford, 1976)
122. Ziman, J.M., *Electrons and Phonons* (Clarendon Press, Oxford, 1962)
123. Callaway, J., Phys. Rev. **113**, 1046 (1959)
124. Holland, M.G., Phys. Rev. **132**, 2461 (1963); Thermal Conductivity, in Willardson, R.K., Weber, E.R. (eds.), *Thermal Conductivity in Semiconductors and Semimetals* (Academic Press, New York, Vol. 2)
125. Herring, C., Phys. Rev. **95**, 954 (1954)
126. Geballe, T.H., Hull, G., Phys. Rev. **110**, 773 (1958)
127. Hamilton, R.A.H., Parrot, J.P., Phys. Rev. **178**, 1284 (1969)
128. Zhernov, A.P., ZETF (Moscow) **120**, 1237 (2001)
129. Omini, M., Sparavigna, A., Nuovo Cimento **19** D, 1537 (1997); Sparavigna, A., Phys. Rev. B **65**, 064305-1 (2002)
130. Anthony, T.R., Banholzer, W.F., Fleisher, J.F., Wei, L.-H., Kuo, P.K., Thomas, R.L., Phys. Rev. B **42**, 1104 (1990)
131. Banholzer, W., Anthony, T., Gilmore, R., Properties of Diamond as a Function of Isotope Composition, in New Diamond Science and Technology, *Proc. Second Int. Conf. New Diamond Sci. Technol.*, September 1990, Washington D.C. (Materials Research Society, Pittsburgh, PA, 1991, p. 857)
132. Anthony, T.R., Banholzer, W.F., Diamond Relat. Mater. **1**, 717 (1992)
133. Onn, D.G., Vitek, A., Qiu, Y.Z., Anthony, T.R., Banholzer, W.F., Phys. Rev. Lett. **68**, 2806 (1992)
134. Olson, J.R., Pohl, R.O., Vandersande, J.W., Zoltan, A., Anthony, T.R., Banholzer, W.F., Phys. Rev. B**47**, 14850 ((1993)
135. Graebner, J.E., Reiss, M.E., Seibles, L., Harfnett, T.M., Miller, R.P., Rolinson, C.J., Phys. Rev. B **50**, 3702 (1994)
136. Wilks, J., Wilks, E., *Properties and Applications of Diamond* (Butterworth-Heinemann, Oxford, 1991)
137. Graebner, J.E., Thermal Conductivity of Diamond, in Prelas, M.A., Popovici, G., Bigelow, L.K. (eds.), *Handbook of Industrial Diamonds and Diamond Films* (Marcel Dekker, New York, Basel, Hong Kong, 1998, p. 193)
138. Banholzer, W.F., Anthony, T.R., Thin Solid Films **212**, 1 (1992)
139. Novikov, N.V., Kocherzhinski, Ju.A., Shulman, L.A., *Physical Properties of Diamond*, Handbook (Naukova Dumka, Kiev, 1987) (in Russian)
140. Berman, R., Phys. Rev. B **45**, 5726 (1992)

141. Asen-Palmer, M., Bartkowski, K., Gmelin, E., Cardona, M., Zhernov, A.P., Inyushkin, A.V., Phys. Rev. B **56**, 9431 (1997); Kuleev, I.G., Kuleev, I.I., ZETP (Moscow) **120**, 649 (2000) (in Russian); Kuleev, I.G., Kuleev, I.I., Taldenkov, A.N., Inyushkin, A.V., Ozhogin, V.I., Itoh, K., Haller, E.E., ZETP (Moscow) **123**, 1227 (2003)
142. Jekson, H.E., Walker, C.T., Phys. Rev. B **3**, 1428 (1971)
143. Zhernov, A.P., Zhernov, D.A., ZETF(Moscow) **114**, 1757 (1998); Fiz. Tverd. Tela, St.-Petersburg **40**, 1604 (1998) (in Russian)
144. Capinski, W.S., Maris, H.J., Tamura, S., Phys. Rev. B **59**, 10105 (1999)
145. Ruf, T., Henn, R.W., Asen-Palmer, A.M., Gmelin, E., Cardona, M., Solid State Commun. **115**, 243 (2003)
146. Ruf, T., Henn, R.W., Asen-Palmer, A., Gmelin, E., Cardona, M., Solid State Commun. **127**, 257 (2000); Kuleev, I.G., Kuleev, I.I., ZETP(Moscow) **122**, 558 (2002) (in Russian)
147. Lyding, J.W., Hess, K., Kizilyalli, I.C., Appl. Phys. Lett. **68**, 2526 (1996)
148. London, G., Zs. Phys. Chem. Neue Folge **16**, 302 (1958)
149. London, F., *Macroscopic Theory of Superfluids: Superfluid Helium* (Dover, New York, 1964, Vol. 2)
150. Kogan, V.S., Sov. Phys. Uspekhi **5**, 951 (1963)
151. Anderson, G.L., Nasise, G., Philipson, K., Pretzel, F.E., J. Phys. Chem. Solids **31**, 613 (1970)
152. Smith, D.K., Leider, H.R., J. Appl. Crystallogr. **1**, 246 (1968)
153. Pavone, P., Baroni, S., Solid State Commun. **90**, 295 (1994)
154. Shpilrain, E.E., Yakimovich, K.A., Melnikova, T.N., *Thermal Properties of Lithium Hydride, Deuteride and Tritide and Their Solution with Lithium* (Energoizdat, Moscow, 1983) (in Russian)
155. Buschert, R.C., Merlin, A.E., Pace, S., Rodriguez, S., Grimsditch, M.H., Phys. Rev. B**38**, 5219 (1988)
156. Noya, J.C., Herrero, C.P., Ramirez, R., Phys. Rev. B **56**, 237 (1997)
157. Holloway, H., Hass, K.C., Tamor, M.A., Phys. Rev. B **45**, 6353 (E) (1992)
158. Yamanaka, I., Morimoto, S., Kanda, H., Phys. Rev. B **49**, 9341 (1994)
159. Touloukian, Y.S., Powell, R.W., Ho, C.Y., Klemens, P.G., *Thermal Conductivity-Metallic Elements and Alloys, Thermophysical Properties of Materials* (IFI/Plenum Press, New York, Washington, 1970)
160. Zhernov, A.P., Fiz. Tverd. Tela **41**, 1185 (1999); ZETF **120**, 1237 (2001) (in Russian)
161. Frenkel, J., Phys. Rev. **37**, 11 (1931); **37**, 1276 (1931)
162. Peierls, R.E., Ann. Phys. **31**, 905 (1932)
163. Slater, J.K., Schokley, W.H., Phys. Rev. **50**, 705 (1936)
164. Wannier, G.H., Phys. Rev. **52**, 191 (1937)
165. Mott, N.F., Proc. R. Soc. A **167**, 384 (1938)
166. Davydov, A.S., *Theory of Molecular Excitons* (Science, Moscow, 1968) (in Russian)
167. Knox, R.S., *Theory of Excitons* (Academic Press, New York, 1963)
168. Agranovich, V.M., Ginzburg, V.L., *Crystalloptic and Theory of Excitons* (Science, Moscow, 1979) (in Russian)
169. Gross, E.F., *Selected Papers* (Science, Leningrad, 1976) (in Russian)
170. Pekar, S.I., *Crystal Optics and Additional Light Waves* (Benjamin, Menlo Park, CA, 1983)

171. Haken, H., *Quantum Field Theory of Solids* (North-Holland, Amsterdam, 1976)
172. Dexter, D.L., Knox, R.S., *Excitons* (Wiley, New York, 1965)
173. Elliot, R.J., Theory of Wannier - Mott Excitons, in Kuper, C.G., Whitefield, G.D. (eds.), *Polarons and Excitons* (Plenum Press, New York, 1963)
174. Dimmock, J.O., Theory of Excitons, in Willardson, R.K., Beer, A.G. (eds.), *Semiconductors and Semimetals* (Academic Press, New York, 1967, Vol. 3, p. 259)
175. Dresselhaus, G., J. Phys. Chem. Solids **1**, 14 (1955)
176. Kittel, C., *Introduction in Solid State Physics* (Wiley and Sons, New York, 1963)
177. Bethe, H.A., Salpeter, E., *Quantum Theory of One and Two Electron Atoms* (Academic Press, New York, 1957)
178. Elliott, R.J., Phys. Rev. **108**, 1384 (1957)
179. Plekhanov, V.G., Betenekova, T.A., Pustovarov, V.A., Sov. Phys. Solid State **18**, 1422 (1976)
180. Klochikhin, A.A., Plekhanov, V.G., Soviet Phys. Solid State **22**, 342 (1980)
181. Bir, G.L., Pikus, G.E., *Symmetry and Deformation Effects in Semiconductors* (Science, Moscow, 1972) (in Russian)
182. Ansel'm, A.I., Firsov, Yu.A., ZETF **30**, 719 (1956) (in Russian)
183. Thomas, D.G. (ed.), *II-VI Semiconducting Compounds* (Benjamin, New York, 1967)
184. Firsov, Yu.A. (ed.), *Polarons* (Science, Moscow, 1975) (in Russian)
185. Permogorov, S.A., Exciton - LO Phonon Scattering, in Rashba, E.I., Sturge, M.D. (eds.), *Excitons* (North-Holland, Amsterdam, 1982, Chap. 5)
186. Toyozawa, Y., Progr. Theor. Phys. (Kyoto) **20**, 53 (1958)
187. Frohlich, H., Adv. Phys. **3**, 325 (1954)
188. Klochikhin, A.A., Sov. Phys. Solid State **22**, 986 (1980)
189. Bulyanitza, D.S., Fizika i Technika Poluprovodnikov, St.-Petersburg, **4**, 1273 (1970). (in Russian)
190. Kroger, F.A., Physica, **7**, 1 (1940)
191. Plekhanov, V.G., Altukhov, V.I., Sov. Phys. Solid State **23**, 439 (1981)
192. Zheng, R., Matzuura, M., Phys. Rev. B**59**, 15422 (1999)
193. Matzuura, M., J. Phys. C: Solid State Phys. **10**, 3345 (1977)
194. Wang, X., Liang, X.X., Phys. Rev. B **42**, 8915 (1990)
195. Nicholas, R.J., Strafling, R.A., Ramage, J.C., J. Phys. C: Solid State Phys. **12**, 1641 (1979)
196. Adachi, S., J. Appl. Phys. **58**, R1 (1985)
197. Bach, F., Banhoeffer, K.F., Zs. Phys. Chem. B **23**, 256 (1933)
198. Kapustinsky, A.F., Shamovsky, L.M., Baushkina, K.S., Acta Physicochem. (USSR) **7**, 799 (1937)
199. Rauch, W., Zs. Phys. **111**, 650 (1938/1939)
200. Gavrilov, F.F., Opt. Spectrosk. **7**, 371 (1959) (in Russian)
201. Pretzel, F.E., Ruppert, G.N., Malder, C.L., J. Phys. Chem. Solids **16**, 10 (1960)
202. Gavrilov, F.F., Kalder, K.A., Shulgin, B.V., Fiz. Tverd. Tela **13**, 3107 (1971) (in Russian)
203. Plekhanov, V.G., Betenekova, T.A., Sov. Phys. Solid State **19**, 1926 (1977)
204. Plekhanov, V.G., Resonant Secondry Emission Spectra, *Proc. Int. Conf. Lasers'80* (McClean, VA, 1981, p. 94)

205. Calder, R.S., Cochran, W., Grifith, D., Lowde, R.D., J. Phys. Chem. Solids **23**, 621 (1962)
206. Pauling, L., *The Nature of the Chemical Bond* (Cornell University Press, Ithaca, 1960)
207. Tyutyunnik, O.I., Tyutyunnik, V.I., Shulgin, B.V., Oparin, D.V., Pilipenko, G.I., Gavrilov, F.F., J. Crystal Growth **68**, 741 (1984)
208. Plekhanov, V.G., Emelyanenko, A.V., Grinfelds, A.U., Phys. Lett. A **101**, 291 (1984)
209. Zimmerman, W.B., Phys. Rev. B **5**, 4704 (1972)
210. Plekhanov, V.G., Altukhov, V.I., Free Exciton Luminescence and Exciton-Phonon Interaction Parameters of Wide-Gap Insulators, *Proc. Int. Conf. Lasers'82* (McClean, VA, 1983, p. 292)
211. Plekhanov, V.G., Phys. Rev. B **54**, 3869 (1996)
212. Plekhanov, V.G., Opt. Spectrosk. **69**, 822 (1990) (in Russian)
213. Kink, R.A., Kink, M.F., Soovik, T.A., Nucl. Instrum. Methods Phys. Res. A **261**, 138 (1987)
214. Segall, B., Marple, D.T.F., Phonon-Assisted Emission of Free Excitons in CdS, in Aven, M., Prener, J. (eds.), *Physics and Chemistry of A_2B_6 Compounds* (North-Holland, Amsterdam, 1967, Chaps. 1 and 7)
215. Plekhanov, V.G., Phys. Rev. B **53**, 9558 (1996)
216. Kunz, A.B., Mickish, D.J., Phys. Rev. B **11**, 1700 (1975)
217. Baroni, S., Parravichini, P.G., Pezzica, G., Phys. Rev. B **32**, 4077 (1985)
218. Perrot, F., Phys. Stat. Solidi **77**, 517 (1976)
219. Segall, B., Mahan, G.D., Phys. Rev. **171**, 935 (1968)
220. Agekyan, V.F., Phys. Stat. Solidi (a) **43**, 11 (1977)
221. Herzberg, G., *Molecular Spectra and Molecular Structure. Spectra of Diatomic Molecules* (Van Nostrand, New York, 1945, Vol. 1)
222. Nelson, R.J., Excitons in Semiconductor Alloys, in Rashba,E.I., Sturge, M.D. (eds.), *Excitons* (North-Holland, Amsterdam, 1982, Chap. 8)
223. Nelson, R.J., Holonjak N., Groves, W., Phys. Rev. B **13**, 5415 (1976)
224. Monemar, R., Shih, K.K., Pettit, G.D., J. Appl. Phys. **47**, 2604 (1976)
225. Radautsan, S.I., Tsurkan A.E., Maksimova, A.E., *Complex Semiconductors and Their Physical Properties* (Shtiinza, Kishinev, USSR, 1971) (in Russian)
226. Brodin, M.S., Vitrikhovski, N.I., Kipen, A.A., *Exciton Reflection Spectra of Mixed Crystals:* $Zn_xCd_{1-x}S$ (unpublished, 1982)
227. Kreingol'd, F.I., Lider, K.F., Shabaeva, M.B., Fiz. Tverd. Tela **26**, 3940 (1984) (in Russian)
228. Zhang, J.M., Ruf, T., Lauck R., Cardona, M., Phys. Rev. B **57**, 9716 (1998)
229. Onodera, Y., Toyozawa, Y., J. Phys. Soc. Japan **24**, 341 (1968)
230. Plekhanov, V.G., Plekhanov, N.V., Phys. Lett. **A313**, 231 (2003)
231. Bajaj, K.K., Mater. Sci. Eng. R: Rep. **34**, 59 (2001)
232. Brodsky, M.N., Burstein, E., J. Phys. Chem. Solids **28**, 1655 (1967)
233. Klochikhin, A.A., Permogorov, S.A., Reznitsky, A.N., Fiz. Tverd. Tela **39**, 1170 (1997)
234. Elliott, R.J., Ipatova, I.P. (eds.), *Optical Properties of Mixed Crystals* (North-Holland, Amsterdam, 1988)
235. Parks, V., Ramdas, A.K., Rodriguez, S., Itoh, K.M., Haller, E.E., Phys. Rev. B**49**, 14244 (1994)
236. Sak, J., Phys. Rev. B **6**, 2226 (1972)

237. Mahanti, S.D., Varma, C.M., Phys. Rev. B **6**, 2209 (1972)
238. Kanehisa, V.A., Elliott, R.J., Phys. Rev. B **35**, 2228 (1987)
239. Schwabe, N.F., Elliott, R.J., Phys. Rev. B **53**, 5318 (1996)
240. Mahanti, D., Phys. Rev. B **10**, 1384 (1974)
241. Hama, J., Kawakami, N., Phys. Rev. B **39**, 3351 (1989)
242. Betenekova, T.A., Shabanova, I.N., Gavrilov, F.F., Fiz. Tverd. Tela **20**, 2470 (1978) (in Russian)
243. Ichikawa, K., Susuki, N., Tsutsumi, K., J. Phys. Soc. Japan **50**, 3650 (1981)
244. Chang, R.K., Long, M.B., Optical Multichannel Detection, in Cardona, M., Guntherodt, G. (eds.), *Light Scattering in Solids* (Springer-Verlag, Berlin, 1982) Vol. 2, p. 179
245. Plekhanov, V.G., O'Konnel-Bronin, A.A., JETP Lett. **27**, 387 (1978)
246. Fisher, B., Stolz, H., Appl. Phys. Lett. **40**, 56 (1982)
247. Schultheis, L., Balslev, I., Phys. Rev. B **28**, 2292 (1983)
248. Collins, A.T., Lawson, A.C., Davies, G., Kanda, H., Phys. Rev. Lett. **65**, 891 (1990)
249. Kreingol'd, F.I., Fiz. Tverd. Tela **20**, 3158 (1978) (in Russian)
250. Davies, G., Lightowlerst, E.L., Hui, T.S., Ozhogin, V., Itoh, K., Haller, E.E., Semicond. Sci. Technol. **8**, 2201 (1993)
251. Etchegoin, P., Weber, J., Cardona, M., Solid State Commun. **83**, 843 (1992)
252. Roberts-Austen, W.C., Philos. Trans. R. Soc. (London) A **187**, 404 (1896)
253. Mehl, R.F., Trans. AIME **122**, 11 (1936)
254. Jost, W., *Diffusion in Solids, Liquids, Gas* (Academic Press, New York, 1952)
255. Seith, W., *Diffusion in Metallen* (Springer-Verlag, Berlin, 1955)
256. Slifkin, L., Lazarus, D., Tomizuka, C.T., J. Appl. Phys. **23**, 1032 (1952)
257. Tomizuka, C.T., Experimental Techniques, in Larl-Horrowitz, K., Johnson, V.A. (eds.), *Methods of Experimental Physics* (Academic Press, New York, 1959, Vol. 6, Part A, p. 364)
258. Nowick, A.S., J. Appl. Phys. **22**, 1182 (1951)
259. Mott, N.F., Gurney, R.F., *Electronic Processes in Ionic Crystals* (Clarendon Press, Oxford, 1948)
260. Crawford, Jr, J.H., Slifkin, L.M. (eds.), *Point Defects in Solids* (Plenum Press, London, 1975, Vols. 1–2
261. Hoffman, R.E., Turnbull, D., J. Appl. Phys. **22**, 634 (1951)
262. Cadek, J., Janda, J., Hung. Listy **12**, 1008 (1957) (English transl.: AERE Transl. 840, Harwell)
263. Adda, Y., Philbert, J., *La Diffusion dans les Solides* (Presse Universitaires de France, Paris, 1966)
264. Lundy, T.S., Experimental Techniques, in Rapp, R.A. (ed.), *Techniques of Metals Research* (Wiley, New York, 1970, Vol. 4, Part 2, Chap. 9A)
265. Beniere, F., in Beniere, F., Catlow, C.R.A. (eds.), *Mass Transport in Solids* (Plenum Press, London, 1983)
266. Nowick, A.S., Burton, J.J. (eds.), *Diffusion in Solids: Recent Developments* (Academic Press, New York, 1975)
267. Rothman, S.J., *Thesis* (Stanford University, Stanford, CA, 1954)
268. Rothman, S.J., J. Nucl. Mater. **3**, 77 (1961)
269. Rothman, S.J., Peterson, N.L., Phys. Rev. **154**, 552 (1967)
270. Rothman, S.J., Peterson, N.L., Walter, C.M., Nowicki L.J., J. Appl. Phys. **39**, 5041 (1968)

271. Sigmund, P., Experimental Techniques in the Sputerring, in Behrisch, R. (ed.), *Sputtering by Particle Bombardment* (Springer-Verlag, Berlin, 1981) Chap. 2
272. High-Power Semiconductor Devices, IEEE Trans. Electron Devices ED-**23** (1976)
273. Antoniadis, D.A., Moskowitz, I., J. Appl. Phys. **53,** 9214 (1982)
274. Craven, R.A., Neutron Transmutation Doping of Silicon, in Huff, H.R., Kriegr, R.J., Takeishi, Y. (eds.), *Semiconductor Silicon'81* (The Electrochem. Soc. Pennington, NJ, 1981)
275. Peterson, N.L., Some Peculiarities of Diffusion in Solids, in Nowick, A.S., Burton, J.J. (eds.), *Diffusion in Solids – Recent Developments* (Academic Press, New York, 1975) p. 115
276. Murch, G.E., Nowick, A.S. (eds.), *Diffusion in Crystalline Solids* (Academic Press, New York, 1984)
277. Fick, A., Poggy Ann. **94**, 59 (1855)
278. Manning, J.R., *Diffusion Kinetics of Atoms in Crystals* (Van Nostrand, Princeton, NJ, 1968)
279. Nowick, A.S., Berry, B.S., *Inelastic Relaxation in Crystalline Solids* (Academic Press, New York, 1987, Chaps. 7 and 10)
280. Wolf, D., *Spin-Temperature and Nuclear-Spin Relaxation in Matter: Basic Principles and Applications* (Oxford University Press, London, New York, 1979)
281. Crank, J., *The Mathematics of Diffusion* (Oxford University Press, London, New York, 1975)
282. Carslaw, H.S., Jaeger, J.C., *Conduction of Heat in Solids* (Oxford University Press, London–New York, 1959)
283. Le Claire, A.D., British J. Appl. Phys. **14**, 35 (1963)
284. Carman, P.C., Haul, R.A.W., Proc. R. Soc. (London) A **222**, 109 (1954)
285. Rothman, S.J., The Measurement of Tracer Diffusion Coefficients in Solids, in Murch, G.E., Nowick, A.S. (eds.), *Diffusion in Crystalline Solids* (Academic Press, New York, 1984, Chap. 1)
286. Swanson, M.L., Mehl, R.F., Pound, G.M., Hirth, J.P., Trans AIME **224**, 742 (1962)
287. Seith, W., Kottman, A., Angew. Chem. **64**, 379 (1952)
288. Queisser, H.J., J. Appl. Phys. **32**, 1776 (1961)
289. Ayres, P.S., Winchell, P.G., J. Appl. Phys. **34**, 4820 (1968)
290. Ayres, P.S., Winchell, P.G., J. Appl. Phys. **43**, 816 (1972)
291. Waida, E.S., Acta Metall. **2**, 184 (1954)
292. Frank, W., Gösele, U., Mehrer, H., Seeger, A., Diffusion in Silicon and Germanium, in Murch, G.E., Nowick, A.S. (eds.), *Diffusion in Crystalline Solids* (Academic Press, New York, 1984, Chap. 2)
293. Aboltyn, D.E., Karis, Ya.E., Plekhanov, V.G., Sov. Phys. Solid State **22**, 510 (1980)
294. Ghoshtagore, R.N., Phys. Rev. **155**, 598 (1967)
295. Werner, M., Mehrer, H., Siethoff, H., J. Phys. C: Solid State Phys. **16**, 6185 (1983)
296. Bratter, P., Gobrecht, H., Phys. Stat. Solidi **37**, 869 (1970)
297. Seeger, A., Chik, K.P., Phys. Stat. Solidi **29**, 455 (1968)
298. Gupts, D., Thin Solid Films **25**, 231 (1975)
299. Wutting, M., Scr. Metall. **3**, 175 (1969)

300. Barr, L.W., Blackburn, D.A., Brown, A.F., *Radioisotopes in the Physical Sciences and Industry* (International Atomic Energy Agency, Vienna, 1962, p. 137)
301. Styris, D.L., Tomizuka, C.T., J. Appl. Phys. **34**, 1001 (1963)
302. Barr, L.W., Mundy, J.N., Rowe, A.H., Electronic Properties of Amorphous Materials, in Douglas, R.W., Ellis, B. (eds.), *Amorphous Materials* (Wiley, New York, 1972, p. 243)
303. Kelly, J.E., Tomozawa, N., J. Am. Ceram. Soc. **63**, 478 (1980)
304. Holloway, D.M., J. Vac. Sci. Technol. **12**, 392 (1975)
305. Mayer, J.W., Poate, J.M., in Urli, N.B., Corbett, J.W. (eds.) *Thin Film Interdiffusion and Reactions* (Wiley, New York, 1978, p. 119)
306. Stein, D.F., Joshi, A., Annu. Rev. Mater. Sci. **11**, 485 (1981)
307. Lea, C., Met. Sci. 17, 357 (1983)
308. Lutz, H., Sizman, R., Z. Naturforsch. **19**A, 1079 (1964)
309. Behrisch, R., The Measurement of Radioactive Tracer Diffusion Coefficients in Solids, in Behrisch, R. (ed.), *Sputtering by Particle Bombardment I* (Springer-Verlag, Berlin, New York, 1981);Hofman, S., Rep. Progr. Phys. **61**, 800 (1998)
310. Chapman, B., *Glow Discharge Processes* (Wiley, New York, 1980)
311. Reader, R.D., Kauffman, H.R., J. Vac. Sci. Technol. **12**, 1344 (1975)
312. Mundy, J.N., Rothman, S.J., J. Vac. Sci. Technol. A **1**, 74, (1983)
313. Gupta, D., Tsui, R.R.C., Appl. Phys. Lett. **17**, 294 (1970)
314. Atkinson, A., Taylor, R.I., Thin Solid Films **46**, 291 (1977)
315. Dorner, P., Gust, W., Lodding, A., Ocklins, H., Predel, B., Roll, U., Z. Metallk. **73**, 325 (1982)
316. Liebl, H., J. Vac. Sci. Technol. **12**, 385 (1975)
317. Reuter, W., Baglin, J.E.E., J. Vac. Sci. Technol. **18**, 282 (1981)
318. Seran, J.-L., *Thesis* (Universite de Paris, 1976)
319. Macht, M.-P., Naundorf, V., J. Appl. Phys. **53**, 7551 (1982)
320. Bracht, H., Norseng, M., Haller, E.E., Eberl, K., Cardona, M., Solid State Commun. **112**, 301 (1999); Bracht, H., Physica B: Condensed Matter **273–274**, 981 (1999)
321. Slusser, G.J., Slattery, J.S., J. Vac. Sci. Technol. **18**, 301 (1981)
322. Wilzbach, K.E., Kaplan, L., J. Am. Chem. Soc. **72**, 5795 (1950).
323. Catlett, D.S., Spencer, J.N., Vogt, G.J., J. Chem Phys. **58**, 3432 (1973)
324. Spencer, J.N., Catlett, D.S., Vogt, G.J., J. Chem. Phys. **59**, 1314 (1973)
325. Ptashnik, V.B., Dunaeva, T.Yu., Fiz. Tverd. Tela **19**, 1643 (1977) (in Russian)
326. Funke, K., Richtering, H., Ber. Bunsenges. Phys. Chem. **72**, 619 (1968)
327. Dellin, T.A., Dienes, G.J., Rischer, C.R., Hatcher, R.D., Wilson, W.D., Phys. Rev. B**1**, 1745 (1970)
328. Pretzel, F.E., Vier, D.T., Szklarz, E.G., Lewis, W.B., Los Alamos Scientific Lab. Rep. LA 2463 (1961)
329. Letaw, Jr, H., Portnoy, W.M., Slifkin, L., Phys. Rev. **102**, 636 (1956)
330. Valenta, M.W., Ramasastry, C., Phys. Rev. **106**, 73 (1957)
331. Widmer, H., Grunther-Mohr, G.R., Helv. Phys. Acta **34**, 635 (1961)
332. Campbell, D.R., Phys. Rev. B **12**, 2318 (1975)
333. Vogel, G., Hettich, G., Mehrer, H., J. Phys. C: Solid State Phys. **16**, 6197 (1983)
334. Werner, M., *Ph. D. Thesis* (University of Stuttgart, Stuttgart, FRG, 1984)
335. Gruzin, P.L., Dokl. Akad. Nauk SSSR **15**, 108 (1952) (in Russian)

336. Steigmann, J., Shockley, W., Nix, F.C., Phys. Rev. **56**, 13 (1939)
337. Seeger, A., Frank, W., Appl. Phys. A **27**, 171 (1982)
338. Peart, R.F., Phys. Stat. Solidi **15**, K119 (1966)
339. Ghoshtagore, R.N., Phys. Rev. Lett. **16**, 890 (1966)
340. Masters, B.J., Fairfield, J.M., Appl. Phys. Lett. **8**, 280 (1966)
341. Fairfield, J.M., Masters, B.J., J. Appl. Phys. **38**, 3148 (1967)
342. Mayer, H.J., Mehrer, H., Maier, K., Inst. Phys. Conf. Ser. **31**, 186 (1977)
343. Hirvonen, J., Anthla, A., Appl. Phys. Lett. **35**, 703 (1979)
344. Kalinowski, L., Seguin, R., Appl. Phys. Lett. **35**, 211 (1979)
345. Demond, F.J., Kalbitzer, S., Mannsperger, H., Damjantschitsch, H., Phys. Lett. A**93**, 503 (1983)
346. Fuchs, H.D., Walukiewicz, W., Haller, E.E., Dondl, W., Schorer, R., Abstreite, G., Phys. Rev. B **51**, 16817 (1995)
347. Haller, E.E., Semicond. Sci. Technol. **5**, 319 (1990)
348. Spitzer, J., Ruf, T., Cardona, M., Dondl, W., Schorer, R., Abstreiter, G., Haller, E.E., Phys. Rev. Lett. **72**, 1565 (1994)
349. Berezin, A.A., J. Phys. Chem. Solids **48**, 853 (1987); **50**, 5 (1989)
350. Compaan, K., Haven, Y., Trans. Faraday Soc. **52**, 786 (1956)
351. Bourgoin, J.C., Lanno, M., Radiat. Effects **46**, 157 (1980)
352. Mullenn, J.G., Phys. Rev. **121**, 1649 (1961)
353. LeClaire, A.D., Philos. Mag. **14**, 1271 (1966)
354. Itoh, K., Hansen, W.L., Haller, E.E., Farmer, J.W., Ozhogin, V.I., Neutron Transmutation Doping of Isotopically Controlled Ge, in *Proc. 5th Int. Conf. Shallow Levels Semicond.*, Kobe, Japan, 1992; Fuchs, H.D., Itoh, K.M., Haller, E.E., Philos. Mag. B **70**, 661 (1994)
355. Peterson, N.L., J. Nucl. Mater. **69–70**, 3 (1978)
356. Mehrer, H. (ed.), *Diffusion in Solid Metals and Alloys* (Landolt-Bornstein New Series, Group III, Springer-Verlag, Berlin, 1990, Vol. 26)
357. Madelung, O. (ed.), *Intrinsic Properties of Group IV Elements and III-V and I-VII Compounds* (Landolt-Bornstein, New Series, Group III, Springer-Verlag, Berlin, 1989, Vol. 22, Pt. a)
358. Tan, T.Y., Gösele, U.M., Yu, S., Cri. Rev. Solid State Phys. **17**, 47 (1991)
359. Goldstein, B., Phys. Rev. **121**, 1305 (1960)
360. Palfrey, H.D., Brown, M., Willoughby, A.F.W., J. Electrochem. Soc. **128**, 2224 (1981)
361. Tan, T.Y., You, H.M., Yu. S., Gösele, U.M., J. Appl. Phys. **72**, 5206 (1992)
362. Deppe, D.G., Holonyak, Jr., N., J. Appl. Phys. **64**, R93 (1988)
363. Wang, L., Hsu, L., Haller, E.E., Erickson, J.W., Fischer, A., Eberl, K., Cardona, M., Phys. Rev. Lett. **76**, 2342 (1996)
364. Wang, L., Wolk, J.A., Hsu, L., Haller, E.E., Appl. Phys. Lett. **70**, 1831 (1997)
365. Krause-Rehberg, R., Polity, A., Siegel, W., Kuhnel, G., Semicond. Sci. Technol. **8**, 290 (1993)
366. O'Brien, S., Bour, B.P., Shealy, J.R., Appl. Phys. Lett. **53**. 1859 (1988)
367. Beernik, K.J., Sun, D., Treat, D.W., Bour, B.P., Appl. Phys. Lett. **66**, 3597 (1995)
368. Bracht, H., Haller, E.E., Eberl, K., Cardona, M., Appl. Phys. Lett. **74**, 49 (1999)
369. Wee, S.F., Chai, M.K., Homewood, K.P., Gilin, W.P., J. Appl. Phys. **82**, 4842 (1997)

370. Olmsted, B.L., Houde-Walter, S.N., Appl. Phys. Lett. **63**, 530 (1993) and references therein
371. Walukiewicz, W., Mater. Res. Soc. Symp. **300**, 421 (1993)
372. Schokley, W., Moll, J.L., Phys. Rev. **119**, 1480 (1960)
373. Kawabe, M., Matzuura, N., Shimizu, N., Hasegawa, F., Nannichi, Y., Jpn. J. Appl. Phys. **23**, L632 (1984)
374. Beernik, K.J., Thornton, R.L., Anderson, G.B., Emanuel, M.A., Appl. Phys. Lett. **66**, 2422 (1995)
375. Coleman, J.J., Dapkus, P.D., Kirkpatric, G.G., Camras, M.D., Holonyak, Jr., N., Appl. Phys. Lett. **40**, 904 (1982)
376. Kobayashi, J., Nakajima, M., Bamba, Y., Fukunaga, T., Jpn. J. Appl. Phys. **25**, L385 (1986)
377. Kawabe, M., Shimizu, N., Hasegawa, F., Nannichi, Y., Appl. Phys. Lett. **46**, 849 (1985)
378. Kobayashi, J., Nakajima, M., Fukunaga, T., Takamori, T., Ishida, K., Nakashima, H., Jpn. J. Appl. Phys. **25**, L376 (1986)
379. Tan, T.Y., Gösele, U., J. Appl. Phys. **61**, 1841 (1987)
380. Baraff, G.A., Schlütter, M., Phys. Rev. Lett. **55**, 1327 (1988)
381. Tan, T.Y., Gösele, U., Appl. Phys. Lett. **52**, 1240 (1988)
382. Mei, P., Schwarz, S.A., Venkatesan, T., Schwartz, C.L., Colas, E., Appl. Phys. Lett. **50**, 1823 (1987)
383. Mei, P., Schwarz, S.A., Venkatesan, T., Schwartz, C.L., Colas, E., J. Appl. Phys. **65**, 2165 (1989)
384. Li, W.M., Cohen, R.M., Simons, D.S., Chi, P.H., Appl. Phys. Lett. **70**, 3392 (1997)
385. Muraki, K., Horikoshi, Y., J. Crystal Growth **175/176**, 162 (1997)
386. Muraki, K., Horikoshi, Y., Inst. Phys. Conf. Ser. **145**, 547 (1996).
387. Bracht, H., Norseng, M., Haller, E.E., Eberl, K., Cardona, M., Solid State Commun. **112**, 301 (1999)
388. Bracht, H., Norseng, M., Haller, E.E., Eberl, K., Cardona, M., Clark-Phelps, R., Mat. Res. Soc. Symp. Proc. **527**, 335 (1998)
389. Bockstedte, M., Scheffer, M., Zs. Phys. Chemie **200**, 195 (1997)
390. Blakemore, J.S., J. Appl. Phys. **53**, R123 (1982)
391. Tan, T.Y., Mater. Sci. Eng. B**10**, 227 (1991)
392. Tan, T.Y., You, H.-M., Gösele, U.M., Appl. Phys. A **56**, 249 (1993)
393. Lahiri, I., Nolte, D.D., Melloch, M.R., Woodall, J.M., Walukiewicz, W., Appl. Phys. Lett. **69**, 236 (1996)
394. Meese, J.M., The NTD Process - A New Reactor Technology, in Meese, J.M. (ed.), *Neutron Transmutation Doping in Semiconductors* (Plenum Press, New York–London, 1979, p. 1)
395. Meese, J.M. (ed.), *Neutron Transmutation Doping in Semiconductors* (Plenum Press, New York–London, 1979)
396. Cullen, D.E., Hlavac, P.J., *ENDF/B Cross Sections* (Brookhaven National Laboratory, Upton, NY, 1972)
397. Smirnov, L.S. (ed.), *Semiconductors Doped by Nuclear Reactions* (Science, Novosibirsk, 1981) (in Russian)
398. De Soete, D., *Neutron Activation Analysis* (John Wiley, New York, 1971)
399. Mukhin, K.N., *Introduction in Nuclear Physics* (Atomizdat, Moscow, 1965) (in Russian)

400. Tanenbaum, M., Mils, A.D.; J. Electrochem. Soc. **108**, 171 (1961)
401. Snöller, M.S., IEEE Trans. Electron Devices **ED-21**, 313 (1974); Haas, W., Snöller, M., J. Electron. Mater. **5**, 57 (1976)
402. Billington, D.S., Crawford, Jr., J.H., *Radiation Damage in Solids* (Princeton University Press, Princeton, NJ, 1961, Chap. 2)
403. Chukichev, M.V., Vavilov, V.S., Sov. Phys. Solid State **3**, 1103 (1961); Vavilov, V.S., *Influence of Radiation on the Semiconductors* (Science, Moscow, 1963) (in Russian)
404. Smith, T.G.G., Neutron Doping of Silicon in the Harwell Research Reactor, in [395], p. 157
405. Smith, T.G.G., Future Reactor Capacity for the Irradiation of Silicon, in Larrabee, R.D. (ed.), *Neutron Transmutation Doping of Semiconductor Material* (Plenum Press, New York–London, 1984, p. 83)
406. Bourdon, J.L., Restelli, G., An Automated Irradiation Facility for Neutron, in [395], p. 181
407. Morrisey, J.E., Tillinghast, T., Perro, A.P., Baliga, B.J., General Electric Test Reactor NTD Silicon, in [395], p. 171
408. Baliga, B.J., Neutron Transmutation Doped Silicon for Power Semiconductor Devices, in Larrabee, R.D. (ed.), *Neutron Transmutation Doping of Semiconductor Materials* (Plenum Press, New York–London, 1984, p. 167)
409. Bickford, N.A., Fleming, R.F., Silicon Irradiation Facilities at the NBS Reactor, in [395], p. 165
410. Becker, D.A., Lafleur, P.D., J. Radioanalytical Chemistry **19**, 149 (1974)
411. Kirk, M.A., Greenwood, L.R., Determination of the Neutron Flux and Energy Spectrum, in [395], p. 143
412. Breant, P., Irradiation of Single Silicon Crystals with Diameters in the 3 - to 5 - Inch Range in French Reactors, in Larrabee, R.D. (ed.), *Neutron Transmutation Doping of Semiconductor Materials* (Plenum Press, New York–London, 1984, p. 103)
413. Robinson, M.T., Theory of Neutron Transmutation Doping, in Hall, J.L., Maple, J.H.C. (eds.), *Nuclear Fision Reactors* (British Nuclear Energy Society, London, 1970, p. 364)
414. Lindhard, J., Nielsen, V., Scharff, M., Thomsen, P.V., Mat. Fys. Medd. Dan. Vid. Selsk. **33**, N10 (1963)
415. Kaltenborn, N., Malmros, O., Experiences with the Norwegian Research Reactor JEEP II in Neutron Transmutation Doping, in Larrabee, R.D. (ed.), *Neutron Transmutation Doping of Semiconductor Materials* (Plenum Press, New York, London, 1984, p. 139)
416. Hansen, K., Stendal, K., Andersen, K., Heydorn K., An Automatic Contolled, Heavy Water Cppled Facility, in [413], p. 91
417. Rossin, A.D., published by T.H. Blewitt and T.J. Koppenaal, Neutron Radiation Damage, in Sheely, W.F. (ed.), *Radiation Effects* (Gordon and Breach, New York, 1966, p. 561)
418. Horak, J.A., Blewitt, T.H., Phys. Stat. Soidi **9**, 721 (1972)
419. Brown, B.S., Blewitt, T.H., Scott, T.L., Klank, A.C., J. Nucl. Mater. **52**, 215 (1974)
420. Parma, Jr., E.J., Hart, R.R., Measurements of the Gamma Abundance of Silicon - 31, in Larrabee, R.D. (ed.), *Neutron Transmutation Doping of Semiconductor Materials* (Plenum Press, New York, London, 1984, p. 127)

421. Vlasov, N.A., *Neutrons* (Science, Moscow, 1971) (in Russian)
422. Arifov, U.A., Sinukov, V.A., Korostelev, Yu.A., in *Crystallization of Thin Films* (Fan, Tashkent, 1970, p. 133) (in Russian)
423. Plekhanov, V.G., Physics-Uspekhi (Moscow) **43**, 1147 (2000)
424. Clairon, P.J., Meese, J.M., Isochronal Annealing of Resistivity in Floar Zone and Czochralski NTD Silicon, in [395], p. 291
425. Brugger, R.M., Yellow, W., Neutron Scattering in Semiconductor, in Moon, R.M. (ed.), *Proc. Conf. Neutron Scattering* (1976) Vol. 11, p. 1117
426. Lark-Horowitz, K., Process of Transmutation, in Henish, H.K. (ed.), *Proc. Conf. on Semicond. Mater.* (Butterworth, London, 1951, p. 47)
427. Schweinler, H.C., J. Appl. Phys. **30**, 1125 (1959)
428. Haller, E.E., J. Appl. Phys. **77**, 2857 (1995)
429. Mirianashwili, Sh.M., Nanobashwili, D.I., Fiz. i Techn. Poluprovodnikov **10**, 1879 (1970) (in Russian)
430. Young, M.H., Hunter, A.T., Baron, R., Marsch, O.I., Neutron Transmutation Doping of p-Type Czochralski - Grown Gallium Arsenide, in Larrabee, R.D. (ed.), *Neutron Transmutation Doping of Semiconductor Materials* (Plenum Press, New York, London, 1984, p. 1)
431. Uematsu, M., Wada, K., Appl. Phys. Lett. **58**, 2015 (1991).
432. Galushka, A.P., Konosemko, I.D., Atom Energy **13**, 277 (1962).
433. Meese, J.M., Clairon, P.J., Resistivuty Fluctuations in Highly Compensated NTD Silicon, in [395], p. 109
434. Klahr, C.N., Cohen, M., Nucleonics **22**, 62 (1964)
435. Kharchenko, V.A., Solov'ev, S.P., Fiz. i Techn. Poluprovodnikov **11**, 1641 (1971) (in Russian)
436. Guldberg, J. (ed.), *Neutron-Transmutation Doped Silicon* (Plenum Press, New York, 1981)
437. Certis, L., *Introduction in Neutron Physics* (Atomizdat, Moscow, 1965) (in Russian)
438. Kikoin, I.K. (ed.), *Handbook of Physical Constants* (Atomizdat, Moscow, 1976) (in Russian)
439. James, H.M., Malmros, O., IEEE Trans. Electron Devices, ED-**23**, 797 (1976)
440. Cleland, J.W., Defects in Neutron Irradiation Crystals, in Billington, D.S. (ed.), *Proc. Int. School of Physics, Enrico Fermi – Course XVIII – Radiation Damage in Solids* (Academic Press, New York, 1962, p. 384)
441. Stein, H.J., Atomic Displacement Effects in Neutron Transmutation Doping, in Meese, J.M. (ed.), *Neutron Transmutations Doping in Semiconductors* (Plenum Press, New York, London, 1979, p. 229)
442. van Lint, V.A.J., Leadon, R.E., Colwell, J.F., IEEE Trans. Nucl. Sci. NS – **19**, 181 (1972)
443. Yoshida, M., J. Phys. Soc. Japan **16**, 44 (1961)
444. Brice, D.K., Radiat. Effects **11**, 227 (1971)
445. Bertolotti, M., Papa, T., Sette, D., Grasso, V., Vitali, G., IL Nuovo Cimento **29**, 4310 (1963)
446. Bertolotti, M., Sette, D., Vitali, G., J. Appl. Phys. **38**, 2645 (1967)
447. Swanson, M.L., Parsons, J.R., Hoelke, C.W., Defects in Semiconductors, in Corbett, J.W., Watkins, G.D. (eds.), *Albany Conf. Radiat. Effects Semiconductors* (Plenum Press, New York, 1971, p. 359)
448. den Onden, G., Philos. Mag. **19**, 321 (1969)

449. Nelson, R.S., Radiat. Effects **32**, 19 (1977)
450. Gossick, B.R., J. Appl. Phys. **30**, 1214 (1959)
451. Crawford, Jr., J.H., Cleland, J.W., J. Appl. Phys. **30**, 1204 (1959)
452. Gregory, B.L., IEEE Trans. Nucl. Sci. NS - **16**, 53 (1969).
453. Stein, H.J., ibid NS - **15**, 69 (1968)
454. Stein, H.J., Appl. Phys. Lett. **15**, 61 (1969)
455. Haller, E.E., Palaio, N.P., Rodder, M., Hansen, W.L., Kreysa, E., NTD Germanium: A Novel Material for Low Temperature Bolometers, in Larrabee, R.L. (ed.), *Neutron Transmutation Doping of Semiconductor Materials* (Plenum Press, 1984, p. 21); Haller, E.E., Infrared Phys. Technol. **35**, 127 (1994)
456. Ootuka, Y., Matsuoka, H., Kobayashi, S., Theory of Anderson Model, in Ando, T., Fukuyama, H. (eds.), *Anderson Localization* (Springer-Verlag, Berlin, 1988, p. 40)
457. Itoh, K.M., Haller, E.E., Beeman, J.W., Hansen, W.L., Phys. Rev. Lett. **77**, 4058 (1996)
458. Schlimak, I., Kaveh, M., Ussyshkin, R., Phys. Rev. Lett. **77**, 1103 (1996)
459. Schlimak, I., Fiz. Tverd. Tela **41**, 794 (1999) (in Russian)
460. Rentzsch, R., Ionov, A.N., Reich, Ch., Ginodman, V., Schlimak, I., Fiz. Tverd. Tela **41**, 837 (1999) (in Russian); Rentzsch, R., Ionov, A.N., Phil. Mag. B**81**, 1065 (2001)
461. Ionov, A.N., Matveev, M.N., Shmik, D.V., J. Tech. Phys. (St.-Petersburg) **59**, 169 (1989) (in Russian)
462. Schlimak, I., Ionov, A.N., Rentzsch, R., Lazebnik, J.M., Semicond. Sci. Technol. **11**, 1826 (1996)
463. Lederer, C.M., Hollander, J.M., Perlman, I. (eds.), *Table of Isotopes*, 6th ed. (Wiley, New York, 1967)
464. Shklovskii, B.I., Efros, A.L., Electronic Properties of Doped Semiconductors, in *Electronic Properties of Doped Semiconductors* (Solid State Series, Vol. 45, Springer-Verlag, Berlin, 1984)
465. Haller, E.E., Hansen, W.L., Goulding, F.S., Adv. Phys. **30**, 93 (1981)
466. Fritzsche, H., *The Metal Non-Metal Transition in Disordered Systems* (L.R. Friedman and D.P. Tunstall, Scotish Universities Summer School in Physics, St. Andrews, Scotland, 1978)
467. Mott, N.F., *Metal-Insulator Transition*, 2nd ed. (Taylor and Francis, London, 1990)
468. Lee, P.A., Ramakrishnan, T.V., Rev. Mod. Phys. **57**, 287 (1985)
469. Rentzsch, R., Ionov, A.N., Reich, Ch., Müller, A., Phys. Stat. Solidi (b) **205**, 269 (1998)
470. Rosenbaum, T.F., Andres, K., Thomas, G.A., Bhatt, F.N., Phys. Rev. Lett. **45**, 1723 (1980)
471. Newman, P.F., Holcomb, D.F., Phys. Rev. B **28**, 638 (1983); Phys. Rev. Lett. **51**, 2144 (1983)
472. Shafarman, W.N., Koon, D.W., Castner, T.G., Phys. Rev. B**40**, 1216 (1989)
473. Ionov, A.N., Lea, M.J., Rentzsch, R., JETP Lett. (Moscow) **54**, 473 (1991)
474. Dai, P., Zhang, Y., Sarachik, M.P., Phys. Rev. Lett. **66**, 1914 (1991)
475. Thomas, G.A., Ootuka, Y., Katsumoto, S., Kobayashi, S., Sasaki, W., Phys. Rev. B **25**, 4288 (1982)
476. Hirsch, M.J., Thomanschefsky, U., Holcomb, D.F., Phys. Rev. B **37**, 8257 (1988)

477. Zabrodskii, A.G., Zinov'eva, K.N., Sov. Phys. JETP **59**, 425 (1984)
478. Rohde M., Micklitz, H., Phys. Rev. B **36**, 7572 (1987)
479. Hertel, G., Bishop, D.J., Spencer, E.G., Dynes, R.C., Phys. Rev. Lett. **50**, 743 (1983)
480. McMillan, W.L., Mochel, J., Phys. Rev. Lett. **46**, 556 (1981)
481. Zint, Th., Rohde, M., Micklitz, H., Phys. Rev. B**41**, 4831 (1990)
482. Hass, E.W., Schnöller, M.S., Phosphorus Doping of Silicon by means of Neutron Irradiation, in [272], p. 803
483. Mott, N.F., Proc. Phys. Soc. (London) A **62**, 416 (1949)
484. Anderson, P.W., Phys. Rev. **109**, 1492 (1958)
485. MacKinnon, A., Kramer, B., Phys. Rev. Lett. **47**, 1546 (1981); Henneke, M., Kramer B., Ohtsuki, T., Europhys. Lett. **27**, 389 (1994)
486. Hafstetter, E., Schreiber, M., Phys. Rev. Lett. **73**, 3137 (1994)
487. Kawarabayashi, T., Ohtsuki, T., Slevin K., Ono, Y., Phys. Rev. Lett. **77**, 3593 (1996)
488. Chayes, J., Chayes, L., Fisher, D.S., Spencer, T., Phys. Rev. Lett. **54**, 2375 (1986)
489. Larrabee, R.D. (ed.), *Neutron Transmutation Doping of Semiconductor Materials* (Plenum Press, New York, London, 1984)
490. Ionov, A.N., Shlimak, I.S., Matveev, M.N., Solid State Commun. **47**, 763 (1983)
491. Chattopadhyay, D., Queisser, H.J., Rev. Mod. Phys. **53**, 745 (1981)
492. von Ammon, W., Nucl. Instrum. Methods B **63**, 95 (1992)
493. Itoh, K.M., Walukiewicz, W., Fuchs, H.D., Beeman, J.W., Haller, E.E., Ozhogin, V.I., Phys. Rev. B**50**, 16995 (1994)
494. Itoh, K., Hansen, W.L., Haller, E.E., Farmer J.W., Ozhogin, V.I., Mater. Sci. Forum **117–118**, 117 (1993)
495. Itoh, K.M., Hansen, W.L., Haller, E.E., Farmer J.W., Ozhogin, V.I., J. Mater. Res. **8**, 127 (1993)
496. Erginsoy, C., Phys. Rev. **79**, 1013 (1950)
497. Ansel'm, A.I., Zh. Eksp. Teor. Fiz. (Moscow) **24**, 85 (1953) (in Russian)
498. Sclar, N., Phys. Rev. **104**, 1559 (1956); **104**, 1548 (1956)
499. McGill, T.C., Baron, R., Phys. Rev. B**11**, 5208 (1975)
500. Ridley, B.K., *Quantum Process in Semiconductors*, 3rd ed. (Clarendon Press, Oxford, 1993)
501. Dingle, R.D., Philos. Mag. **46**, 831 (1955)
502. Brooks, H., Adv. Electron Phys. **7**, 85 (1955)
503. Shockley, W., *Electrons and Holes in Semiconductors* (Van Nostrand Reinhold, Princeton, 1950)
504. Blakemore, J., *Semiconductor Statistics*, 2nd ed. (Dover, New York, 1985)
505. Dzhakeli, V.G., Kachlishvili, Z.S., Sov. Phys. Semicond. **18**, 926 (1984)
506. Farmer J.W., Nugent, J.C., Transit Current Spectroscopy of Neutron Irradiated Silicon, in [489], p. 225
507. Hill, M.J., van Iseghem, P.M., Zimmerman, W., Preparation and Application of Neutron Transmutation Doped Silicon for Power Device Research, in [272], p. 809
508. Muhlbauer, A., Seldak, F., Voss, P., J. Electrochem. Soc. **122**, 1113 (1975)
509. Messier, J., le Corroler Y., Flores, J.M., IEEE Trans. Nucl. Sci. NS - **11**, 276 (1964)

510. van Iseghem, P.V., IEEE Trans. Electron Devices ED-**23**, 823 (1976)
511. *Regulations for the Safe Transport of Radioactive Materials*, rev. ed. (International Atomic Energy Agency, 1973, p. 8)
512. Hamanaka, H., Kuriyama, K., Yahagi, M., Appl. Phys. Lett. **45**, 786 (1984)
513. Koon, D.W., Castner, T.G., Phys. Rev. Lett. **60**, 1755 (1988)
514. Dai, P., Zhang, Y., Sarchik, M.P., Phys. Rev. B **49**, 14039 (1994)
515. Grussbach, H., Schreiber, M., Phys. Rev. B **51**, 663 (1995).
516. Dai, P., Zhang, Y., Sarchik, M.P., Phys. Rev. Lett. **70**, 1968 (1993)
517. Field, S.B., Rosenbaum, T.F., Phys. Rev. Lett. **55**, 522 (1985).
518. Rentzsch, R., Friedland, K.J., Ionov, A.N., Phys. Stat. Solidi (b) **146**, 199 (1988)
519. Magerle, R., Burchard, A., Deicher, M., Kerle, T., Phys. Rev. Lett. **75**, 1594 (1995)
520. Kuriyama, K., Sakai, K., Phys. Rev. B **53**, 987 (1996)
521. Kuriyama, K., Miyamoto, Y., Koyama, T., Ogawa, O., J. Appl. Phys. **86**, 2352 (1999)
522. Kuriyama, K., Ohbora, K., Okada, M., Solid State Commun. **113**, 415 (2000)
523. Sze, S.M., *Physics of Semiconductor Devices* (Wiley, New York, 1969)
524. Winston, H., Kimura, H., Young, M.H., Marsh, O.J., Neutron Transmutation Doping of Semiconducting Crystals, in *AAGG/West 6 Conf. Crystal Growth* (Fallen Leaf Lake, CA, 1982)
525. Lederer, C.M., Shirley, V.S. (eds.), *Table of Isotopes* (Wiley, New York, 1978)
526. Pons, D., Bourgoin, J.C., J. Phys. C: Solid State Phys. **18**, 3839 (1985)
527. *Properties of Gallium Arsenide*, EMIS Data Reviews Series No 2 (INSPEC, The Institute of Electrical Engineers, London, 1990)
528. Pavesi, L., Guzzi, M.J., Appl. Phys. **75**, 4779 (1994)
529. Mooradian, A., MsWorter, A.L., Phys. Rev. Lett. **19**, 849 (1967)
530. Sekine, T., Uchinokura, K., Matzuura, E., J. Phys. Chem. Solids **48**, 109 (1977)
531. Kawakubo, T., Okada, M., J. Appl. Phys. **67**, 3111 (1990)
532. Kuriyama, K., Miyamoto, Y., Okada, M., J. Appl. Phys. **85**, 3499 (1999)
533. Dean, P.J., Inter-Impurity Recombinations in Semiconductors, in McCaldin, J.O., Somorjai, G., (eds.), *Progress in Solid State Chemistry* (Pergamon Press, Oxford, 1973, Vol. 8, p. 1)
534. Alawadhi, H., Vogelgesang, R., Chin, T.P., Woodall, J.M., J. Appl. Phys. **82**, 4331 (1997)
535. Klick, C.C., Shulman, J.H., The Color Centers in Ionic Crystals, in Seitz, F., Turnbull, D. (eds.), *Solid State Physics* (Academic Press, New York, 1957, Vol. 5, p. 100)
536. Kressel, H. (ed.), *Semiconductor Devices for Optical Communications* (Topics in Applied Physice, Vol. 39, Springer-Verlag, Berlin, Heidelberg, New York, 1982)
537. Becker, P.C., de Barras, M.R.X., Optical Fibers for Telecommunications: Transmission and Amplification, in Quilec, M. (ed.), *Materials for Optoelectronics* (Kluwer Academic Publishers, Boston, Dordrecht, London, 1996)
538. Snyder, A.W., Love, J.D.,*Optical Waveguide Theory* (Chapman and Hall Medical, London, New York, Melbourne, 1996)
539. Marcuse, D., *Light Transmission Optics* (Van Nostrand, New York, 1972)
540. Allan, W.B., *Fibre Optics Theory and Practice* (Plenum Press, New York, 1973)

541. Kapany, N.S., *Fiber Optics* (Academic Press, New York, London, 1967)
542. Arnaud, J.A., *Beam and Fibre Optics* (Academic Press, New York, 1976)
543. Marcuse, D., *Theory of Dielectric Optical Waveguides* (Academic Press, New York, 1974)
544. Midwinter, J.E., *Optical Fibers for Transmission* (Wiley and Sons, New York, Toronto, 1979)
545. Tamm, I.E., *Theory of Electricity* (Science, Moscow, 1957) (in Russian)
546. Born M., Wolf, E., *Principles of Optics* (Pergamon Press, Oxford, 1970)
547. Collin, R.E., *Field Theory of Guided Waves* (McGraw-Hill, New York, 1960)
548. Snitzer, E., J. Opt. Soc. Am. **51**, 491 (1961)
549. Gloge, D., Appl. Opt., **10**, 2252 (1971)
550. Stratton, J.A., *Electromagnetic Theory* (McGraw-Hill, New York, 1941)
551. Moon, P., Spencer, D.E., *Field Theory Handbook* (Springer-Verlag, Berlin, 1961)
552. Personick, S.D., Bell Syst. Tech. J. **52**, 843 (1973)
553. Olshansky, R., Keck, D.B., Appl. Opt. **15**, 483 (1976)
554. Olshansky, R., Keck, D.B., in *Technical Digest of OSA Topical Meeting on Optical Fiber Transmission* (Optical Society of America, Washington DC, 1975)
555. Quillec, M. (ed.), *Materials for Optoelectronics* (Kluwer Academic, Boston, Dordrecht, London, 1996)
556. Beck, Th., Reng, N., Weber, H., Opt. Lasers Eng. **34**, 255 (2000); Heitman, W., J. Opt. Commun. **8**, 3 (1987)
557. Philipp, H.R., J. Phys. Chem. Solids **32**, 1935 (1971)
558. Pantelides, S.T., Harrison, W., Phys. Rev. B**13**, 2667 (1976)
559. Mott, N.F., in Pantelides, S.T. (ed.), *Physics of SiO_2 and Its Interfaces* (Proc. Int. Conf., New York, 1978) p. 1
560. Griscom, D.L., Non-Crystal. J., Solids **40**, 211 (1980); Griscom, D.L., Friebele, E.L., Long, K.J., J. Appl. Phys. **54**, 3743 (1983)
561. Grinfelds, A.U., Aboltyn, D.E., Plekhanov, V.G., Sov. Phys. Solid State **26**, 1075 (1984); Phys. Stat. Solidi (a) **81**, K23 (1984)
562. Gupta, R.P., Phys. Rev. B**32**, 8278 (1985)
563. Horiguchi, M., Electron. Lett. **12**, 311 (1976)
564. Osonai, H., Shioda, T., Moriyama, T., Araki, S., Horiguchi, M., Izawa, T., Takate, H., Electron. Lett. **12**, 550 (1976)
565. Pinnow, D.A., Rich, T.C., Ostermayer, F.W., DiDomenico, Jr., M., Appl. Phys. Lett. **22**, 527 (1973)
566. Schroeder, J., Mohr, R., Macedo, P.B., Montrose, C.J., J. Am. Ceram. Soc. **56**, 510 (1973)
567. Beales, K.J., Midwinter, J.E., Newns, G.R., Day, C.R., Post Off. Elec. Eng. J. **67**, 80 (1974)
568. Scott, B., Rawson, H., Glass Technol. **14**, 115 (1973)
569. Beales, K.J., Day, C.R., Dunkan, W.J., Midwinter, J.E., Newns, G.R., Proc. IEEE **123**, 591 (1976)
570. Inoue, T., Koizumi, K., Ikeda, Y., ibid. **123**, 577 (1976)
571. Zheltikov, A.M., Uspekhi Fiz. Nauk (Moscow) **170**, 1203 (2000) (in Russian)
572. Thyagarajan, K., Ghatak, A.K. (eds.), *Lasers Theory and Applications* (Plenum Press, New York, 1982)
573. Svelto, O., *Principles of Lasers*, 2nd ed. (Plenum Press, New York, 1982)
574. Tarasov, L.V., *Laser Physics* (Mir, Moscow, 1983)

575. Karlov, N.V., *Lectures on Quantum Electronics* (Science, Moscow, 1983) (in Russian)
576. Einstein, A., Mitt. Phys. Ges. (Zurich) **16**, (18) 47 (1916)
577. Reif, F., *Statistical Thermal Physics* (McGraw-Hill, New York, 1965)
578. Brodin, M.S., Reznitchenko, A V.Ya., Interaction of the Intensity Laser Radiation with A_2B_6 Semiconductors, in Georgabiani, A.N., Sheinkman, M.K. (eds.) *Physics A_2B_6 Compounds* (Science, Moscow, 1986, p. 184) (in Russian)
579. Taylor, A.J., Erskine, D.J., Tang, C.L., J. Opt. Soc. Am. (B) **2**, 663 (1985)
580. Kash, J.A., Tsang, J.C., Light Scattering and Other Secondary Emission Studies of Dynamic Process in Semiconductors, in Cardona, M., Güntherodt, G. (eds.), *Light Scattering in Solids* (Springer-Verlag, Berlin, 1991, Vol. 6, p. 423)
581. Dumke, W.P., Phys. Rev. **127**, 1559 (1962)
582. Nahory, R.E., Shakley, K.L., Leheny, R.F., Logan, R.A., Phys. Rev. Lett. **27**, 1647 (1971)
583. Kressel, H., Semiconducror Lasers: Devices, in Arechi, F.T., Schulz-Dubois, E.O. (eds.), *Laser Handbook* (North-Holland, Amsterdam, 1972, Chap. B5)
584. Stern, F., Semiconductor Lasers: Theory, ibid., Chap. B4
585. Klingshirn, C., Semiconductor Emission at the High Density Excitation, in *Spectr. Solid-State Laser Type Matter* (Proc. Course Enrico Fermi, Erice, New York, London, 1987)
586. Klein, C.A., Appl. Opt., **5**, 1922 (1966); IEEE QE - **4**, 186 (1968)
587. Lasher, G., Stern, F., Phys. Rev. **133**, A553 (1964)
588. Haug, H., J. Appl. Phys. **39**, 4687 (1968); Advances in Solid State Physics, Vol. XXII, 149 (1982)
589. van Roosbroeck, W., Shokley, W., Phys. Rev. **94**, 1558 (1954)
590. Hwang, C.J., Phys. Rev. B**2**, 4126 (1970)
591. Haug, H., Koch, S., Phys. Stat. Solidi (b) **82**, 531 (1977)
592. Liu, K.C., Liboff, R.L., J. Appl. Phys. **54**, 5633 (1983)
593. Anderson, P.W., *Concepts in Solids* (Benjamin, Reading, MA, 1963)
594. Haken, H., *Handbuch der Physik* (Springer-Verlag, Berlin, 1970, Vol. XXV/2c)
595. Haken, H., Laser Theory, in Kay, S.M., Maitland, A. (eds.), *Quantum Optics* (Academic Press, New York, 1970)
596. Liboff, R.L., Int. J. Theor. Phys. **18**, 185 (1979)
597. Schawlow, A.W., Townes, C.H., Phys. Rev. **112**, 1940 (1958)
598. Basov, N.G., Bogdankevich, O.V., Devyatkov, A.G., Sov. Phys. JETP **20**, 1902 (1964); Sov. Phys. Solid State **8**, 1221 (1966)
599. Packard, J.R., Campbell, D.A., Tait, W.C., J. Appl. Phys. **38**, 5255 (1967)
600. Benoit a la Guilaume, C., Debever, J.M., Salvan F., High Intensity Effects in Semiconductors, in [183] p. 609; Phys. Rev. **169**, 567 (1969)
601. Kulevsky,L.A., Prokhorov, A.M., IEEE, QE - **2**, 584 (1966)
602. Brodin, M.S., Dmitrenko, K.A., Shevel, S.G., Taranenko, L.V., *Proc. Int. Conf. Lasers'82* (STS Press, USA, 1983, p. 287)
603. Brodin, M.S., Zakrevski, S.V., Mashkevich, V.S., Reznitchenko, V.Ya., Sov. Phys.-Semicond. **1**, 495 (1967)
604. Takiyama, K., Abd-Elrahman, M.I., Fujita, T., Okada, T., Solid State Commun. **99**, 793 (1996)
605. C. Klingshirn and H. Haug, Phys. Reports **70**, 315 (1981).
606. Basov, N.G., Danilychev, V.A., Popov, Yu.M., JETP Lett. **12**, 473 (1970) (in Russian)

607. Schwenter, N., Dossel, O., Nahme, N., Stimulated Vacuum Ultraviolet Emission from Rare Gas Crystals, in *Laser Techniques for Extreme Ultraviolet Spectroscopy* (APTP, New York, 1982, p. 163)
608. Plekhanov, V.G., Experimental Manifestation of Nonlinear Interaction and Excitonic Effects in the Wide-Gap Insulators, in *Proc. Int. Conf. NLO'88*, Dublin, 1988; Progress in Solid State Chemistry **29**, 71 (2001)
609. Ruf, T., Cardona, M., Thonke, K., Pavone, P., Anthony, T.R., Solid State Commun. **105**, 311 (1988)
610. Plekhanov, V.G., Quantum Electronics (Moscow) **16**, 2156 (1989) (in Russian)
611. Plekhanov, V.G., Betenekova, T.A., Gavrilov, F.F., Sov. Phys, Solid State **25**, 159 (1983)
612. Roessler, R.W., Walker, W.C., J. Phys. Chem. Solids **28**, 1507 (1967)
613. Piacentini, M., Solid State Commun. **17**, 697 (1975)
614. Plekhanov, V.G., Wide-Gap Ionic Insulators Excitonic Nonlinearity and Its Potential Applications in Solid State Lasers, in *Proc. Int. Conf. Advances Solid-State Lasers*, March, 1990, Salt Lake City, UT, SOQUE, 1990; J. Materials Science **38**, 3341 (2003)
615. Agekyan, V.F., Alexandrov, B.G., Stepanov, Yu.A., Fiz. Tekh. Poluprovod. **22**, 2240 (1988) (in Russian)
616. Berezin, A.A., Solid State Commun. **65**, 819 (1988)
617. Berezin, A.A., Chang, J.S., Ibrahim, A., Chemtronix (UK) **3**, 116 (1988)
618. Werheit, H., Some Properties of Composite Crystal Structure, in Madelung, O. (ed.), *Landolt-Börnstein* (Springer-Verlag, Berlin, 1983, Vol. 17^e, p. 9)
619. Berezin, A.A., J. Chem. Phys. **80**, 1241 (1984); Kybernetes **15**, 15 (1986)
620. Knoff, H.W., Mueller, M.H., Heaton, L., Acta Crystallographia **23**, 549 (1967)
621. Hidaka, T., Oka, K., Phys. Rev. B **35**, 8502 (1987)
622. Antonov, A.V., Galanov, N.V., Issakov, A.I., Sov. Phys. Tech. Phys. **31**, 942 (1986)
623. Belyakov, V.A., Semenov, S.V., JETP **90**, 290 (2000)
624. Kane, B.E., Nature **393**, 133 (1998)
625. Nielsen, M.A., Chuang, I.L., *Quantum Computation and Quantum Information* (Cambridge University Press, New York, 2000)

Index

Springer Series in

MATERIALS SCIENCE

Editors: R. Hull R. M. Osgood, Jr. J. Parisi

10 **Computer Simulation of Ion-Solid Interactions**
By W. Eckstein

11 **Mechanisms of High Temperature Superconductivity**
Editors: H. Kamimura and A. Oshiyama

12 **Dislocation Dynamics and Plasticity**
By T. Suzuki, S. Takeuchi, and H. Yoshinaga

13 **Semiconductor Silicon**
Materials Science and Technology
Editors: G. Harbeke and M. J. Schulz

14 **Graphite Intercalation Compounds I**
Structure and Dynamics
Editors: H. Zabel and S. A. Solin

15 **Crystal Chemistry of High-T_c Superconducting Copper Oxides**
By B. Raveau, C. Michel, M. Hervieu, and D. Groult

16 **Hydrogen in Semiconductors**
By S. J. Pearton, M. Stavola, and J. W. Corbett

17 **Ordering at Surfaces and Interfaces**
Editors: A. Yoshimori, T. Shinjo, and H. Watanabe

18 **Graphite Intercalation Compounds II**
Editors: S. A. Solin and H. Zabel

19 **Laser-Assisted Microtechnology**
By S. M. Metev and V. P. Veiko
2nd Edition

20 **Microcluster Physics**
By S. Sugano and H. Koizumi
2nd Edition

21 **The Metal-Hydrogen System**
By Y. Fukai

22 **Ion Implantation in Diamond, Graphite and Related Materials**
By M. S. Dresselhaus and R. Kalish

23 **The Real Structure of High-T_c Superconductors**
Editor: V. Sh. Shekhtman

24 **Metal Impurities in Silicon-Device Fabrication**
By K. Graff 2nd Edition

25 **Optical Properties of Metal Clusters**
By U. Kreibig and M. Vollmer

26 **Gas Source Molecular Beam Epitaxy**
Growth and Properties of Phosphorus Containing III–V Heterostructures
By M. B. Panish and H. Temkin

27 **Physics of New Materials**
Editor: F. E. Fujita 2nd Edition

28 **Laser Ablation**
Principles and Applications
Editor: J. C. Miller

29 **Elements of Rapid Solidification**
Fundamentals and Applications
Editor: M. A. Otooni

30 **Process Technology for Semiconductor Lasers**
Crystal Growth and Microprocesses
By K. Iga and S. Kinoshita

31 **Nanostructures and Quantum Effects**
By H. Sakaki and H. Noge

32 **Nitride Semiconductors and Devices**
By H. Morkoç

33 **Supercarbon**
Synthesis, Properties and Applications
Editors: S. Yoshimura and R. P. H. Chang

34 **Computational Materials Design**
Editor: T. Saito

35 **Macromolecular Science and Engineering**
New Aspects
Editor: Y. Tanabe

36 **Ceramics**
Mechanical Properties, Failure Behaviour, Materials Selection
By D. Munz and T. Fett

37 **Technology and Applications of Amorphous Silicon**
Editor: R. A. Street

38 **Fullerene Polymers and Fullerene Polymer Composites**
Editors: P. C. Eklund and A. M. Rao

Springer Series in
MATERIALS SCIENCE

Editors: R. Hull R. M. Osgood, Jr. J. Parisi

39 **Semiconducting Silicides**
Editor: V. E. Borisenko

40 **Reference Materials in Analytical Chemistry**
A Guide for Selection and Use
Editor: A. Zschunke

41 **Organic Electronic Materials**
Conjugated Polymers and Low Molecular Weight Organic Solids
Editors: R. Farchioni and G. Grosso

42 **Raman Scattering in Materials Science**
Editors: W. H. Weber and R. Merlin

43 **The Atomistic Nature of Crystal Growth**
By B. Mutaftschiev

44 **Thermodynamic Basis of Crystal Growth**
P–T–X Phase Equilibrium and Non-Stoichiometry
By J. Greenberg

45 **Thermoelectrics**
Basic Principles and New Materials Developments
By G. S. Nolas, J. Sharp, and H. J. Goldsmid

46 **Fundamental Aspects of Silicon Oxidation**
Editor: Y. J. Chabal

47 **Disorder and Order in Strongly Nonstoichiometric Compounds**
Transition Metal Carbides, Nitrides and Oxides
By A. I. Gusev, A. A. Rempel, and A. J. Magerl

48 **The Glass Transition**
Relaxation Dynamics in Liquids and Disordered Materials
By E. Donth

49 **Alkali Halides**
A Handbook of Physical Properties
By D. B. Sirdeshmukh, L. Sirdeshmukh, and K. G. Subhadra

50 **High-Resolution Imaging and Spectrometry of Materials**
Editors: F. Ernst and M. Rühle

51 **Point Defects in Semiconductors and Insulators**
Determination of Atomic and Electronic Structure from Paramagnetic Hyperfine Interactions
By J.-M. Spaeth and H. Overhof

52 **Polymer Films with Embedded Metal Nanoparticles**
By A. Heilmann

53 **Nanocrystalline Ceramics**
Synthesis and Structure
By M. Winterer

54 **Electronic Structure and Magnetism of Complex Materials**
Editors: D.J. Singh and D. A. Papaconstantopoulos

55 **Quasicrystals**
An Introduction to Structure, Physical Properties and Applications
Editors: J.-B. Suck, M. Schreiber, and P. Häussler

56 **SiO_2 in Si Microdevices**
By M. Itsumi

57 **Radiation Effects in Advanced Semiconductor Materials and Devices**
By C. Claeys and E. Simoen

58 **Functional Thin Films and Functional Materials**
New Concepts and Technologies
Editor: D. Shi

59 **Dielectric Properties of Porous Media**
By S.O. Gladkov

60 **Organic Photovoltaics**
Concepts and Realization
Editors: C. Brabec, V. Dyakonov, J. Parisi and N. Sariciftci

Printing: Saladruck, Berlin
Binding: Stein+Lehmann, Berlin